Linux架构（基础篇）

韩艳威◎著

人民邮电出版社

北京

图书在版编目（CIP）数据

跟老韩学Linux架构. 基础篇 / 韩艳威著. —— 北京：人民邮电出版社，2022.10（2024.4重印）
ISBN 978-7-115-56160-2

Ⅰ．①跟… Ⅱ．①韩… Ⅲ．①Linux操作系统 Ⅳ．①TP316.85

中国版本图书馆CIP数据核字（2021）第048669号

内 容 提 要

本书全面、系统地介绍了 Linux 基础架构服务。主要内容包括 DNS 服务、DHCP 服务器、vsftpd 服务、rsync 服务、SFTP 服务、Samba 服务和 LAMP 基础架构等，能为读者后续学习 Linux 操作系统、高可用分布式文件系统等高级应用打下坚实的基础。

本书既适合 Linux 系统管理员、系统架构师、Linux 相关技术从业人员学习，也适合软件开发人员、软件测试人员、数据库管理人员参考，还可作为高等院校计算机及相关专业、计算机培训机构师生的教材或参考书。

◆ 著　　韩艳威
　　责任编辑　张　涛
　　责任印制　王　郁　焦志炜

◆ 人民邮电出版社出版发行　北京市丰台区成寿寺路 11 号
　　邮编　100164　电子邮件　315@ptpress.com.cn
　　网址　https://www.ptpress.com.cn
　　北京盛通印刷股份有限公司印刷

◆ 开本：787×1092　1/16
　　印张：23.75　　　　　　　　　　2022 年 10 月第 1 版
　　字数：681 千字　　　　　　　　 2024 年 4 月北京第 2 次印刷

定价：99.80 元

读者服务热线：（010）81055410　印装质量热线：（010）81055316
反盗版热线：（010）81055315
广告经营许可证：京东市监广登字 20170147 号

前　　言

编写背景

从 Linux 基础指令的使用到 Linux 基础服务的搭建和维护，再到 Shell 编程自动化运维，这是 Linux 系统工程师的学习路线。

本书着重讲解 Linux 基础架构服务，书中所有案例及经验总结均来自笔者近 10 年的一线 Linux 运维实战。因此，可作为一线 Linux 运维人员的实战参考书。

本书特色

本书特色主要有以下几点。

（1）理论丰富，注重实战。

本书的核心理念是理论丰富，注重实战。例如书中的 DNS 服务器部署和运维，读者学完这些内容后可以直接将其应用到企业 DNS 服务器的运维工作中。

（2）结构合理，图文并茂。

一图胜千言，书中对一些晦涩难懂的知识点采用图文并茂的方式讲解，让读者轻松掌握实战知识。

（3）答疑服务。

读者遇到问题时，可以通过本书提供的公众号联系笔者。笔者将在线为读者提供 Linux 系统和架构方面的答疑服务。

本书内容

第 1 章介绍 DNS 的基础知识，以及如何使用 BIND 软件构建 DNS 服务器，可为读者后续的学习打下坚实的基础。

第 2 章结合案例讲解如何使用 DNS 作为负载均衡器，并基于 BIND 软件构建 DNS 主从服务器。

第 3 章以案例的方式讲解 DNS 中的视图应用和如何构建智能 DNS 服务器。

第 4 章详细介绍 DHCP 的工作过程和原理，以及 DHCP 服务的数据报文格式和如何构建 DHCP 服务器。

第 5 章讲解 vsftpd 的基础知识、FTP 架构选型，以及虚拟用户等高级应用。

第 6 章以案例的方式讲解 rsync 指令的具体应用。

第 7 章以案例的方式讲解如何构建安全的文件传输服务。

第 8 章讲解 Samba 服务中 user 和 share 级别的具体配置，以及如何将共享目录映射为 Windows 操作系统的驱动器。

第 9 章详细讲解 LAMP 架构的基本知识和 LAMP 指令。

建议和反馈

读者在学习的过程中若遇到问题,既可以扫描下方二维码,关注笔者的微信公众号,也可以发送邮件到笔者邮箱 3128743705@qq.com、编辑邮箱 zhangtao@ptpress.com.cn,就本书的相关知识与我们交流。

欢迎读者提出宝贵的意见和建议,我们将不胜感激。

致谢

感谢父母给了我生命,让我体会到作为子女的快乐;感谢我的一对儿女,让我体会到做父亲的欣喜和骄傲,是你们让我有了更为强劲的工作动力。

感谢老师和各位朋友给了我非常多的鼓励和大量的写作指导,没有你们的支持和帮助,就没有本书的出版。

注:本书中的一些网址是虚拟的,只起到演示作用,不影响读者阅读。读者学习中有任何问题,可通过前面提供的联系方式,与我们进行交流。

韩艳威

目　　录

第1章　DNS 服务入门 ··· 1
　1.1　网站、域名及 IP 地址基础 ··· 2
　　1.1.1　访问网站的基本流程 ··· 2
　　1.1.2　网站和域名的基础知识 ·· 4
　1.2　hosts 文件 ·· 5
　　1.2.1　hosts 文件概述 ·· 5
　　1.2.2　hosts 文件解析原理 ·· 5
　　1.2.3　Linux 操作系统 hosts 文件详解 ·· 6
　　1.2.4　Linux 操作系统 hosts 文件解析主机名实验 ······································ 7
　　1.2.5　Windows 操作系统配置 hosts 文件详解 ·· 12
　　1.2.6　hosts 文件域名解析实战 ·· 14
　1.3　DNS 服务基础知识 ··· 17
　　1.3.1　DNS 的演变和作用 ··· 17
　　1.3.2　为什么要设置 DNS 服务器地址才能上网 ······································ 18
　　1.3.3　DNS 服务基础概念 ·· 18
　　1.3.4　DNS 服务器作用 ··· 20
　　1.3.5　DNS 服务器组织架构 ··· 21
　　1.3.6　DNS 服务器类型 ··· 23
　　1.3.7　DNS 查询过程 ·· 24
　　1.3.8　DNS 服务器解析类型 ··· 26
　　1.3.9　DNS 服务器各资源记录 ·· 27
　　1.3.10　TTL 值详解 ·· 30
　　1.3.11　DNS 解析配置文件 ··· 30
　　1.3.12　JVM 设定 DNS 缓存时间 ··· 31
　1.4　DNS 服务器部署实战 ·· 31
　　1.4.1　实验环境介绍 ··· 31
　　1.4.2　DNS 服务器安装 BIND 软件 ·· 32
　　1.4.3　启动 BIND 服务器 ··· 34
　　1.4.4　配置 BIND 服务器 ··· 34
　　1.4.5　BIND 主配置文件详解 ·· 38
　　1.4.6　测试主 DNS 服务器 ··· 44
　　1.4.7　保存并备份配置文件 ·· 52
　　1.4.8　named.conf 配置文件详解 ·· 52
　　1.4.9　BIND 服务资源记录 ··· 53
　　1.4.10　BIND 配置 DNS 服务器总结 ·· 53

第2章　DNS 服务进阶 ··· 57
　2.1　BIND 实现网站负载均衡实战 ·· 58
　　2.1.1　主流负载均衡器介绍 ·· 58
　　2.1.2　BIND 实现轮询基础知识 ··· 59

2 | 目录

- 2.1.3 BIND 实现 Web 服务器负载均衡 ·········· 61
- 2.1.4 BIND 实现 Web 服务器负载均衡总结 ·········· 67
- 2.1.5 BIND 实现 DNS 轮询探讨 ·········· 68
- 2.1.6 BIND 实现网站负载均衡深入探讨 ·········· 68
- 2.2 DNS 服务器部署实战 ·········· 72
  - 2.2.1 从 DNS 服务器应用场景 ·········· 72
  - 2.2.2 DNS 主从同步原理 ·········· 72
  - 2.2.3 DNS 主从同步架构选型 ·········· 73
  - 2.2.4 DNS 主从实验环境介绍 ·········· 73
  - 2.2.5 主 DNS 服务器设置 ·········· 73
  - 2.2.6 从 DNS 服务器设置 ·········· 77
  - 2.2.7 主从同步数据的安全性 ·········· 83
  - 2.2.8 DNS 主从配置优化 ·········· 84
  - 2.2.9 DNS 主从搭建总结 ·········· 84
- 2.3 DNS 服务常用分析指令 ·········· 85
  - 2.3.1 DNS 服务查询基础指令 ·········· 85
  - 2.3.2 DNS 高级查询指令之 dig ·········· 86
  - 2.3.3 查询 DNS 服务器记录类型 ·········· 87
  - 2.3.4 DNS 迭代查询的具体流程 ·········· 92
  - 2.3.5 DNS 查询指令之 host 进阶 ·········· 95
  - 2.3.6 DNS 查询指令之 nslookup 进阶 ·········· 96
  - 2.3.7 DNS 服务类型查询指令总结 ·········· 97
- 2.4 用 BIND 实现子域授权和区域转发 ·········· 98
  - 2.4.1 实现 DNS 服务器子域授权 ·········· 98
  - 2.4.2 实现 DNS 服务器域名解析转发 ·········· 103
- 2.5 用 BIND 实现域名解析 ·········· 105
  - 2.5.1 直接域名、泛域名及子域 ·········· 105
  - 2.5.2 直接域名解析实例 ·········· 106
  - 2.5.3 泛域名解析实例 ·········· 112

第 3 章 DNS 服务器核心应用与运维管理 ·········· 117

- 3.1 构建企业级缓存 DNS 服务器 ·········· 118
  - 3.1.1 BIND 缓存基本实现 ·········· 118
  - 3.1.2 DNS 转发器工作原理 ·········· 118
  - 3.1.3 使用 BIND 搭建缓存 DNS 服务器 ·········· 119
- 3.2 BIND 实现智能 DNS 服务器 ·········· 127
  - 3.2.1 智能 DNS 服务器基础知识 ·········· 127
  - 3.2.2 构建智能 DNS 服务器基础环境 ·········· 128
  - 3.2.3 智能 DNS 服务器实现核心步骤 ·········· 129
  - 3.2.4 智能 DNS 服务器核心构建步骤 ·········· 130
  - 3.2.5 测试 BIND 视图 ·········· 137
- 3.3 BIND 日志配置 ·········· 137
  - 3.3.1 BIND 日志概念 ·········· 137
  - 3.3.2 logging 语句 ·········· 138
  - 3.3.3 配置实例 ·········· 139
- 3.4 DNS 与 CDN 企业级缓存架构 ·········· 143
  - 3.4.1 DNS 安全问题 ·········· 143
  - 3.4.2 CDN 基础知识 ·········· 144

3.5　DNS 服务运维技巧 145
　　3.5.1　CNAME 记录和 A 记录 145
　　3.5.2　CNAME 解析运维技巧 146
3.6　DNS 管理工具之 rndc 146
　　3.6.1　rndc 基本环境描述 146
　　3.6.2　配置 rndc 147
　　3.6.3　配置 rndc 本地管理 148
　　3.6.4　配置 rndc 远程管理 149
　　3.6.5　rndc 管理工具常用选项和指令 152
　　3.6.6　管理 DNS 注意事项 152
3.7　TTL 值配置 153
　　3.7.1　TTL 值基础知识 153
　　3.7.2　TTL 值最佳配置实战 153

第 4 章　DHCP 服务器运维实战 155

4.1　DHCP 服务器详解 156
　　4.1.1　DHCP 服务器基础 156
　　4.1.2　DHCP 运行机制 157
　　4.1.3　DHCP 服务器工作原理 158
4.2　DHCP 服务应用场景 162
　　4.2.1　网络与 IP 地址基本管理理念 162
　　4.2.2　DHCP 应用场景解析 163
4.3　DHCP 数据包格式 164
　　4.3.1　DHCP 的封装 164
　　4.3.2　DHCP 数据包本身的报文格式 165
　　4.3.3　DHCP 报文类型简析 168
4.4　DHCP 服务器部署规划 169
　　4.4.1　准备 DHCP 服务器基础环境 169
　　4.4.2　配置网络环境与防火墙 170
　　4.4.3　配置 DHCP 客户端环境信息 174
4.5　CentOS 搭建 DHCP 服务器实战 175
　　4.5.1　DHCP 服务器基本配置 175
　　4.5.2　DHCP 服务器常用操作 179
4.6　DHCP 客户端测试 181
　　4.6.1　DHCP 客户端测试注意事项 181
　　4.6.2　DHCP 客户端测试步骤 182
　　4.6.3　DHCP 运维常用文件/程序/脚本 183

第 5 章　vsftpd 服务 184

5.1　FTP 基础知识 185
　　5.1.1　FTP 服务主动模式 185
　　5.1.2　FTP 服务被动模式 185
　　5.1.3　FTP 软件种类 186
　　5.1.4　FTP 服务器与客户端选型 186
5.2　搭建 vsftpd 服务器 188
　　5.2.1　初始化 vsftpd 服务器运行环境 188
　　5.2.2　安装 vsftpd 软件 190
　　5.2.3　访问 vsftpd 服务 192

- 5.2.4 vsftpd iptables 设置 ... 193
- 5.3 vsftpd 配置文件和日志配置 ... 195
  - 5.3.1 vsftpd 配置文件详解 ... 195
  - 5.3.2 配置 vsftpd 日志 ... 199
- 5.4 vsftpd 匿名用户配置案例 ... 200
  - 5.4.1 vsftpd 服务匿名用户基础配置 ... 200
  - 5.4.2 配置匿名用户上传、下载案例 ... 202
  - 5.4.3 配置匿名用户仅有上传权限案例 ... 203
- 5.5 vsftpd 本地用户 ... 205
  - 5.5.1 本地用户案例 ... 205
  - 5.5.2 配置本地用户经验谈 ... 207
- 5.6 vsftpd 虚拟用户配置案例 ... 208
  - 5.6.1 配置 vsftpd 虚拟用户 ... 208
  - 5.6.2 创建虚拟用户目录 ... 211
  - 5.6.3 验证 vsftpd 服务 ... 213

第 6 章 rsync 服务 ... 216

- 6.1 rsync 基础知识 ... 217
  - 6.1.1 rsync 快速入门 ... 217
  - 6.1.2 rsync 特性和核心算法 ... 221
  - 6.1.3 rsync 基础运维实例 ... 222
- 6.2 rsync 配置文件和选项规则 ... 231
  - 6.2.1 rsync 配置文件 ... 231
  - 6.2.2 rsync 排除和包含文件规则 ... 233
  - 6.2.3 rsync 镜像同步 ... 235
- 6.3 搭建企业级 rsync 备份服务器 ... 244
  - 6.3.1 为什么需要搭建备份服务器 ... 244
  - 6.3.2 rsync 服务端初始化 ... 246
  - 6.3.3 rsync 客户端配置 ... 248
- 6.4 搭建 rsync+inotify 实时备份服务器 ... 255
  - 6.4.1 企业级主流实时同步工具比较 ... 255
  - 6.4.2 rsync+inotify 组合基础知识 ... 256
  - 6.4.3 inotifywait 实时同步企业级案例 ... 257
- 6.5 Lsyncd 实时同步详解 ... 262
  - 6.5.1 安装 Lsyncd ... 262
  - 6.5.2 配置 Lsyncd ... 264
  - 6.5.3 本机同步设置 ... 265
  - 6.5.4 远程同步设置 ... 267

第 7 章 SFTP 服务 ... 271

- 7.1 构建 SFTP 服务运行环境 ... 272
  - 7.1.1 初始化 SFTP 服务器 ... 272
  - 7.1.2 初始化 SFTP 用户运行环境 ... 275
- 7.2 搭建 SFTP 服务 ... 276
  - 7.2.1 基本配置 ... 276
  - 7.2.2 安全设置 ... 277
  - 7.2.3 验证 SFTP 环境 ... 279
  - 7.2.4 开启 SFTP 服务日志记录 ... 280

	7.2.5	SFTP 服务基础环境初始化	281
	7.2.6	192.168.2.172 搭建 SFTP 服务	283
	7.2.7	创建 SFTP 服务的用户和组	285
	7.2.8	配置双机互信	288
7.3	SFTP 服务配置文件对比		289
	7.3.1	192.168.2.171 配置文件	289
	7.3.2	192.168.2.172 配置文件	291
	7.3.3	192.168.2.173 配置文件	294

第 8 章 Samba 服务 298

8.1	搭建基本的 Samba 服务器		299
	8.1.1	Samba 简介	299
	8.1.2	构建 Samba 服务器环境	299
	8.1.3	Samba 服务器组件说明	303
	8.1.4	配置 Samba 服务器	304
	8.1.5	用户权限与配置文件	308
	8.1.6	Windows 客户端访问 Samba 服务器	312
8.2	Samba 服务之 user 配置案例		314
	8.2.1	案例需求及其分析	314
	8.2.2	初始化 Samba 服务器	314
	8.2.3	配置 Samba 服务器	316
8.3	Samba 服务之 share 配置案例		318
	8.3.1	Samba 服务需求及分析	318
	8.3.2	初始化 Samba 服务器	318
	8.3.3	配置 Samba 服务器	320

第 9 章 网站架构之 LAMP 323

9.1	LAMP 架构安装前基本规划		324
	9.1.1	LAMP 基础知识	324
	9.1.2	LAMP 架构数据流	324
9.2	安装 LAMP		325
	9.2.1	环境规划	325
	9.2.2	安装 httpd	326
	9.2.3	安装 PHP	327
	9.2.4	安装 MariaDB	329
	9.2.5	LAMP 常用运维指令	335
9.3	优化编译安装 LAMP 架构		336
	9.3.1	配置 LAMP 运行环境	336
	9.3.2	为什么要编译 LAMP	336
9.4	高标准编译安装 Apache		337
	9.4.1	彻底隐藏 Apache 版本	337
	9.4.2	安装 httpd 依赖包	339
	9.4.3	Apache 2.4 编译参数详解	343
	9.4.4	编译安装 Apache 2.4	346
9.5	高标准安装 MySQL Percona		349
	9.5.1	为什么要使用 Percona 版本	349
	9.5.2	优化 Percona 5.6.28 运行环境	350
	9.5.3	初始化 MySQL	352

  9.5.4　导出 MySQL 头文件和库文件 ······353
  9.5.5　安装 MySQL 总结 ······353
9.6　高标准编译安装 PHP ······356
  9.6.1　构建 PHP 基础环境 ······356
  9.6.2　配置 PHP ······359
9.7　使用 WordPress 搭建企业级站点优化建站环境 ······364

第 1 章
DNS 服务入门

1.1 网站、域名及 IP 地址基础

在互联网通信技术领域中，计算机网络中的各主机节点间通信是通过 IP 地址实现的，IP 地址通常由 4 组数字加"."组成。Linux 系统管理员管理服务器时，很难让每台主机都接收或长期记忆几十甚至几百个 IP 地址。随着技术的不断更新和发展，网络上出现了一套针对此问题的解决方案，即域名系统（Domain Name System，DNS）。它能够将数以千万计的 IP 地址通过查询数据库，轻松地转化成与之对应的名称字符串（域名），相对于 IP 地址而言，用英文字母组成的域名的结构形式更符合人类的记忆和阅读习惯。

IPv4 虽然是互联网协议（Internet Protocol，IP）的第 4 版，却是第一个被广泛使用并成为当今互联网技术基石之一的协议。它的作用是为每一个网络和每一台主机分配一个 IP 地址。IP 地址是一个 32 位的二进制数，共有 2^{32} 个地址，看起来好像非常多。但是在 5G 时代，智能手机、智能手环、智能手表、智能运动相机、路由器等智能终端设备都会占据一个 IP 地址，甚至网络电视机、智能冰箱、智能洗衣机、智能空调、智能门锁、智能监控系统等能接入互联网的智能家电和家具，也都需要一个 IP 地址。

2019 年 11 月 26 日，所有 IPv4 地址已经分配完毕，这意味着没有更多的 IPv4 地址可以分配给因特网服务提供方（Internet Service Provider，ISP）或其他大型网络基础设施提供商。

国际互联网工程任务组为了应对 IPv4 地址分配完毕的局面，推出了 IPv4 的升级版——IPv6。海量的 IPv6 地址除了能解决分配不足的问题，还有利于防止扫描攻击。网络攻击者通过地址扫描识别用户的地理位置，发现漏洞并入侵。目前的技术可实现在 45min 内扫描 IPv4 的全部地址空间。而每一个 IPv6 地址有 128 位，假设网络前缀有 64 位，那么在一个子网中就会存在 2^{64} 个地址。如果网络攻击者以每秒百万个地址的速度扫描，需要 50 万年才能遍历所有的地址，这无疑将显著增强网站与用户终端设备的安全性。

1.1.1 访问网站的基本流程

用户访问一个网站其实是访问该网站的服务器，服务器一般被各个公司托管在 IDC 机房或自建机房内。当用户通过浏览器或 App 访问对应的网站服务器时，首先需要知道该网站的地址。该地址本质上是一串数字，它对应的专业术语是 IP 地址。当用户需要访问网站内容时，只需输入该网站的 IP 地址就可以浏览网站内容。使用 IP 地址访问网站，如图 1-1 所示。

用户使用 IP 地址访问对应的网站，而 IP 地址不方便记忆。试想，如果用户经常需要访问的网站有 100 个，那么用户就需要记住 100 个 IP 地址，非专业管理人员很难记住这么多数字，所以就有了网站的域名。用户只需在浏览器地址栏中输入要访问的网站域名，就可以访问相应的网站服务器，如图 1-2 所示。

当用户通过浏览器输入域名访问网站时，浏览器如何知道用户输入的域名对应的 IP 地址呢？计算机中的 hosts 文件相当于手机中的联系人功能，记录了网站**域名对应的 IP 地址映射关系**。当用户访问 ptpress.com 网站时，计算机首先会在 hosts 文件里面查找是否有该网站域名对应的 IP 地址信息。如果有域名对应的 IP 地址信息，那么用户在浏览器地址栏中直接输入 ptpress.com 即可访问该网站，而不需要输入一长串难以记忆的数字。

1.1 网站、域名及 IP 地址基础 | 3

图 1-1　使用 IP 地址访问网站　　　　图 1-2　使用域名访问网站

随着互联网技术的不断发展，出现了图 1-3 所示的**多域名体系结构**，即不同的域名对应不同的 IP 地址。

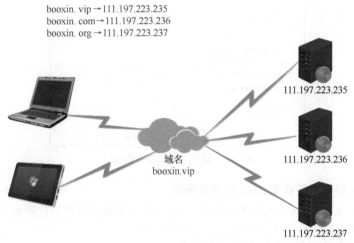

图 1-3　多域名体系结构

网站的域名越来越多，导致计算机本地的 hosts 文件解析条目随之增加，hosts 文件变得越来越庞大。全世界网站的个数在不断增加，而本地的 hosts 文件解析记录数据库中，域名对应的 IP 地址不一定能及时更新到最新状态。就像我们不可能把全世界所有电话号码都存到手机联系人里，并实时更新联系人一样。因此**分布式 DNS 服务器**应运而生。

当用户访问一个网站时，如果用户计算机的本地 hosts 文件里面没有该网站的域名和 IP 地址的映射关系，则会将该请求转到本地 DNS 服务器，查询到该网站域名对应的 IP 地址后就可以访问该网站了。就像我们不知道店铺 A 的电话号码，发现手机联系人（hosts 文件）里面也没有留存，然后拨打 114 查号台（DNS 服务器）查询一样，最后通过 DNS 服务器**递归查询**帮助我们查找与域名对应的 IP 地址。

如果用户在家或在公司上网时，遇到莫名跳转到其他网站或者弹出广告之类的情况，有可能是攻击者对**上游**的 **DNS** 动了手脚。比如，一个网站域名对应的 IP 地址如果被攻击者篡改成其他的 IP 地址，当用户输入该域名时就会遇到打开的网页是其他网页的情况，该情况是 **DNS 劫持**的一种表现。

如何避免 DNS 劫持的情况产生？当通过本地浏览器访问对应的网站内容时，首先需要访问本地运营商的 DNS 服务器。因此，在遇到 DNS 劫持等问题时，除向客服反映之外，还可以使用一些"靠谱"的第三方 DNS 服务器。下面将介绍如何设置第三方 DNS 服务器地址。

1. 为 Windows 操作系统设置 DNS 服务器地址

以 Windows 10 操作系统为例，先打开"网络和共享中心"，单击"更改适配器设置"，然后右击正在使用的连接，单击"属性"，找到"Internet 协议版本 4(TCP/IPv4)"，双击该项目或者单击"属性"，选择"使用下面的 DNS 服务器地址"单选按钮，填入第三方 DNS 服务器地址，单击"确定"按钮即可，如图 1-4 所示。

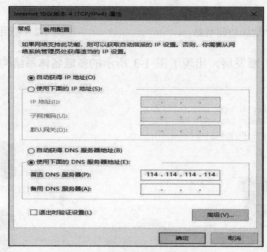

图 1-4　设置 DNS 服务器地址

2. 为 Linux 操作系统设置 DNS 服务器地址

以 CentOS 6 或 CentOS 7 操作系统为例，设置 DNS 服务器地址的代码如下。

```
1 [root@www.booxin*.com ~]#cat /etc/resolv.conf
2 ; generated by /sbin/dhclient-script
3 nameserver 114.114.114.114
4 nameserver 202.106.195.68
```

上述代码第 3、4 行自定义 DNS 服务器地址。开发人员可根据当前的运营商网络环境设置适合自己的 DNS 服务器地址。

1.1.2　网站和域名的基础知识

1. 网站基础

网站是指在互联网上，根据一定的规则和标准，使用 HTML、CSS、JavaScript 等制作的用于展示特定内容的相关网页的集合。人们可以通过网站来发布自己想要公开的信息，或者通过网站来提供相关的网上服务，也可以通过浏览器来访问网站，获取自己需要的信息或者享受共享服务。

一个完整的网站由 HTML 页面、网站程序、域名、IP 地址、服务器等元素组成，其中网站程序、服务器、域名和 IP 地址是网站的重要元素，缺一不可。

目前以 HTML5 和 CSS3 等 W3C 标准技术创建的网站更注重用户体验和产品体验，功能更全面、更强大，无论是从网站的响应速度、还是从布局设计、视觉效果等方面来看，都越来越完善。

现在的网站更注重用户交互和用户体验，如在网站加入智能机器人、小视频、VR 技术等，以便让用户更快地了解某种产品或功能，提供更丰富的交互功能和改善用户体验。

2. 域名基础

域名（Domain Name）是互联网上的一个服务器或一个网络系统的名字。

在全世界范围内，不允许有重复的域名。域名由若干个英文字母、数字和"−"组成，由"."分隔成几部分。

如果网站是一座房子，那么域名就是这座房子的门牌号。若要创建一个网站，第一步就是买一个域名，让这座房子有一个属于自己的门牌号。所以域名的申请、购买，以及备案是建立一个网站的必需条件。域名是网站在互联网世界里的一张名片。企业拥有易于记忆和阅读的域名，是建立良好品牌的重要步骤，因此域名的商业价值对于企业非常重要。

1.2 hosts 文件

1.2.1 hosts 文件概述

hosts 文件是早期各操作系统（macOS、Windows、Linux 等）中解析域名和 IP 地址对应关系的主要工具。

hosts 文件的主要作用是定义 **IP 地址**和**域名之间**的对应关系，即 IP 地址和域名的一对一或一对多的关系。用文本处理工具可以打开和编辑 hosts 文件。

当用户在浏览器地址栏中输入一个需要访问的网站域名时，系统会首先从 hosts 文件中查找域名对应的 IP 地址。若找到对应的解析记录，浏览器会立即打开对应网页；若没有找到解析记录，浏览器会将域名提交到 DNS 服务器并进行域名解析，这也是快速打开网页的方法。

技术层面上，hosts 文件是本地的 IP 地址和域名的解析文件（本地局域网的 DNS 解析文件）。

1.2.2 hosts 文件解析原理

计算机中的 hosts 文件是一个用于存储计算机网络中各节点信息的**文本文件**，该文件负责将域名对应到相应的 IP 地址。

hosts 文件通常用于**补充**或**取代**网络中 DNS 服务器的功能。与分布式多域名体系结构不同的是，用户可以**直接对本地计算机中的 hosts 文件**进行**控制操作**（增、删、改、查）。为了方便用户记忆，系统管理员将 IP 地址与域名对应表保存到 hosts 文件中。当用户通过浏览器输入域名访问网站时，需要先将域名解析成 IP 地址。

DNS 服务器的作用是把域名解析为对应的 IP 地址，而 hosts 文件可以**提高解析效率**。用户通过浏览器向 DNS 服务器发出请求之前，系统会检查 hosts 文件中是否有该 IP 地址与域名的对应关系。如果有对应关系，则将域名对应到 IP 地址；如果没有对应关系，则需要向已知的 DNS 服务器发出域名解析请求，因为 hosts 文件的**请求级别比 DNS 服务器的高**。

当计算机中的 hosts 文件中有域名和 IP 地址的对应关系时，浏览器就会直接访问 hosts 文件中域名对应的 IP 地址，而不再查找 DNS 服务器中的记录。

hosts 文件域名解析的优先级最高（默认值），读者可以根据需要进行修改，但一般不建议读者对域名解析的优先级配置文件进行修改。

1.2.3　Linux 操作系统 hosts 文件详解

1．IP 地址基础

无论是在公网环境中还是在局域网环境中，每台主机都会有一个或多个 IP 地址，IP 地址是互联网体系结构中主机的"门牌号"。

公网 IP 地址不方便记忆，所以就出现了域名。全世界都可以访问的域名只有公网中才存在，每个域名对应一个公网 IP 地址，但一个 IP 地址又可以对应多个域名。

在计算机网络通信环境中，每台计算机都有一个主机名，系统管理员通常为每台计算机设置主机名，便于计算机间相互访问。我们在局域网中可以根据每台计算机的功能为其命名，根据 Linux 操作系统版本的不同，域名相关的配置文件分别在/etc/hosts 文件（如 CentOS 6 系列操作系统）和/etc/sysconfig/network 文件（如 CentOS 7 系列操作系统）中进行配置。

2．/etc/hosts 文件的格式

Linux 操作系统中，/etc/hosts 文件可用来把主机名对应到 IP 地址，记录的是本地主机的对应关系。每台计算机都是独立的，所有的计算机都不能通过主机名来互相访问，/etc/hosts 文件的格式如下。

```
127.0.0.1    localhost localhost.localdomain localhost4 localhost4.localdomain4
127.0.0.1    localhost.localdomain localhost
192.168.1.100 www.booxin*.vip
```

/etc/hosts 文件中每行为一个主机条目信息，由两部分内容组成，每个部分由空格隔开。以"#"开头的行为注释，不被系统解释和执行。各部分介绍如下。

- 第一部分：IP 地址。
- 第二部分：主机名.域名。注意主机名和域名之间的"."。

假设公司有 3 台服务器，我们按照给每台服务器分配任务的不同，将它们分为 FTP 服务器、Web 服务器、MySQL 服务器，建立 IP 地址与主机名的数据库解析记录信息的代码如下。

```
127.0.0.1 localhost.localdomain localhost
192.168.1.2 ftp.localdomain ftp
192.168.1.3 www.localdomain www
192.168.1.4 mysql.localdomin mysql
```

将上述代码写入每台服务器的/etc/hosts 文件中，这样这 3 台局域网内的服务器就可以通过主机名来互相访问。

3．主机名和域名的区别

主机名（Host Name）是计算机本身的名字，域名需要被解析为具体的 IP 地址。在局域网中，主机名是可以解析为 IP 地址的。比较重要的一点是：主机名一般用于**局域网**内部的名称解析；而域名用于对应**公网**的 IP 地址，一般属于分布式解析，因为 DNS 本身属于分布式文件系统。

（1）要在 Linux 操作系统下显示主机名，代码如下。

```
[root@www.booxin*.com ~]#hostname
www.booxin*.com
```

该服务器的主机名是 www.booxin*.com，hostname 指令默认不添加任何参数来显示当前操作的服务器的主机名。

（2）临时设置主机名。

使用 hostname 指令后接要设置的主机名，可以设置当前服务器的主机名。要设置主机名为

booxin*.com，代码如下。

```
[root@www*.booxin*.com ~]#hostname booxin*.com
[root@www*.booxin*.com ~]#logout
```

通过 hostname 指令设置主机名只是临时生效，下次重启服务器后，临时设定的主机名将会失效，因此建议永久修改主机名，这需要在/etc/hosts 配置文件中进行设定。

（3）显示主机 IP 地址。

显示当前主机的所有 IP 地址，代码如下。

```
[root@booxin*.com ~]#hostname -i
127.0.0.1 192.168.2.30
```

上述代码表示当前主机的 IP 地址分别是 127.0.0.1 和 192.168.2.30。

4．/etc/hosts 文件作用

企业级办公环境中开发人员或工程师经常需要使用非生产机器程序运行环境来模拟生产机器程序运行环境。如需要在办公网环境中部署测试和研发环境，通信双方在同一个局域网内想通过局域网内的域名相互访问，只需修改自己的/etc/hosts 文件内容即可，该方法一般适用于测试和准生产环境，生产环境中建议使用 DNS。

1.2.4　Linux 操作系统 hosts 文件解析主机名实验

1．设定集群服务器主机名

任务是为集群内的服务器设定主机名。

利用 hosts 文件来解析自己的集群中所有的主机名，CentOS 7 系列操作系统可以使用 hostnamectl set-hostname 指令设定主机名，因此需要先准备两台虚拟机，如图 1-5 所示。

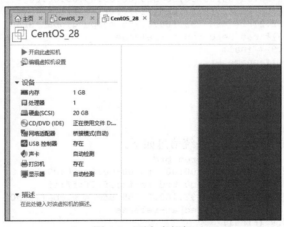

图 1-5　两台虚拟机

这里采用 CentOS 6.9 操作系统，hosts 主机设定架构规划如表 1-1 所示。

表 1-1　hosts 主机设定架构规划

操作系统	主机名	IP 地址	防火墙和 SELinux 是否关闭
CentOS 6.9 x86_64	CentOS_27	192.168.2.27	关闭
CentOS 6.9 x86_64	CentOS_28	192.168.2.28	关闭

2．架构

基于 hosts 解析网络文件系统（Network File System，NFS）服务主机名实验架构如图 1-6

所示。

（1）基本架构说明。

NFS 服务器 IP 地址为 192.168.2.27，NFS 客户端 IP 地址为 192.168.2.28。

（2）实验目的。

192.168.2.27 通过主机名挂载 192.168.2.28 共享的目录。

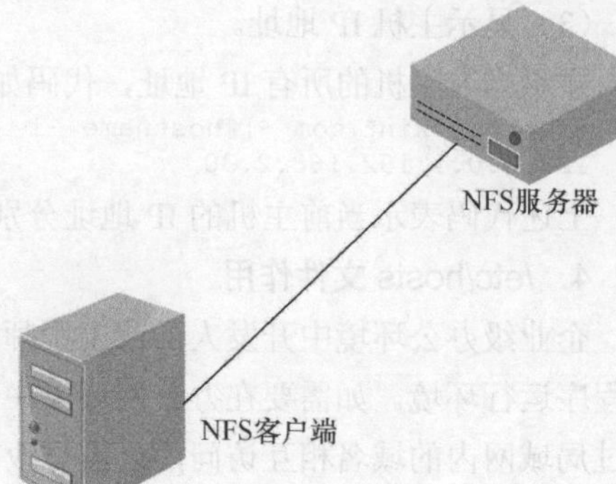

3. 确认服务器信息

NFS 服务器（192.168.2.27）环境设置如下。

要使用定时任务加入时间同步，代码如下。

```
[root@CentOS_27 ~]# crontab -l
00 00 * * * ntpdate ntp6.aliyun.com > /dev/null 2>&1
[root@CentOS_28 ~]# crontab -l
00 00 * * * ntpdate ntp5.aliyun.com > /dev/null 2>&1
```

要使用阿里云时间服务器进行同步，代码如下。

ntp1.aliyun.com
ntp2.aliyun.com
ntp3.aliyun.com
ntp4.aliyun.com
ntp5.aliyun.com
ntp6.aliyun.com
ntp7.aliyun.com

图 1-6　基于 hosts 解析 NFS 服务主机名实验架构

192.168.2.27 服务器显示的操作系统信息如下。

```
[root@CentOS_27 ~]# ip addr |grep brd
    link/loopback 00:00:00:00:00:00 brd 00:00:00:00:00:00
    link/ether 00:0c:29:22:f9:f7 brd ff:ff:ff:ff:ff:ff
    inet 192.168.2.27/24 brd 192.168.2.255 scope global eth0
[root@CentOS_27 ~]# cat /etc/redhat-release
CentOS release 6.9 (Final)
[root@CentOS_27 ~]# uname -r
2.6.32-696.el6.x86_64
[root@CentOS_27 ~]# uname -m
x86_64
[root@CentOS_27 ~]# uname -n
CentOS_27
```

192.168.2.28 客户端显示的操作系统信息如下。

```
[root@CentOS_28 ~]# ip addr |grep brd
    link/loopback 00:00:00:00:00:00 brd 00:00:00:00:00:00
    link/ether 00:0c:29:d5:00:b0 brd ff:ff:ff:ff:ff:ff
    inet 192.168.2.28/24 brd 192.168.2.255 scope global eth0
[root@CentOS_28 ~]# cat /etc/redhat-release
CentOS release 6.9 (Final)
[root@CentOS_28 ~]# uname -r
2.6.32-696.el6.x86_64
[root@CentOS_28 ~]# uname -n
CentOS_28
[root@CentOS_28 ~]# uname -m
x86_64
```

4. 设置其他信息

设置服务器主机名、防火墙等信息，具体代码如下。

```
1   [root@localhost ~]# hostname CentOS_27
2   [root@localhost ~]# logout
3   Connection closing...Socket close.
4
5   Connection closed by foreign host.
```

```
 6
 7  Disconnected from remote host(CentOS_127) at 16:13:54.
 8
 9  Type 'help' to learn how to use Xshell prompt.
10  [D:\~]$
11
12  Connecting to 192.168.2.27:22...
13  Connection established.
14  To escape to local shell, press 'Ctrl+Alt+]'.
15
16  Last login: Sun Jun 30 16:09:33 2019 from 192.168.2.3
17  [root@CentOS_27 ~]# sed -i 's#HOSTNAME=*.*#HOSTNAME=CentOS_27#' /etc/
    sysconfig/network
18  [root@CentOS_27 ~]# grep HOSTNAME /etc/sysconfig/network
19  HOSTNAME=CentOS_127
20
21  [root@CentOS_27 ~]# setenforce  0
22  [root@CentOS_27 ~]# sed -i 's#SELINUX=enforcing#SELINUX=disabled#' /etc/
    SELINUX/config
23  [root@CentOS_27 ~]# /etc/init.d/iptables stop
24  iptables: ACCEPT: filter                                    [确定]
25  iptables: 清除防火墙规则：                                     [确定]
26  iptables: 正在卸载模块：                                       [确定]
27  [root@CentOS_27 ~]# chkconfig iptables off
28  [root@CentOS_27 ~]# reboot
29
30  Broadcast message from root@CentOS_27
31        (/dev/pts/0) at 16:22 ...
32
33  The system is going down for reboot NOW!
34  [root@CentOS_27 ~]# Connection closing...Socket close.
35
36  Connection closed by foreign host.
37
38  Disconnected from remote host(CentOS_27) at 16:22:59.
39
40  Type 'help' to learn how to use Xshell prompt.
41  [D:\~]$
42
43  Connecting to 192.168.2.27:22...
44  Connection established.
45  To escape to local shell, press 'Ctrl+Alt+]'.
46
47  Last login: Sun Jun 30 16:21:17 2019 from 192.168.2.3
```

上述代码中，第 1 行设置临时主机名，重启操作系统后会失效。第 17 行使用 sed 指令修改配置文件，即永久修改主机名。第 21～22 行修改配置文件，永久关闭 SELinux。第 23 和 27 行临时关闭防火墙。第 28 行使用 reboot 指令重启操作系统。

CentOS 重启后，要确认防火墙、主机名、SELinux 等设置是否正确，代码如下。

```
[root@localhost ~]# hostname CentOS_28
[root@localhost ~]# logout
Connection closing...Socket close.

Connection closed by foreign host.

Disconnected from remote host(CentOS_28) at 16:14:10.

Type 'help' to learn how to use Xshell prompt.
[D:\~]$

Connecting to 192.168.2.28:22...
Connection established.
To escape to local shell, press 'Ctrl+Alt+]'.
```

```
Last login: Sun Jun 30 16:09:06 2019 from 192.168.2.3
[root@CentOS_28 ~]# sed -i 's#HOSTNAME=*.*#HOSTNAME=CentOS_28#' /etc/sysconfig/network
[root@CentOS_28 ~]# cat /etc/sysconfig/network
NETWORKING=yes
HOSTNAME=CentOS_28
[root@CentOS_28 ~]# sed -i 's#SELINUX=enforcing#SELINUX=disabled#' /etc/SELINUX/config
[root@CentOS_28 ~]# /etc/init.d/iptables stop
iptables：将链设置为政策 ACCEPT: filter                  [确定]
iptables：清除防火墙规则：                              [确定]
iptables：正在卸载模块：                                [确定]
[root@CentOS_28 ~]# chkconfig iptables off
[root@CentOS_28 ~]# reboot

Broadcast message from root@CentOS_28
    (/dev/pts/0) at 16:24 ...

The system is going down for reboot NOW!
[root@CentOS_28 ~]# Connection closing...Socket close.

Connection closed by foreign host.

Disconnected from remote host(CentOS_28) at 16:24:58.

Type 'help' to learn how to use Xshell prompt.
```

从上述代码中可以看到，两台虚拟机的主机名和防火墙等已经设置完毕。

5. 将主机名写入配置文件

更改/etc/hosts 文件让两台虚拟机可以互相识别主机名。在 CentOS_27 主机上更改文件内容，代码如下。

```
[root@CentOS_27 ~]#echo " " >>/etc/hosts
[root@CentOS_27 ~]#echo "#2018/12/23 hanyanwei" >>/etc/hosts
[root@CentOS_27 ~]#echo "192.168.2.27 CentOS_27" >>/etc/hosts
[root@CentOS_27 ~]#echo "192.168.2.28 CentOS_28" >>/etc/hosts
[root@CentOS_27 ~]#tail -5 /etc/hosts
::1        localhost localhost.localdomain localhost6 localhost6.localdomain6

#2018/12/23 hanyanwei
192.168.2.27 CentOS_27
192.168.2.28 CentOS_28
```

上述代码使用重定向的方式修改主机名。在 CentOS_28 主机上更改文件内容的代码如下。

```
[root@CentOS_28 ~]#echo " " >>/etc/hosts
[root@CentOS_28 ~]#echo "#2018/12/23 hanyanwei" >>/etc/hosts
[root@CentOS_28 ~]#echo "192.168.2.27 CentOS_27" >>/etc/hosts
[root@CentOS_28 ~]#echo "192.168.2.28 CentOS_28" >>/etc/hosts
[root@CentOS_28 ~]#tail -5 /etc/hosts
::1        localhost localhost.localdomain localhost6 localhost6.localdomain6

#2018/12/23 hanyanwei
192.168.2.27 CentOS_127
192.168.2.28 CentOS_128
```

6. 使用 ping 指令验证主机名设置是否正确

在 192.168.2.27 主机上验证 CentOS_28 主机名解析是否正确，代码与验证结果如下。

```
[root@CentOS_27 ~]# ping CentOS_28 -c 3
PING CentOS_28 (192.168.2.28) 56(84) bytes of data.
64 bytes from CentOS_28 (192.168.2.28): icmp_seq=1 ttl=64 time=0.184 ms
64 bytes from CentOS_28 (192.168.2.28): icmp_seq=2 ttl=64 time=0.527 ms
64 bytes from CentOS_28 (192.168.2.28): icmp_seq=3 ttl=64 time=0.623 ms

--- CentOS_28 ping statistics ---
3 packets transmitted, 3 received, 0% packet loss, time 2002ms
```

```
rtt min/avg/max/mdev = 0.184/0.444/0.623/0.190 ms
```
在 192.168.2.28 主机上验证 CentOS_27 主机名解析是否正确，代码与验证结果如下。
```
[root@CentOS_28 ~]# ping CentOS_27 -c 3
PING CentOS_27 (192.168.2.27) 56(84) bytes of data.
64 bytes from CentOS_27 (192.168.2.27): icmp_seq=1 ttl=64 time=0.181 ms
64 bytes from CentOS_27 (192.168.2.27): icmp_seq=2 ttl=64 time=0.487 ms
64 bytes from CentOS_27 (192.168.2.27): icmp_seq=3 ttl=64 time=0.645 ms

--- CentOS_27 ping statistics ---
3 packets transmitted, 3 received, 0% packet loss, time 2002ms
rtt min/avg/max/mdev = 0.181/0.437/0.645/0.194 ms
```
上述代码使用 ping 指令验证主机名设置是否正确。

7. 设置 NFS 服务器

（1）安装 NFS 软件包，代码如下。
```
[root@CentOS_28 ~]# yum -y install nfs-utils nfs portmap rpcBind >/dev/null 2>&1
```
（2）配置共享文件，代码如下。
```
[root@CentOS_28 ~]# cat /etc/exports
/hanyanwei     192.168.2.0/24(rw,sync)
```
（3）创建共享目录，代码如下。
```
[root@CentOS_28 ~]# mkdir -pv /hanyanwei
mkdir: 已创建目录 "/hanyanwei"
```
（4）重启服务，代码如下。
```
[root@CentOS_28 ~]# /etc/init.d/rpcBind restart
停止 rpcBind：                                           [确定]
正在启动 rpcBind：                                       [确定]
[root@CentOS_28 ~]# /etc/init.d/nfs restart
关闭 NFS 守护进程：                                      [确定]
关闭 NFS mountd：                                        [确定]
关闭 NFS 服务：                                          [确定]
Shutting down RPC idmapd：                               [确定]
启动 NFS 服务：                                          [确定]
启动 NFS mountd：                                        [确定]
启动 NFS 守护进程：                                      [确定]
正在启动 RPC idmapd：                                    [确定]
```

注意：先重启 RPCBind 服务，再重启 NFS 服务。

8. NFS 客户端设置

（1）安装 NFS、RPCBind 软件，代码如下。
```
[root@CentOS_27 ~]# yum install nfs-utils rpcbind -y >/dev/null 2>&1
```
（2）启动 RPCBind 服务，代码如下。
```
[root@CentOS_27 ~]# /etc/init.d/rpcbind start
[root@CentOS_27 ~]# showmount -e 192.168.2.28
Export list for 192.168.2.28:
/hanyanwei 192.168.2.0/24
```
CentOS 6 系列使用/etc/init.d/rpcbind 启动 RPCBind 服务。

9. 使用主机名将 CentOS_28 的 NFS 目录重新挂载

（1）使用 IP 地址验证挂载的远程目录。
```
[root@CentOS_27 ~]#mount 192.168.2.28:/hanyanwei /test/
[root@CentOS_27 ~]#df -Th
Filesystem         Type    Size  Used Avail Use% Mounted on
/dev/mapper/vg_www-lv_root
                   ext4    8.3G  801M  7.1G  10% /
tmpfs              tmpfs   931M     0  931M   0% /dev/shm
/dev/sda1          ext4    477M   28M  425M   7% /boot
192.168.2.28:/hanyanwei
                   nfs     8.3G  801M  7.1G  10% /test
```

```
[root@CentOS_27 ~]#umount /test/
```
上述代码验证挂载目录是否正确。

（2）使用主机名验证挂载的远程目录。
```
[root@CentOS_27 ~]#mount CentOS_28:/hanyanwei /test/
[root@CentOS_27 ~]#df -Th
Filesystem           Type    Size  Used Avail Use% Mounted on
/dev/mapper/vg_www-lv_root
                     ext4    8.3G  801M  7.1G  10% /
tmpfs                tmpfs   931M     0  931M   0% /dev/shm
/dev/sda1            ext4    477M   28M  425M   7% /boot
CentOS_28:/hanyanwei
                     nfs     8.3G  801M  7.1G  10% /test
```
从上述代码中可以看到，使用 IP 地址和使用主机名的效果是一样的。

10. 注意事项

为 NFS 客户端 192.168.2.28 设置 NFS 开机启动，代码如下。
```
[root@CentOS_28 ~]#chkconfig nfs --level 3 on
```
确认 NFS 服务开机启动设置是否生效，代码如下。
```
[root@CentOS_28 ~]#chkconfig --list |grep nfs
nfs                0:off   1:off   2:on    3:on    4:on    5:on    6:off
nfslock            0:off   1:off   2:off   3:on    4:on    5:on    6:off
```
为 NFS 客户端设置 NFS 挂载开机启动，代码如下。
```
[root@CentOS_27 ~]#echo "mount 192.168.2.28:/hanyanwei /test/" >>/etc/rc.local
[root@CentOS_27 ~]#tail -1 /etc/rc.local
mount 192.168.2.28:/hanyanwei /test/
```
查看 NFS 客户端设置 NFS 开机启动是否成功，代码如下。
```
[root@CentOS_27 ~]#shutdown -r 1

Broadcast message from root@CentOS_27
    (/dev/pts/0) at 22:30 ...

The system is going down for reboot in 1 minute!
[root@CentOS_27 ~]#df -Th
Filesystem           Type    Size  Used Avail Use% Mounted on
/dev/mapper/vg_www-lv_root
                     ext4    8.3G  799M  7.1G  10% /
tmpfs                tmpfs   931M     0  931M   0% /dev/shm
/dev/sda1            ext4    477M   28M  425M   7% /boot
192.168.2.28:/hanyanwei
                     nfs     8.3G  799M  7.1G  10% /test
```
从上述代码的输出结果中可以看到，NFS 客户端挂载开机启动设置已经被成功执行。

11. 实验总结

通过上述实验，得到如下结果。

当服务器使用两台或者多台主机作为测试环境部署服务时，可使用主机名解析对应的 IP 地址。相对于 IP 地址，域名更易于用户阅读和记忆，且当服务器的 IP 地址发生变动时，不需要对客户端做任何修改，只需要在服务器中写入主机名对应的新 IP 地址即可。

1.2.5 Windows 操作系统配置 hosts 文件详解

1. Windows 操作系统 hosts 文件基础

在 Windows 操作系统中设置本地主机名解析时，首先在网络环境正常的条件下找到本地计算机的 hosts 文件。该文件的路径信息代码如下。
```
C:\Windows\System32\drivers\etc
```

详细信息如图 1-7 和图 1-8 所示。

图 1-7　Windows 操作系统 hosts 文件

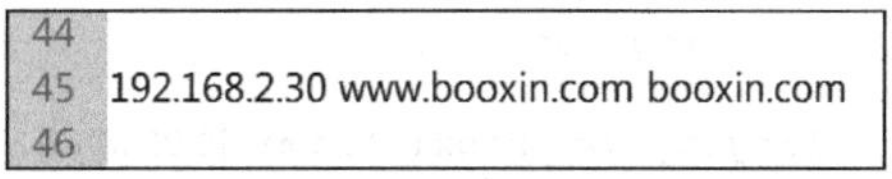

图 1-8　Windows 操作系统 hosts 文件内容

用户通过浏览器访问 www.booxin*.com 网站时，如果在 hosts 文件中找到 www.booxin*.com 域名的对应 IP 地址，就把真实的用户请求发送到该 IP 地址对应的服务器；如果 hosts 文件中没有该域名的解析信息，则向外部的 DNS 服务器发起查询请求。如果该域名在供应商的 DNS 服务器注册过，则 DNS 服务器返回该域名的对应 IP 地址。

在局域网中设计集群架构时，可以先在每台服务器上的 hosts 文件中写入每台服务器对应的主机名和 IP 地址的对应关系，然后可以通过主机名相互访问，不需要记住每台计算机的域名。

2．如何配置 hosts 文件

（1）hosts 文件在 Windows 操作系统上的路径是 C:\Windows\System32\drivers\etc，我们在该路径中找到 hosts 文件并用记事本打开它。

（2）按照"**IP 地址　域名**"的格式添加单独的一行记录，记录的内容如下。

```
192.168.2.30 www.booxin*.com booxin*.com
```

设置主机名注意事项如下。

- IP 地址前面不要有空格，IP 地址和域名之间，要有至少 1 个空格。
- 修改后，一定要记得保存文件使配置生效。

（3）确认主机名解析配置是否生效。

在安装有 Windows 操作系统的计算机上，徽标键与"R"键组合使用，可以快速打开计算机"运行"窗口，然后输入 cmd 指令，即可进入 Windows 操作系统的运行和命令行模式，如图 1-9 和图 1-10 所示。

图 1-9　进入 Windows 操作系统的运行模式

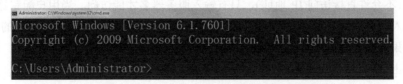

图 1-10　进入 Windows 操作系统的命令行模式

进入 Windows 操作系统的命令行模式后，使用 ping www.booxin*.com 指令，按"Enter"键，查看结果，显示结果代码如下。

```
C:\>ping www.booxin*.com

Pinging www.booxin*.com [192.168.2.30] with 32 bytes of data:

Reply from 192.168.2.30: bytes=32 time<1ms TTL=64
Reply from 192.168.2.30: bytes=32 time<1ms TTL=64
```

```
Reply from 192.168.2.30: bytes=32 time<1ms TTL=64
Reply from 192.168.2.30: bytes=32 time<1ms TTL=64

Ping statistics for 192.168.2.30:
    Packets: Sent = 4, Received = 4, Lost = 0 (0% loss),
Approximate round trip times in milli-seconds:
    Minimum = 0ms, Maximum = 0ms, Average = 0ms

C:\>ping booxin*.com

Pinging www.booxin*.com [192.168.2.30] with 32 bytes of data:

Reply from 192.168.2.30: bytes=32 time<1ms TTL=64
Reply from 192.168.2.30: bytes=32 time<1ms TTL=64
Reply from 192.168.2.30: bytes=32 time<1ms TTL=64
Reply from 192.168.2.30: bytes=32 time<1ms TTL=64

Ping statistics for 192.168.2.30:
    Packets: Sent = 4, Received = 4, Lost = 0 (0% loss),
Approximate round trip times in milli-seconds:
    Minimum = 0ms, Maximum = 0ms, Average = 0ms
```

上述代码中，192.168.2.30 就是域名 booxin*.com 对应的 IP 地址。

3. 配置 hosts 文件的意义

（1）加快域名解析。

对于经常访问的网站，可以先在 hosts 文件中配置域名和 IP 地址的对应关系。当输入域名请求后，计算机可以在本地局域网中很快解析出 IP 地址，而不用向网络上的 DNS 服务器发出请求，减少网络交互请求，加快访问网站的速度。

（2）屏蔽网站。

有些网站不经过用户同意就将各种各样的插件安装到用户的计算机中，用户很难判断其是木马还是病毒。对于此类网站可以利用 hosts 文件把该网站的域名对应到错误的 IP 地址或自己计算机的 IP 地址，以达到屏蔽访问的目的。如不想访问 www.duoduo*.com 网站时，可以在 hosts 文件中写入以下内容，代码如下。

```
127.0.0.1 www.duoduo*.com #屏蔽的网站
0.0.0.0 www.duoduo*.com #屏蔽的网站
```

这样计算机解析域名时就解析到不存在的 IP 地址或错误的 IP 地址，以达到屏蔽访问的目的。

（3）顺利连接系统。

对于特定的网站服务器和一些数据库服务器，在访问时如果直接输入 IP 地址是不能访问的（安全策略，不允许直接访问 IP 地址），只有输入服务器名才能访问。当用户配置好 hosts 文件后，只需输入域名就可以访问服务器了。注意：hosts 文件配置的对应关系是静态的，不支持动态更新，如果网络上的计算机（服务器）更新了 IP 地址，请及时更新 hosts 文件，否则将导致某些服务器上的服务无法访问。

（4）强制指定某域名对应某个 IP 地址。

如果新的网站代码文件已经上传到服务器，网站部分功能需要调试，网站暂时不对外提供服务或者不想被爬虫抓取到时，可以修改本地的 hosts 文件。这样只有内网中特定的计算机才能通过域名来访问该网站，以达到安全调试的目的。

1.2.6 hosts 文件域名解析实战

公司内网需要搭建一台 Web 服务器，定期发布一些公司新闻。由于公司内部未搭建 DNS

服务器，暂时采用 hosts 文件来做此 Web 服务器的解析，Web 服务器架构规划如表 1-2 所示。

表 1-2　　　　　　　　　　　　Web 服务器架构规划

架构组成	域名记录	IP 地址
Web 服务器（CentOS）	www.booxin*.com booxin*.com	192.168.2.30
客户端（Windows 10 操作系统）	无	192.168.2.31
域名解析器	无	192.168.2.31

定时任务时间同步计划加入服务器，代码如下。

```
[root@booxin* ~]# crontab -l
00 00 * * * ntpdate ntp3.aliyun.com >/dev/null 2>&1
[root@booxin* ~]# ntpdate ntp3.aliyun.com >/dev/null 2>&1
```

上述代码设置服务器做定时任务时间同步。

1．查看服务器基本信息

确认 SELinux 安全策略组件的运行状态为 disabled，代码如下。

```
1  [root@CentOS_30 ~]# sestatus
2  SELINUX status:                disabled
3  [root@CentOS_30 ~]# /etc/init.d/iptables status
4  iptables：未运行防火墙。
```

上述代码第 2 行中的 disabled 关键字和第 4 行的未运行防火墙关键字，表示 iptables 防火墙和 SELinux 安全策略均已被关闭。

2．设置主机名

通过 hostname 指令设置临时主机名，并通过修改主机名配置文件永久修改主机名，具体代码如下。

```
[root@booxin* ~]# hostname
booxin*.com
[root@booxin* ~]# cat /etc/sysconfig/network
NETWORKING=yes
HOSTNAME=booxin*.com
```

上述代码设置的主机名为 booxin*.com。

3．在 192.168.2.30 主机安装 httpd 服务器

使用 yum 指令安装 Apache 软件，并使用 rpm 指令查询该软件是否安装成功，具体代码如下。

```
[root@booxin* ~]# yum -y install httpd >/dev/null 2>&1
[root@booxin* ~]# rpm -q httpd
httpd-2.2.15-69.el6.centos.x86_64
```

使用 yum 指令安装 Apache Web 服务器软件，方便测试。

4．创建独立的 Web 项目解析文件

新建 www.booxin*.com 域名的区域解析文件，具体代码如下。

```
cat >/etc/httpd/conf.d/www.booxin*.com.conf<<q
<VirtualHost *:80>
  DocumentRoot /data/Web/www.booxin*.com
  ServerName   www.booxin*.com
  ServerAlias  booxin*.com
  DirectoryIndex index.html
</VirtualHost>
q
```

5．创建 Web 项目网站目录

使用 mkdir 指令中的 –p 选项递归创建 Web 项目网站目录，–v 选项显示创建目录的过程信息，代码如下。

```
[root@www.bj-apache-1.com conf.d]#mkdir -pv /data/Web/www.booxin*.com
mkdir: created directory '/data'
mkdir: created directory '/data/Web'
mkdir: created directory '/data/Web/www.booxin*.com'
^C[root@booxin* ~]# yum install tree -y  >/dev/null  2>&1
[root@booxin* ~]# tree /data/Web/
/data/Web/
└── www.booxin*.com

1 directory, 0 files
```

创建 www.booxin*.com 项目网站目录,用于存放网站项目工程文件。

6. Windows 10 设置 IP 地址等

Windows 10 客户端设置 IP 地址、子网掩码、默认网关等信息,如图 1-11 所示。

7. 在 Windows 10 客户端设置 hosts 文件域名解析

```
192.168.2.30 www.booxin*.com booxin*.com
```

在 Windows 10 客户端 hosts 文件中写入图 1-11 所示的配置信息,输入 192.168.2.30 地址对应解析 www.booxin*.com、booxin*.com 两个域名。另外还需要确定主机名与 IP 地址的对应解析是否正确,可以使用 ping 指令或者 nslookup 指令测试是否正确,测试过程及结果代码如下。

图 1-11 Windows 10 客户端设置 IP 地址等

```
C:\>ping -n 2 www.booxin*.com

Pinging www.booxin*.com [192.168.2.30] with 32 bytes of data:

Reply from 192.168.2.30: bytes=32 time<1ms TTL=64
Reply from 192.168.2.30: bytes=32 time<1ms TTL=64

Ping statistics for 192.168.2.30:
    Packets: Sent = 2, Received = 2, Lost = 0 (0% loss),
Approximate round trip times in milli-seconds:
    Minimum = 0ms, Maximum = 0ms, Average = 0ms


C:\>ping -n 2 booxin*.com

Pinging www.booxin*.com [192.168.2.30] with 32 bytes of data:

Reply from 192.168.2.30: bytes=32 time<1ms TTL=64
Reply from 192.168.2.30: bytes=32 time<1ms TTL=64

Ping statistics for 192.168.2.30:
    Packets: Sent = 2, Received = 2, Lost = 0 (0% loss),
Approximate round trip times in milli-seconds:
    Minimum = 0ms, Maximum = 0ms, Average = 0ms
```

上述代码中,ping 指令执行后返回的 hosts 文件域名解析结果准确无误。

8. 服务器端创建测试文件

(1) 写入静态网页测试文件。

写入 HTML 静态网页测试文件,方便后续测试使用,代码如下。

```
echo "<h1 style="color:pink">test-page</h1>">/data/Web/www.booxin*.com/index.html
[root@booxin* ~]# echo "<h1 style="color:pink">test-page</h1>">/data/Web/www.booxin*.com/index.html
[root@booxin* ~]# tree /data/Web/
```

```
      /data/Web/
      └── www.booxin*.com
              └── index.html

1 directory, 1 file
```

（2）重启 httpd 服务。

重启 httpd 服务，并确认端口和进程是否存在，代码如下。

```
[root@booxin* ~]# /etc/init.d/httpd restart
停止 httpd:                                                [确定]
正在启动 httpd:
httpd: Could not reliably determine the server's fully qualified domain name, using
 booxin*.com for ServerName
                                                            [确定]

[root@booxin* ~]# netstat -ntpl |grep 80
tcp        0      0 :::80                 :::*            LISTEN      1600/httpd
[root@booxin* ~]# ps -ef |grep httpd
root      1600     1  0 21:19 ?        00:00:00 /usr/sbin/httpd
Apache    1602  1600  0 21:19 ?        00:00:00 /usr/sbin/httpd
Apache    1603  1600  0 21:19 ?        00:00:00 /usr/sbin/httpd
Apache    1604  1600  0 21:19 ?        00:00:00 /usr/sbin/httpd
Apache    1605  1600  0 21:19 ?        00:00:00 /usr/sbin/httpd
Apache    1606  1600  0 21:19 ?        00:00:00 /usr/sbin/httpd
Apache    1607  1600  0 21:19 ?        00:00:00 /usr/sbin/httpd
Apache    1608  1600  0 21:19 ?        00:00:00 /usr/sbin/httpd
Apache    1609  1600  0 21:19 ?        00:00:00 /usr/sbin/httpd
root      1613  1467  0 21:21 pts/0    00:00:00 grep httpd
```

上述代码中，httpd 服务已经重启，80 端口和 httpd 进程已经存在。

（3）浏览器访问测试。

浏览器访问测试效果如图 1-12 所示。

9. hosts 文件解析域名小结

- 使用两台虚拟机，角色分别为 Apache 服务器和 Windows 10 客户端。
- 设置 Web 服务器的 ServerName。
- 虚拟机各自创建网站目录，并建立静态网页测试文件。

图 1-12 浏览器访问测试效果

- 在 Windows 10 客户端设置 hosts 文件并对 www.booxin*.com、booxin*.com 域名设置解析记录。
- 测试 www.booxin*.com、booxin*.com 域名是否可以被正确访问。

10. hosts 文件缺点

- 无法测试配置文件的语法是否错误。
- 无法对单个域名设置并实现负载均衡功能（单 IP 地址对应多域名解析记录）。

1.3 DNS 服务基础知识

1.3.1 DNS 的演变和作用

1. DNS 的演变

用户在上网时，通常需要在浏览器中输入 www.ziroom*.com 一类的网址，此类网址即是一个域名。互联网发展早期计算机使用 IP 地址进行通信，各个计算节点之间只有用 IP 地址才能相互识别。

后来 UNIX 操作系统中出现了名为 hosts 的文件（Linux 和 Windows 操作系统均继承和保留了这个文件），此文件中记录着主机名与 IP 地址的对应关系。用户通过浏览器访问某个网站时，只要输入主机名，系统就会加载 hosts 文件并查找主机名与 IP 地址的对应关系，找到对应的 IP 地址后，就可以访问这个 IP 地址的主机，进而访问该主机提供的服务。

随着互联网的发展，主机越来越多，无法保证所有主机都能获取统一的、最新的 hosts 文件，于是就出现了在**文件服务器上集中存放的 hosts 文件**，以供下载使用。

随着互联网规模的进一步扩大，这种方式也不再适合，而且把所有地址解析记录形成的文件都同步到所有的客户端似乎也不是一个好办法。再者，当读者通过浏览器向 Web 服务器请求 Web 页面时，可以在浏览器中输入网址或者是相应的 IP 地址。例如浏览自如网，可在浏览器的地址栏中输入 www.ziroom*.com，也可输入类似 119.254.108.123 的 IP 地址进行访问，但是 IP 地址不方便人类记忆。

互联网上的网站成千上万，如果登录每个网站都需要记住一大串数字，是特别不方便的，因此，DNS 技术诞生了。它的作用是将 119.254.108.123 解析为 www.ziroom*.com，类似的域名人们更容易记住。当下次访问自如网时，只需在浏览器中直接输入 www.ziroom*.com 这个域名，就可以了。随着解析规模的继续扩大，DNS 也在不断演化，发展成现今的多层分布式体系架构。

2. DNS 的作用

将全限定域名（Full Qualified Domain Name，FQDN）解析为 IP 地址，IP 地址解析成主机名的数据解析过程，称为域名解析（Name Resolving）。DNS 服务解析示例代码如下。

```
FQDN ==> IP （正向解析） www.mingze*.com. --> 192.168.1.190
IP ==> FQDN （反向解析） 192.168.1.190 --> www.mingze*.com.
```

邮件服务器邮件解析也需要用到 DNS 服务，如在 DNS 服务器中设置 MX 邮件解析记录。

1.3.2 为什么要设置 DNS 服务器地址才能上网

为什么一定要设置 DNS 服务器地址才能上网？有些读者可能会发现，为什么我可以登录 QQ，但却打不开网页呢？原因是 DNS 服务器故障。

DNS 服务器地址是唯一的，一般是运营商提供给终端用户用来解析 IP 地址和域名的关系。如果不设置 DNS 服务器地址，那么就**无法查询 IP 地址的去向**，自然也就打不开网页。而 QQ 等即时聊天软件，为了提高服务器吞吐能力，很多服务器采用用户数据报协议（User Datagram Protocol，UDP），既不靠传输协议，也无须提供 DNS 服务器地址，用户同样可以登录 QQ。因此就会出现虽然可以登录 QQ，但无法打开网页的情况。

1.3.3 DNS 服务基础概念

1. DNS 基础概念

DNS 是因特网的一项核心服务。它作为一个可以将域名和 IP 地址相互对应的**分布式数据库**，能够使人们更方便地访问互联网，而不用去记忆服务器端的 IP 地址。域名是由"."分开的一串单词或缩写组成的，每一个域名都对应一个唯一的 IP 地址，这一命名的方法或这样管理域名的系统叫作**域名管理系统**。通过主机名，最终得到该主机名对应的 IP 地址的过程叫作**域名解析**（或主机名解析）。

DNS 使用传输控制协议（Transmission Control Protocol，TCP）和 **UDP** 的 **53** 端口进行数据传输。实际使用中，根据场景需要选择 UDP 的 53 端口。使用 UDP 会为服务器减小很多压力。DNS 的分布式数据库是以**域名**为**索引**的，每个域名实际上就是一棵很大的逆向树中的路径，这棵逆向树称为**域名空间**（Domain Name Space）。

整个因特网的规模是很大的，单台 DNS 服务器无法满足服务的需求。所以，很早的时候，DNS 就开始使用**分布式模型设计**，并采用了**分层次**、树状的结构命名方法。

2. 关于域名的基础概念

请读者思考一个问题：www.booxin*.com 是一个域名吗？

答案是否定的，技术专业人员把类似 www.booxin*.com 格式的名称叫作**全限定域名**，即 FQDN，其中（www.booxin*.com 代表的是一台主机，而 booxin*.com 代表的是域），FQDN 是域加计算机名（主机名）的总称。

3. DNS 解析过程

DNS 解析名称的过程，即实现"FQDN↔IP 地址"的转换（双向转换，但这种双向是由两种完全不同的机制实现的），代码如下。

```
172.16.0.1-->www.test*.com
172.16.0.2-->mail.test*.com
```

其他名称（用户名、服务名称）解析的过程如下。

- 用户名↔ID（如 Linux 系统 root 用户的 UID 和 GID 默认均为 0）。
- 服务名称↔端口号（如 http 服务默认使用 80 端口）。

实现名称解析的机制有很多，因为需要对这些名称解析的机制进行统一管理，所以出现了统一的名称解析平台，即 nsswitch。

4. nsswitch 运行机制

/etc/nsswitch.conf 文件，即名称服务切换（name service switch，nsswitch）配置文件，它规定通过哪些途径，以及按照什么顺序来查找特定类型的信息。nsswitch 是为多种需要实现名称解析的机制提供名称解析的平台，需要注意的是 nsswitch 只提供平台，并不负责实际的名称解析，其架构如图 1-13 所示。

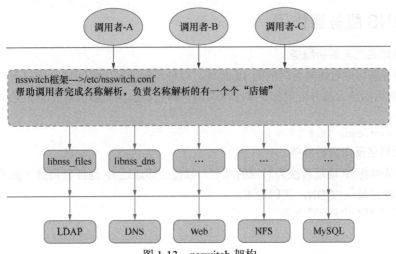

图 1-13　nsswitch 架构

能够将主机名转换成 IP 地址的解析机制有两个，分别为 libnss_files、libnss_dns。这两个都是库文件，库是需要被调用的，调用库的对象就是图 1-13 中平台下的对应的"店铺"（LDAP、DNS、Web、NFS、MySQL），而它们之间建立关联关系就是通过 nsswitch 这个框架实现的。nsswitch 对我们而言的展现形式就是库文件（/etc/nsswitch.conf），/etc/nsswitch.conf 文件格式如下。

```
hosts:   files   DNS
```

- files：/etc/hosts 文件的位置。
- DNS：指 DNS 服务。

用户通过浏览器访问主机名时，无法与主机名直接建立联系，因此需要调用一个库文件完成主机名到 IP 地址的转换。此机制在本机叫作 stub resolver（最原始的名称解析器，它本质上就是一个程序），stub resolver 会通过调用某个库来完成对应关系的转换。

stub resolver 先去寻找/etc/nsswitch.conf 中指定的配置，并根据此处的次序找出 files 指定的文件，然后再从这个文件中查找访问的主机名有没有对应的 IP 地址。如果有，解析成功。当系统管理员使用 ping www.baidu.com 指令去测试域名解析记录时，由于 www.baidu.com 是一个主机名，无法直接与之建立联系，需要将它转换成 IP 地址，因此 ping 指令就会借助本地的 stub resolver 来完成名称解析。

stub resolver 先去找/etc/hosts 中有没有 www.baidu.com 对应的 IP 地址，如果没有，继续向 DNS 服务器发起域名请求解析，CentOS 中/etc/hosts 的文件格式如下。

```
IPADDR         FQDN              Aliases（主机别名）
172.16.0.1     www.test*.com     www
```

5. ICANN 等机构的诞生

为什么要用到/etc/hosts 文件呢？互联网诞生之初，主机数量很少，通过每一台主机的 hosts 文件记录每一台主机的 IP 地址和它的主机名的对应关系即可。

在小规模的网络中，hosts 文件用于实现名称解析，它有效地工作了很长一段时间，但随着互联网规模的不断扩大，若每一台主机都随意地给自己起名，那么彼此之间的通信会很麻烦。因此出现了对应的域名注册管理机构，域名注册管理相关的 3 个机构分别是 ICANN、InterNIC、CNNIC。

1.3.4 DNS 服务器作用

1. 提供域名正向解析服务

DNS 服务器将域名转换为对应的 IP 地址，以便能够通过主机的域名访问到对应的服务器主机，互联网内的绝大部分应用都是基于域名的正向解析。正向解析格式为 FQDN→IP 地址，代码如下。

```
www.duoduo*.com.  16.8.12.9
```

2. 提供域名反向解析服务

DNS 服务器将 IP 地址转换为对应的域名，以便能够通过 IP 地址查询到主机的域名，反向解析格式为 IP 地址→FQDN，代码如下。

```
16.8.12.9 www.duoduo*.com.
```

1.3.5 DNS 服务器组织架构

1．DNS 服务系统的组成

DNS 服务系统由以下两部分组成。

（1）域名服务器：提供域名解析的软件，默认监听 UDP 与 TCP 的 53 端口。

（2）解析器：访问域名服务器的客户端，它负责获取域名服务器响应后解析，或显示结果，并把结果返回给调用它的主机或服务器。

2．DNS 服务体系是一个庞大的分布式数据库

在庞大的互联网世界里，DNS 不是一个或者少数几个服务器，而是由大量的服务器共同构成的一个服务体系。每一个 DNS 服务器都会保存它所负责解析的区域文件的**数据库，区域文件分布式存储在互联网的不同 DNS 服务器上**，所有的 DNS 服务器区域文件都是以数据库的形式保存的，以加快传输速度。

3．DNS 服务采用树形层次结构

DNS 树形层次结构如图 1-14 所示。

根域：位于域名层次结构最顶层的域。

DNS 是有层级关系的，整个层级关系的表现形式就像一棵大树，每一层均被称为域。最顶层的域叫作根域，用点号表示；第二层通常用 com、net、edu、gov、mil、org 等表示，该层叫作顶级域；往下第三层，即是经常见到的域名了，如 51cto*.com，这一层的域叫作二级域；当二级域加上 www 后，就是第四层结构了，叫作三级域……域名通常是由小到大、从左往右书写的，用"."连接，最后加"."表示根域。

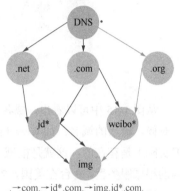

图 1-14 DNS 树形层次结构

并不是所有的 DNS 服务器地位都是平等的，其有严格的**层次结构**。比如 DNS 根域负责顶级域的 DNS 服务器数据信息，因此所有顶级域的相关解析信息由根域负责解析，二级域和三级域就不一定由根域负责解析了，其特点如下。

- 一级域负责管理和维护二级域的 DNS 服务器数据信息。
- 二级域负责管理和维护三级域的 DNS 服务器数据信息。

在互联网上，各个 DNS 服务器之间是相互**协作**的，有严格的层次关系。

4．全世界的 DNS 服务器具有相同的根域

因为 DNS 服务器在全世界只有一个**根**，所以**根域**实际上是不存在的。但是要表示出来，则采用"."来表示。

5．对域名的查询是分层次进行的

对域名 www.yahoo*.com.cn 的解析需要依次经过如下 DNS 服务器。

- 根域"."的 DNS 服务器。
- "cn."域的 DNS 服务器。
- "com.cn."域的 DNS 服务器。
- "yahoo*.com.cn."域的 DNS 服务器。

每一个 DNS 服务器只响应自身已知的数据，若查询的结果未知则向其他服务器继续转发该请求，查询的范围逐步缩小，查询的结果越来越准确，这种查询方式叫作**迭代（DNS 服务器之**

间的互相查询）。递归是客户端发出 DNS 请求到上层 DNS 服务器，若上层 DNS 服务器没有返回结果，则客户端继续等待，直到上层 DNS 服务器返回客户端请求的**最终结果**，DNS 域名层次结构如图 1-15 所示。

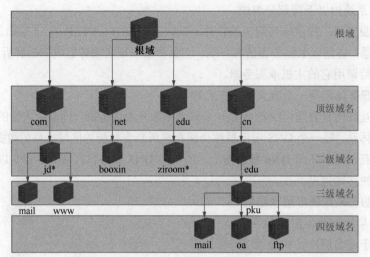

图 1-15　DNS 域名层次结构

从图 1-15 中可以看出，**域名层次结构中最高级的是根域**，使用"."表示。读者在访问网站的时候，输入的域名，如 www.baidu.com，其实最后还有一个"."，只是不需要手动输入而已，但实际上是存在的，这就是根域。根域服务器中，除了一台主根域服务器（美国），还有 12 台辅助根域服务器，9 台在美国，2 台在欧洲（英国、瑞典各 1 台），1 台在亚洲（日本）。

第二层是顶级域名，顶级域名分为三大类。

（1）组织顶级域名：com 一般代表公司、企业，net 代表网络机构，edu 代表教育机构，gov 代表非军事的政府机构等。

（2）国家或地区级顶级域名：根据 ISO 3166，cn 代表中国，us 代表美国，uk 代表英国，jp 代表日本等。

（3）基础结构顶级域名：这种顶级域名只有一个，是 arpa，用于反向解析。

第三层是二级域名，即企业或个人可以注册的域名，如 baidu.com 是注册后的二级域名。域名被注册后，则该域名下面的主机名，可以根据需求自定义三级、四级、五级，甚至是六级域名等。

DNS 相关协议规定，域名中的标号都由英文和数字组成，每一个标号不超过 63 个字符（为了记忆方便，一般不会超过 12 个字符），也不区分大小写字母。标号中除"-"以外，不能使用其他的标点符号。

级别最低的域名写在最左边，级别最高的写在最右边。由多个标号组成的完整域名总共不超过 255 个字符。DNS 既不规定一个域名需要包含多少个下级域名，也不规定每一级域名代表什么意思。各级域名由其上一级的域名管理机构管理，而顶级域名则由 ICANN 管理。用这种方法可使每一级域名在整个互联网范围内是唯一的，并且也便于设计出一种查找域名的机制。

国家或地区都被分配了根域下的顶级域名，常用国家或地区顶级域名后缀如表 1-3 所示。

表 1-3　　　　　　　　　　　常用国家或地区顶级域名后缀

域名后缀	表示
.us	代表美国
.cn	代表中国
.jp	代表日本

国家或地区级顶级域名的特点是只有两位。

按照机构职能的不同划分，可以使用不同的域名后缀，常用机构的域名后缀如表 1-4 所示。

表 1-4　　　　　　　　　　　常用机构的域名后缀

域名后缀	表示
.gov	代表政府机构
.com	代表公司、企业
.edu	代表教育机构、高校
.net	代表网络机构
.org	代表非商业组织

6．域是分层管理的

第一层：根域。

第二层：顶级域。

反向域：arpa，这是反向解析的特殊顶级域。

第三层及以下：顶级域之后即是普通的域，公司或个人在互联网上注册的域名一般都是这些普通的域，如 ziroom*.com 域名即是二级域名。

1.3.6　DNS 服务器类型

1．缓存 DNS 服务器

缓存 DNS 服务器又被称为"唯一高速缓存服务器"，还被称为"纯缓存 DNS 服务器"，其主要功能是提供域名解析的缓存，**默认情况下不进行任何 DNS 服务的解析**时，也可以正常工作。当 DNS 客户端发出查询请求到缓存 DNS 服务器，缓存 DNS 服务器查询不到解析结果的时候会向**根域 DNS 服务器发**出**请求**。经过一段时间的积累以后，它的缓存里面保存了大量域名的解析结果，缓存 DNS 服务器就是靠此来进行工作的，即缓存 DNS 服务器**全靠缓存来工作**。

2．转发 DNS 服务器

转发 DNS 服务器靠缓存进行工作，但是它和缓存 DNS 服务器不同的是它有**转发器**，可指向一个或者多个转发器。当有客户端向转发 DNS 服务器发出查询请求时，转发 DNS 服务器会转发客户端请求到指定的转发器，通过另外的 DNS 服务器进行解析，转发器充当了一个代理客户端请求的角色。

转发 DNS 服务器向指定 DNS 服务器发出请求，寻求结果。DNS 服务器包括主、从两种，两种服务器都有缓存。缓存极大地提高了互联网 DNS 解析的工作效率，一个域名一旦被解析过之后，如果缓存的结果被保留，再次解析时就不需要来回递归和迭代，可直接从高速缓存里面获取结果返回给客户端。

3. 主 DNS 服务器

主 DNS 服务器是特定域所有信息的权威性信息源，对于某个指定域，主 DNS 服务器是唯一的存在；主 DNS 服务器最大的特点是**保存了指定域的区域数据文件（正向解析记录、反向解析记录等）**，主 DNS 服务器必定有某个区域的区域数据文件，这是主 DNS 服务器和从 DNS 服务器的一个不同点。

4. 从 DNS 服务器

从 DNS 服务器没有区域数据文件，它的所有的区域数据文件都是从它所指向的主 DNS 服务器复制过来的。为什么要分为主、从 DNS 服务器呢？主要用于故障转移。当主 DNS 服务器出现问题时，可及时将请求分配到从 DNS 服务器，不会造成业务中断，进而减轻维护的工作量。

1.3.7 DNS 查询过程

1. 域名查询基本过程

域名的查询是分层进行的，图 1-16 所示为 DNS 服务器查询流程。

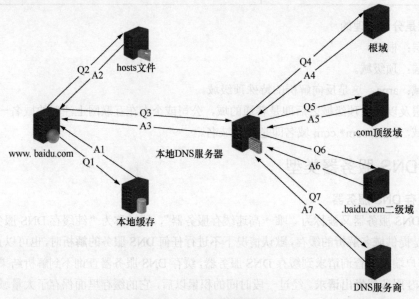

图 1-16　DNS 服务器查询流程

以查询 www.baidu.com 为例，域名查询的基本流程如下。

（1）在客户端打开浏览器，输入 www.baidu.com 并按"Enter"键。

（2）首先查询本地缓存是否有记录。如果有，则直接返回结果，解析完成。

（3）如果没有，则查询本地 hosts 文件是否有记录。如果有，则直接返回结果，解析完成。

（4）如果没有，则向本地 DNS 服务器（TCP/IP 配置指定的 DNS 服务器地址）查询，DNS 服务器查找此域名是否在自己的资源记录中。若在，直接返回结果，解析完成。

（5）如果查询的域名不在当前 DNS 服务器的资源记录中，则向根域服务器查询（所有 DNS 服务器都知道 13 台根域服务器的 IP 地址）。

（6）根域回复。如果当前根域服务器没有该域名的区域文件，请向 .com 顶级域的 DNS 服务器进行查询，查询到该域名对应的区域文件的 IP 地址是 1.1.1.1。

（7）本地 DNS 服务器向 1.1.1.1 查询。

（8）1.1.1.1 服务器回复，此处没有你要查询的记录，请去.baidu.com 二级域的 DNS 服务器查询，它的 IP 地址是 2.2.2.2。

（9）DNS 服务器又向 2.2.2.2 查询。

（10）2.2.2.2 回复，此处没有你要查询的记录，请去 www.baidu.com 的 DNS 服务商查询，它的 IP 地址是 3.3.3.3。

（11）3.3.3.3 回复，www.baidu.com 域名的 IP 地址是 119.75.217.109。

（12）本地 DNS 服务器回复给客户端，www.baidu.com 的 IP 地址是 119.75.217.109。

（13）计算机将 www.baidu.com 的 IP 地址存储到计算机的本地缓存中，以便下次调用。

（14）浏览器请求服务器端网页资源。

（15）网站服务器回复客户端请求页面内容，客户端并在客户端的浏览器中显示。

2. 递归查询

递归（客户端访问 DNS 服务器的方式）：请求（查询）者发送一次请求就能得到结果，如图 1-17 所示。

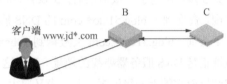

图 1-17 DNS 递归查询

客户端有一道问题不知道答案，它去问 B，而 B 也不知道答案，它去问 C，C 把答案返回给 B，B 再把答案返回给客户端，这个过程就是递归。（在这个过程当中，客户端只发送了一次请求。）

3. 迭代查询

迭代（DNS 服务器查询得到结果的方式）：请求（查询）者发送多次请求才能得到结果，如图 1-18 所示。

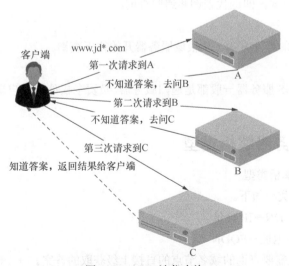

图 1-18 DNS 迭代查询

客户端第一次请求到 A，A 不知道答案，它告诉客户端去问 B，B 也不知道答案，但 B 知道 C 知道答案，于是 B 告诉客户端可以去问 C，于是客户端从 C 处获取了答案，这个过程就是迭代。（在这个过程当中，客户端发送了多次请求。）

> 注意：为安全和效率考虑，根域服务器不能与任何主机递归，互联网中的 DNS 查询是两段式的，先递归后迭代。

4．DNS 服务器工作原理

以访问 hanyanwei.blog.51cto*.com 网站为例，具体过程如下。

第 1 步：当我们使用 DNS 客户端请求查询此域名的 A 记录时，DNS 服务器首先会判断所要**查询域名**是属于哪个区域，是否在自己的管辖范围内（假设此域名不在管辖范围内）。

第 2 步：为了确定当前域名是在哪个级别，DNS 服务器要向根域服务器寻求帮助，根域服务器会向 DNS 服务器返回域名所在的顶级域的 DNS 服务器地址，也就是根域的下一级的 DNS 服务器地址。此时 DNS 服务器根据这个地址再次查询，此次查询获得了 51cto*.com 的 DNS 服务器地址，终于确定了 hanyanwei.blog.51cto*.com 的区域所在的 DNS 服务器地址。DNS 服务器通过查询一个 DNS 服务器确定一个地址，并使用该地址去查询下一个 DNS 服务器的查询方式称为迭代查询。

第 3 步：确定了区域地址后，51cto*.com 为了向 DNS 服务器返回结果，需要一级一级地往下查询。查询到了 blog.51cto*.com 的 DNS 服务器地址，于是 blog.51cto*.com 的 DNS 服务器继续往下查询。blog.51cto*.com 又查询到了 hanyanwei.blog.51cto*.com 的 DNS 服务器地址，此时的地址正是 DNS 服务器要找的地址。到了这一步，hanyanwei.blog.51cto*.com 的 DNS 服务器则会把结果向它的上一级汇报，上一级又向上一级汇报，最终结果回到了最初的 DNS 服务器，此时客户端才能获得最终的 IP 地址。像这样 DNS 服务器的数据逐级遍历和逐级返回的过程，称为递归查询。

如果查询的域名正好和当前查询使用的 DNS 服务器匹配，则 DNS 服务器会先查询本地缓存是否存在。如果存在，则返回缓存数据；如果不存在，则去数据文件中读取数据并返回结果。

5．DNS 服务器查询小结

查询域名有两种方式，即迭代查询和递归查询。

（1）迭代查询。

DNS 服务器之间的查询方式，从根域服务器开始往下查询。

（2）递归查询。

客户端使用的 DNS 服务器一般都是递归服务器，负责全权处理客户端的查询请求直至返回结果。

1.3.8　DNS 服务器解析类型

1．DNS 服务器解析类型

DNS 服务器解析类型如下。

- 正向解析：FQDN→IP 地址。
- 反向解析：IP 地址→FQDN。
- 权威答案：从需要查询的域名节点的直接上级获取的答案。
- 非权威答案：任何不是从其直接上级获取的答案，如从本地缓存 DNS 服务器获取的答案。

DNS 通过**分布式机制**将查询请求一层层地划分给更小的管理机构，使得某一个域内的管理只需要由一个服务器即可完成。若一个公司的域内的主机很少，单独建立一个解析域名的服务器意义不大，也可以支持一个服务器管理多个域。一个 IP 地址可以对应多个主机名（如一台主机既作为 www 服务器又作为 FTP 服务器且共用同一 IP 地址时），一个主机名也可以对应多个 IP 地址（负载均衡，DNS 高级功能，但效果比较差），这是 DNS 服务器的分布式数据库特色，总结如下。

- 上级仅知道其直接下级。
- 下级只知道根域的位置。

2．DNS 服务器工作职责

DNS 服务器工作职责（按请求及范围划分）如下。

（1）接收本地客户端请求（此过程是递归的）。

（2）接收外部客户端请求：请求权威答案。

其中，权威答案的结果如下。

- 肯定答案（有 TTL 值）。
- 否定答案（有 TTL 值）。

（3）接收外部客户端请求：请求非权威答案。

站在 DNS 架构的角度进行分析，**域（Domain）**是**逻辑概念**，**区域（Zone）**是一个**物理概念**，区域的基本概念如图 1-19 所示。

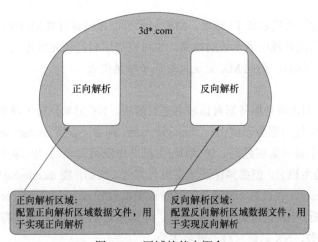

图 1-19　区域的基本概念

域是逻辑概念，如图 1-19 中的.com 域，由于其本身不存储任务数据，即域（这里以.com 域为例）是不会真正存储数据的，因此它是一个逻辑概念。而区域是真正存储下来的数据文件，如图 1-19 中的 3d*.com 域名会存储 3d*域的正向解析和反向解析区域数据文件，所以它是物理概念。

1.3.9　DNS 服务器各资源记录

为客户端提供查询、存储 DNS 数据服务的主机称为 DNS 服务器，DNS 服务器中记录着大量的数据资源信息。网络上为了方便域名的注册、管理和解析，将域名对应到特定类型的资源信息，我们称之为资源记录。DNS 资源记录是分类型的，如表 1-5 所示。

表 1-5　　　　　　　　　　　　　　DNS 资源记录类型

记录类型	说明	补充
SOA	起始授权	有一个且必须是第一个
NS	域名服务器	DNS 服务器域名地址
MX	邮件服务器	域内的邮件服务器
A	域名解析到 IPv4 地址	正向解析
PTR	IP 地址解析到域名	反向解析
CNAME	别名记录	记录域名的别名

能提供域名解析的服务器，上面的记录类型可以是 A 记录、CNAME 记录、MX 记录、NS 记录等。

1．A 记录

A 记录表明 DNS 服务器指向目标主机地址，只能使用 IP 地址。

2．CNAME 记录

CNAME 记录的**目标主机地址只能使用主机名**，不能使用 IP 地址，主机名前不能有任何其他前缀，如 http:// 等是不被允许的。**A 记录优先于 CNAME 记录**，即如果一个主机地址同时存在 A 记录和 CNAME 记录，则 CNAME 记录不生效。

3．MX 记录

MX 记录可以使用主机名或 IP 地址，MX 记录可以通过设置优先级实现主、从 DNS 服务器设置。优先级中的**数字越小表示级别越高**，也可以使用相同优先级达到负载均衡的目的。如果在主机名中填入子域名，则此 MX 记录只对该子域名生效。

4．NS 记录

NS 记录用来表明由哪台服务器对该域名进行解析，NS 记录只对子域名生效。例如用户希望由 119.254.108.108 这台服务器解析 ns.duoduo*.com，则需要设置 ns.duoduo*.com 的 NS 记录，优先级中的数字越小表示级别越高。IP 地址/主机名中既可以填写 IP 地址，也可以填写类似 ns.mydomain.com 的主机名，但必须保证该主机名有效。例如，将 ns.duoduo*.com 的 NS 记录指向到 ns.mydomain.com，在设置 NS 记录的同时还需要设置 ns.mydomain.com 的指向，否则 NS 记录将无法正常解析。**NS 记录优先于 A 记录**，即如果一个主机地址同时存在 **NS 记录和 A 记录**，则 **A 记录不生效**，NS 记录只对子域名生效。

5．负载均衡服务基础知识补充

负载均衡（Server Load Balancing，SLB）是指在一系列资源上智能地分布网络负载。

负载均衡可以减少网络拥塞、提高整体网络性能、提高系统自愈性，并确保企业关键应用的可用性。当相同子域有多个目标地址，或域名的 MX 记录有多个目标地址且优先级相同时，表示轮循，可以达到负载均衡的目的，但需要虚拟机和邮箱服务商支持。

在实际应用中，负载均衡一般是结合高可用服务使用的。因为一旦负载均衡器出现故障，自身将无法进行故障转移。所以，还需要结合高可用软件对外提供服务，如 Nginx+Keepalived 的组合使用、BIND+LVS+Keepalived 的组合使用。

6．其他基础概念

（1）泛域名与泛解析。泛域名是指在一个域名下，以 *.duoduo*.com 的形式，表示该域名所有未建立的子域名。泛解析是把 *.duoduo*.com 的 A 记录解析到某个 IP 地址上，通过访问任意的前缀.duoduo.com 都能访问到解析的站点（一般是负载均衡器）。

（2）域名绑定是指将域名指向服务器 IP 地址的操作。

（3）域名转向又称为域名指向或域名转发。当你在用户地址栏中输入域名时，将会自动跳转到指定的另一个域名，一般是使用短的、易记的域名转向复杂难记的域名。

7. SOA 记录详解

BIND 的 SOA 记录：每个区域仅有一个 SOA 记录。SOA 记录包括区域的名称，一个技术联系人和各种不同的超时值，代码如下。

```
$TTL 3600
$ORIGIN hanyanwei*.com.
@ IN SOA DNS.qqDNS.com. domainadmin.DNSpod.com.(
1247134024 ;Serial
3600 ; Refresh ( seconds )
1800 ; Retry ( seconds )
1209600 ; Expire ( seconds )
3600 );Minimum TTL for zone ( seconds )
;
@ IN NS DNS.qqDNS.com.
@ IN NS DNS.ggDNS.com.
www 3600 IN A 63.220.7.134
www 3600 IN A 63.220.7.134
;
;generated @ 2009-07-09 18:07:12
;end
```

（1）Serial。

数值 Serial 代表这个区域的序列号。

作用：供从 DNS 服务器判断是否从主 DNS 服务器获取新数据。

每次区域文件更新时，都需要修改 Serial 数值，RFC 1912 2.2 建议的格式为 YYYYMMDDnn，其中 nn 为版本修订号。

（2）Refresh。

数值 Refresh 设置从 DNS 服务器多长时间与主 DNS 服务器进行 Serial 核对，目前 BIND 的 notify 参数可设置每次主 DNS 服务器更新时都会主动通知从 DNS 服务器更新，Refresh 参数主要用于 notify 参数关闭时。

（3）Retry。

数值 Retry 用于设置当从 DNS 服务器试图获取主 DNS 服务器的 Serial 参数时，如果主 DNS 服务器未响应，多长时间重新进行检查。

（4）Expire。

数值 Expire 决定从 DNS 服务器在没有主 DNS 服务器的情况下，权威地提供域名解析服务的时间长短。

（5）Minimum。

BIND 8.2 之前，由于没有独立的$TTL 指令，因此通过 SOA 最后一个字段来实现。但由于 BIND 8.2 之后出现了$TTL 指令，该部分功能就不再由 SOA 的最后一个字段来负责。

8. PTR

PTR 是 A 记录的逆向记录，又称作 IP 反查记录或指针记录，负责将 IP 地址反向解析为对应的域名。

1.3.10 TTL 值详解

1. 基本定义

生存时间（Time-To-Live，TTL）原理：TTL 是 IP 包中的一个值，它告诉网络路由器，包在网络中的时间是否太长而应被丢弃。具体的 TTL 值定义，请见本书 3.7.1 小节。

2. TTL 值设置的应用

（1）增大 TTL 值。

增大 TTL 值在很大程度上可以节约域名解析时间，提高网站访问速度。

一般情况下，域名的各种记录是极少更改的，很可能几个月、几年内都不会有什么变化，因此系统管理员可以增大域名记录的 TTL 值，让记录在各地 DNS 服务器中缓存的时间加长。这样在较长的一段时间内，用户访问同一个网站时，本地 ISP 的 DNS 服务器就不需要向负责解析该域名的 DNS 服务器发出解析请求，可以直接从缓存中返回域名解析记录。

（2）减小 TTL 值。

减小 TTL 值是为了减少更换空间记录时的不可访问时间，更换空间记录时，有 99.9%的概率会遇到 DNS 记录更改的问题（因为缓存问题）。

新的域名记录在有的地方可能生效了，但在有的地方可能等几天甚至更久才会生效，这会导致有的用户访问到新服务器，有的用户访问到旧服务器。如果仅是客户端访问，这也不是什么大问题，但如果涉及邮件发送，说不定哪封重要信件就会被发送到已经停掉的旧服务器上，因此为了尽可能地减小各地服务器的解析时间差，合理的做法如下。

- 先查看域名当前的 TTL 值，比如将 TTL 值过期时间设定为 1 天。
- 修改 TTL 值为可设定的最小值，建议为 1 分钟。
- 等待一天。保证各地的 DNS 服务器缓存都过期并更新了记录。

3. TTL 值使用总结

TTL 值表示 DNS 记录在 DNS 服务器上的缓存时间，设定该值（缓存时间）太长和太短都会引发一系列问题。如果缓存时间设置太长，一旦域名到 IP 地址的解析有变化，会使客户端缓存的域名无法解析到变化后的 IP 地址，导致该域名不能正常解析，这段时间内可能会有一部分用户无法访问网站；如果缓存时间设置太短，会导致用户每次访问网站都要重新解析一次域名，增加服务器间通信压力。

1.3.11 DNS 解析配置文件

3 个与 DNS 服务解析有关的配置文件如下。

- /etc/hosts：此文件在没有出现 DNS 服务器的时候，作用和 DNS 服务器类似。不过随着网络技术的发展，单个文件根本无法满足需求，因此此文件通常用于本地 IP 地址解析。
- /etc/resolv.conf：定义 DNS 服务器 IP 地址，本地解析域名通常使用的都是次数设定的 IP 地址。
- /etc/nsswitch.conf：该文件决定先要使用/etc/hosts 还是/etc/resolv.conf 的设置。

总结：理解 DNS 的树状结构、递归查询及迭代查询的基本原理，可为后面的学习打下良好的基础。

1.3.12 JVM 设定 DNS 缓存时间

Linux 操作系统中，在域名被解析为 IP 地址后，IP 地址的信息被保存在 Java 虚拟机（Java Virtual Machine，JVM）的高速缓存中，域名缓存的有效时间默认永远有效，这样一来域名和 IP 地址的解析变更后，Java 类的应用程序必须重启才能连接到新域名。

```
1  [root@master ~]# java -version
2  java version "1.8.0_251"
3  Java(TM) SE Runtime Environment (build 1.8.0_251-b08)
4  Java HotSpot(TM) 64-Bit Server VM (build 25.251-b08, mixed mode)
5  [root@master ~]# javac -version
6  javac 1.8.0_251
7  [root@master ~]# grep -n ttl ${JAVA_HOME}/jre/lib/security/java.security
8  329:#networkaddress.cache.ttl=-1
9  344:networkaddress.cache.negative.ttl=10
```

上述代码第 2~4 行表示 JDK 的版本及相关信息，第 6 行表示 javac 编译器的版本信息，第 8~9 行表示设置解析失败的域名记录在 JVM 中缓存的有效时间，JVM 默认是 10s，0 表示禁止缓存，-1 表示永远有效。

还有两种方式设置 DNS 缓存。

（1）在 JAVA_OPTS 里设置 DNS 缓存。

```
-Dsun.net.inetaddr.ttl=6
-Dsun.net.inetaddr.negative.ttl=2
```

（2）修改 property。

```
System.setProperty("sun.net.inetaddr.ttl", "6");
System.setProperty("sun.net.inetaddr.negative.ttl", "2");
```

一般情况下我们不需要完全取消 JVM 的 DNS 缓存，只需要调小有效时间，经过一些测试发现如下结论。

- 一个域名对应一个 IP 地址和一个域名对应多个 IP 地址，DNS 查询响应时间差别极小，只不过后者占用 CPU 资源稍高一些。
- 在高并发的应用场景中，不做 DNS 缓存时的 CPU 资源消耗比做了 5s 缓存的 CPU 资源消耗要高 3/4 倍，实时 DNS 请求相当消耗 CPU 资源。
- 经测试，3s 和 30s 缓存有效时间对 DNS 查询响应时间的影响差别不大，CPU 内存占用都比较接近。
- 建议将 DNS 缓存时间设置为 3~6s，进而提升运维效率。

1.4 DNS 服务器部署实战

本节会带领读者熟悉 DNS 服务器部署，核心内容如下。

- 搭建一台 DNS 服务器。
- 解析并设置一个域名。
- 搭建 Nginx 服务器+XP 服务器（测试 DNS 服务器域名解析）。
- 进行 DNS 服务器域名解析与 hosts 文件域名解析的优势对比。

1.4.1 实验环境介绍

DNS 服务器实验环境部署如表 1-6 所示。

第1章 DNS 服务入门

表 1-6　　　　　　　　　　DNS 服务器实验环境部署

角色	操作系统	IP 地址	主机名及域名
DNS 服务器	CentOS 6.9 2.6.32 x86_64	192.168.2.33	DNS.booxin*.com
Web 服务器	CentOS 6.9 2.6.32 x86_64	192.168.2.34	www.booxin*.com
Windows 客户端	Windows 10 x86_64	192.168.2.35	laohan

实验环境架构如图 1-20 所示。

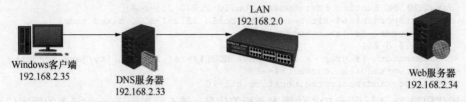

图 1-20　实验环境架构

1.4.2　DNS 服务器安装 BIND 软件

Linux 操作系统通常使用 BIND 软件实现 DNS 服务器的架设。当前 BIND 的稳定版本为 BIND 9，BIND 服务进程名称为 named，监听 53 端口，主要配置文件为 /etc/named.conf。此文件主要用于配置区域解析文件，并指定区域数据文件名称。区域数据文件通常保存于 /var/named/ 目录下，用于存放区域的资源类型文件。首先在 DNS 服务器（192.168.2.33）上安装 BIND 软件，安装过程如下。

1. 设置 DNS 服务器主机名

设置 DNS 服务器主机名和 PS1 环境变量，便于日常管理，代码如下。

```
[root@booxin* ~]# hostname  www.DNS_master_33.com
[root@www.DNS_master_33.com ~]$export PS1="[\u@\H \W]\\$"
[root@www.DNS_master_33.com ~]#tail -1 /etc/profile
export PS1="[\u@\H \W]\\$"
[root@booxin* ~]# hostname  www.DNS_slave_34.com
[root@www.DNS_slave_34.com ~]#export PS1="[\u@\H \W]\\$"
[root@www.DNS_slave_34.com ~]#tail -1 /etc/profile
export PS1="[\u@\H \W]\\$"
```

2. 关闭防火墙和 SELinux

为了使后面的实验顺利，建议读者关闭所有服务器的防火墙和 SELinux，代码如下。

```
[root@www.DNS_master_33.com ~]# service iptables stop
[root@www.DNS_master_33.com ~]# setenforce 0
setenforce: SELINUX is disabled
```

验证上述代码是否执行成功，代码如下。

```
[root@www.DNS_master_33.com ~]# service iptables status
iptables：未运行防火墙。
[root@www.DNS_master_33.com ~]#  sestatus
SELINUX status:                 disabled
```

上述代码表示防火墙和 SELinux 系统已经被关闭。

3. 安装 BIND 软件

BIND 是 Linux 操作系统下的 DNS 服务器程序，安装 BIND 及其组件，代码如下。

```
[root@www.DNS_master_33.com ]# yum install Bind Bind-utils Bind-chroot -y
已加载插件：fastestmirror
设置安装进程
（中间过程略）
```

作为依赖被安装：
　　Bind-libs.x86_64 32:9.8.2-0.62.rc1.el6_9.5　　　　　　　　portreserve.x86_64 0:0.0.4-11.el6

配置 BIND 服务过程中，安装 Bind-chroot 包，就可以将 BIND 服务的运行目录改变，即将 BIND 程序运行目录改为/var/named/chroot 目录（虚拟根目录），则 BIND 的配置文件目录就由原本的/etc 变成了/var/named/chroot/etc，区域文件目录就由原本的/var/named 变成了/var/named/chroot/var/named。当有攻击者入侵 BIND 服务时，因为 BIND 服务的运行目录已经被改变，所以其权限也就在虚拟根目录中，即/var/named/chroot 目录，不会威胁到整个服务器的安全。

> **注意**：CentOS 6 操作系统已经将 chroot 所使用的目录通过 mount-Bind 的功能进行目录链接，所以读者不用再单独配置 chroot，正常配置 BIND 文件即可。

4．查看程序包安装的组件

（1）查看 BIND 程序安装的组件，代码如下。

```
[root@www.DNS_master_33.com ~]#rpm -ql Bind |grep -Ev "doc|DNSsec|sbin|man"
/etc/NetworkManager/dispatcher.d/13-named
/etc/logrotate.d/named
/etc/named
/etc/named.conf
/etc/named.iscdlv.key
/etc/named.rfc1912.zones
/etc/named.root.key
/etc/portreserve/named
/etc/rc.d/init.d/named
/etc/rndc.conf
/etc/rndc.key
/etc/sysconfig/named
/usr/lib64/Bind
/var/log/named.log
/var/named
/var/named/data
/var/named/dynamic
/var/named/named.ca
/var/named/named.empty
/var/named/named.localhost
/var/named/named.loopback
/var/named/slaves
/var/run/named
```

上述代码使用 rpm 指令查看 Bind 程序安装了哪些软件包相关的组件。

（2）查看 Bind-utils 程序安装的组件，代码如下。

```
[root@www.DNS_master_33.com ~]#rpm -ql Bind-utils
/usr/bin/dig
/usr/bin/host
/usr/bin/nslookup
/usr/bin/nsupdate
/usr/share/man/man1/dig.1.gz
/usr/share/man/man1/host.1.gz
/usr/share/man/man1/nslookup.1.gz
/usr/share/man/man1/nsupdate.1.gz
```

上述代码安装了 Bind-utils 软件包，该软件包包含 dig、host、nslookup、nsupdate 等 DNS 解析测试和故障排查实用工具。

（3）查看 Bind-chroot 程序安装的组件，代码如下。

```
[root@www.DNS_master_33.com ~]#rpm -ql Bind-chroot
/var/named/chroot
/var/named/chroot/dev
/var/named/chroot/dev/null
/var/named/chroot/dev/random
```

```
/var/named/chroot/dev/zero
/var/named/chroot/etc
/var/named/chroot/etc/localtime
/var/named/chroot/etc/named
/var/named/chroot/etc/named.conf
/var/named/chroot/etc/pki
/var/named/chroot/etc/pki/DNSsec-keys
/var/named/chroot/lib64
/var/named/chroot/usr
/var/named/chroot/usr/lib64
/var/named/chroot/usr/lib64/Bind
/var/named/chroot/var
/var/named/chroot/var/log
/var/named/chroot/var/named
/var/named/chroot/var/run
/var/named/chroot/var/run/named
/var/named/chroot/var/tmp
```

上述代码中，/var/named/chroot/是 chroot 的主目录，下面包含了 dev、etc、lib64、usr、var 等目录。

1.4.3 启动 BIND 服务器

1. 验证 BIND 服务器是否启动

安装完 BIND 之后，启动 BIND 服务器，验证是否可以正常启动。

```
[root@www.DNS_master_33.com ~]# /etc/init.d/named start
Generating /etc/rndc.key:                                  [确定]
启动 named：                                                [确定]
```

2. 查看 BIND 服务器对应的端口和进程是否存在

查看 BIND 服务器对应的端口和进程是否存在，代码如下。

```
[root@www.DNS_master_33.com ~]# netstat -ntpl
Active Internet connections (only servers)
Proto Recv-Q Send-Q Local Address      Foreign Address  State     PID/Program name
tcp        0      0 127.0.0.1:53       0.0.0.0:*        LISTEN    1436/named
tcp        0      0 0.0.0.0:22         0.0.0.0:*        LISTEN    1108/sshd
tcp        0      0 127.0.0.1:953      0.0.0.0:*        LISTEN    1436/named
tcp        0      0 127.0.0.1:25       0.0.0.0:*        LISTEN    1187/master
tcp        0      0 :::1:53            :::*             LISTEN    1436/named
tcp        0      0 :::22              :::*             LISTEN    1108/sshd
tcp        0      0 :::1:953           :::*             LISTEN    1436/named
tcp        0      0 :::1:25            :::*             LISTEN    1187/master
```

从上述代码的执行结果中可以看到，安装完 BIND 软件之后，默认不做任何配置是可以启动 BIND 服务器的，且 BIND 服务器监听本机的 127 端口，只负责对本机的所有服务进行解析。

3. 将 BIND 服务器加入开机启动

将 BIND 服务器加入开机启动，代码如下。

```
[root@www.DNS_master_33.com ~]#chkconfig named on  --level 3 5
[root@www.DNS_master_33.com ~]#chkconfig --list |grep named
named           0:关闭   1:关闭   2:启用   3:启用   4:启用   5:启用   6:关闭
```

从上述代码中可以看到，BIND 服务器已经加入开机启动了，下次服务器重启时，BIND 服务器会随着操作系统自动启动。

1.4.4 配置 BIND 服务器

1. BIND 服务器主配置文件

/var/named/chroot/etc/named.conf 是 BIND 服务器的主配置文件，其权限信息如下。

```
[root@www.DNS_master_33.com ~]#ll /var/named/chroot/etc/named.conf
-rw-r----- 1 root named 984 Nov 20  2015 /var/named/chroot/etc/named.conf
```
BIND 服务器主配置文件内容如下。
```
[root@www.DNS_master_33.com ~]#cat /var/named/chroot/etc/named.conf
//
// named.conf
//
// Provided by Red Hat Bind package to configure the ISC BIND named(8) DNS
// server as a caching only nameserver (as a localhost DNS resolver only).
//
// See /usr/share/doc/Bind*/sample/ for example named configuration files.
//

options {
    listen-on port 53 { 127.0.0.1; };
    listen-on-v6 port 53 { ::1; };
    directory       "/var/named";
    dump-file       "/var/named/data/cache_dump.db";
        statistics-file "/var/named/data/named_stats.txt";
        memstatistics-file "/var/named/data/named_mem_stats.txt";
    allow-query     { localhost; };
    recursion yes;

    DNSsec-enable yes;
    DNSsec-validation yes;

    /* Path to ISC DLV key */
    Bindkeys-file "/etc/named.iscdlv.key";

    managed-keys-directory "/var/named/dynamic";
};

logging {
    channel default_debug {
        file "data/named.run";
        severity dynamic;
    };
};

zone "." IN {
    type hint;
    file "named.ca";
};

include "/etc/named.rfc1912.zones";
include "/etc/named.root.key";
```

上述配置语句中，以"//"开头的为单行注释内容，以"/*"开头和以"*/"结尾的为多行注释内容。BIND 默认的配置文件使用 named.conf，因此建议对 named.conf 文件进行定期备份，如果 named.conf 文件出错还可以进行恢复。此外还需要修改/var/named/chroot/etc/ named.conf 文件所属的组，避免权限问题导致 named 服务无法启动，修改权限代码如下。

```
chgrp named /var/named/chroot/etc/named.conf
```

重启 named 服务，代码如下。
```
service named restart
```

named.conf 的配置文件路径"很深"，可以为 named.conf 文件创建软链接，代码如下。
```
[root@www.DNS_master_33.com etc]#ln -sv /var/named/chroot/etc/named.conf /etc/
"/etc/named.conf" -> "/var/named/chroot/etc/named.conf"
```

创建完软链接以后，读者可直接使用 vim 指令修改和编辑/ctc/name.conf 文件，等同于使用 vim 指令修改和编辑/var/named/chroot/etc/named.conf 文件。

2. BIND 工作目录

/var/named/chroot/var/named/目录用于保存 BIND 服务器的域名区域文件,需要和主配置文件里面的文件名称一一对应。

3. BIND 启动脚本

BIND 配置文件和区域文件配置完成之后,如果需要启动 BIND 服务器,可以到/etc/init.d/目录中启动。其中 named 可执行文件是 BIND 服务器的启动脚本,用于控制 BIND 服务器的启动和停止,代码如下。

```
[root@www.DNS_master_33.com ~]# /etc/init.d/named
Usage: /etc/init.d/named {start|stop|status|restart|try-restart|reload|force-reload}
```

当读者输入/etc/init.d/named 后,按"Enter"键即可查看相关脚本的启动和停止等脚本参数。

4. 编写 BIND 配置文件

BIND 软件安装完成以后,BIND 配置文件和内容如下。

```
[root@www.DNS_master_33.com ~]#ll /var/named/chroot/etc/named.conf
-rw-r----- 1 root named 984 Nov 20  2015 /var/named/chroot/etc/named.conf
[root@www.DNS_master_33.com ~]#nl /var/named/chroot/etc/named.conf
     1  //
     2  // named.conf
     3  //
     4  // Provided by Red Hat bind package to configure the ISC bIND named(8) DNS
     5  // server as a caching only nameserver (as a localhost DNS resolver only).
     6  //
     7  // See /usr/share/doc/bind*/sample/ for example named configuration files.
     8  //
     9  options {
    10      listen-on port 53 { 127.0.0.1; };
    11      listen-on-v6 port 53 { ::1; };
    12      directory       "/var/named";
    13      dump-file       "/var/named/data/cache_dump.db";
    14          statistics-file "/var/named/data/named_stats.txt";
    15          memstatistics-file "/var/named/data/named_mem_stats.txt";
    16      allow-query     { localhost; };
    17      recursion yes;
    18      DNSsec-enable yes;
    19      DNSsec-validation yes;
    20      /* Path to ISC DLV key */
    21      bindkeys-file "/etc/named.iscdlv.key";
    22      managed-keys-directory "/var/named/dynamic";
    23  };
    24  logging {
    25      channel default_debug {
    26          file "data/named.run";
    27          severity dynamic;
    28      };
    29  };
    30  zone "." IN {
    31      type hint;
    32      file "named.ca";
    33  };
    34   include "/etc/named.rfc1912.zones";
    35   include "/etc/named.root.key";
```

/var/named/chroot/var/named/配置文件如下所示。

```
[root@www.DNS_master_33.com ~]#ll /var/named/chroot/var/named/
total 32
drwxr-x---  / root   named 4096 Nov  2 07:02 chroot
drwxrwx---  2 named  named 4096 Nov  2 07:19 data
drwxrwx---  2 named  named 4096 Nov  2 07:20 dynamic
-rw-r-----  1 root   named 3289 Apr 11  2017 named.ca
-rw-r-----  1 root   named  152 Dec 15  2009 named.empty
```

```
-rw-r----- 1 root    named   152 Jun 21  2007 named.localhost
-rw-r----- 1 root    named   168 Dec 15  2009 named.loopback
drwxrwx--- 2 named   named  4096 Aug 27 23:39 slaves
```

上述代码中，named.ca 是根域的相关配置文件。

5. 配置 DNS 区域解析文件

编辑/var/named/chroot/etc/named.conf 文件，编辑之前先对其备份，可以在因修改失误或误删除该文件后快速复原配置和回滚该文件，代码如下。

```
[root@www.DNS_master_33.com ~]#cd /var/named/chroot/etc/
[root@www.DNS_master_33.com etc]#ll named.conf
-rw-r----- 1 root named 984 Nov 20  2015 named.conf
[root@www.DNS_master_33.com etc]#cp -av named.conf  named.conf_$(date +%Y%m%d)
'named.conf' -> 'named.conf_20181102'
```

使用 Vim 编辑器，打开/var/named/chroot/etc/named.conf，修改监听地址和允许**递归查询**的**白名单**，否则后面测试会有很多麻烦，named.conf 主配置文件内容如下。

```
[root@www.DNS_master_33.com etc]#cat named.conf
//
// named.conf
//
// Provided by Red Hat Bind package to configure the ISC BIND named(8) DNS
// server as a caching only nameserver (as a localhost DNS resolver only).
//
// See /usr/share/doc/Bind*/sample/ for example named configuration files.
//

options {
    listen-on port 53 { 127.0.0.1; };
// 2018-11-02-hanyanwei
    listen-on port 53 { 192.168.2.33; };
// 2018-11-02-hanyanwei
    listen-on-v6 port 53 { ::1; };
    directory       "/var/named";
    dump-file       "/var/named/data/cache_dump.db";
        statistics-file "/var/named/data/named_stats.txt";
        memstatistics-file "/var/named/data/named_mem_stats.txt";
// 2018-11-02-hanyanwei
//  allow-query     { localhost; };
// 2018-11-02-hanyanwei

// 2018-11-02-hanyanwei
    allow-query     { 192.168.2.0/24; };
// 2018-11-02-hanyanwei
    recursion yes;

    DNSsec-enable yes;
    DNSsec-validation yes;

    /* Path to ISC DLV key */
    Bindkeys-file "/etc/named.iscdlv.key";

    managed-keys-directory "/var/named/dynamic";
};

logging {
        channel default_debug {
                file "data/named.run";
                severity dynamic;
        };
};

zone "." IN {
    type hint;
    file "named.ca";
```

```
    };


    // hanyanwei-201811-02
    zone "booxin*.com" IN {
        type master;
        file "booxin*.com.zone";
    };
    zone "1.168.192. in-addr.arpa" IN {
        type master;
        file "booxin*.com.rev";
    };
    // hanyanwei-201811-02

    include "/etc/named.rfc1912.zones";
    include "/etc/named.root.key";
```

检测配置文件语法是否有误,代码如下。

```
[root@www.DNS_master_33.com etc]#named-checkconf
[root@www.DNS_master_33.com etc]#echo $?
0
```

通过上述代码检测可发现,没有返回任何错误,因此语法配置没有问题。

1.4.5 BIND 主配置文件详解

1. named.conf 主配置文件注释

(1) 行注释。

打开 named.conf 主配置文件,以 "//" 开头的内容,其中 "//" 是注释符号,"//" 后面的为注释内容,代表程序忽略这一行,named.conf 里使用 "//" 进行注释。

(2) 段注释。

"/*" 与 "*/" 之间则是表示对某段内容的注释。

(3) named.conf 配置文件特点。

named.conf 配置文件中每一行都以 ";" 结尾,开头的 "{}" 千万不要忘记(初学时容易出现该错误)。

2. 全局配置

(1) options。

BIND 的配置文件以选项(options)开始,其所有关于 options 的设定都包括在 "{}" 内。首先使用 listen-on port 指定监听的端口和 IP 地址,例如 listen-on port 53 { 192.168.1.22; };若要监听服务器的所有 IP 地址,则使用 any,例如,listen-on port { any; }。

(2) 指定工作目录。

使用 directory 指定资源记录文件所在的位置,默认为/var/named 目录,BIND 程序到此目录下查找 DNS 记录。由于此处配置的 directory 为绝对路径,在后面的配置中,就无须再使用绝对路径了,因此资源记录文件一定要放在 directory 指定的目录下。

3. 区域配置

(1) 定义根域。

定义一个区域时,还定义了 ".",即 DNS 的根域。

（2）指定区域数据类型。

使用 type 关键字指定区域数据类型，共有 3 个类型，分别是 master、slave、hint。master 代表主要区域，slave 代表辅助区域，hint 代表根区域。

（3）定义区域记录文件名称。

file 指定区域记录文件的名称，即文件名，如 named.ca。

（4）定义区域。

使用 zone 关键字定义一个区域，名称为 booxin*.com，类型为 master，文件名称为 booxin*.com.zone。

4．反向解析区域

定义名称为 2.168.192.in-addr.arpa 的区域文件，它的作用是把 IP 地址解析到域名，即反向解析区域。假设反向解析 192.168.1.0，先把最后一段（.0）去掉，得到的结果是 192.168.1，再反过来写，是 1.168.192，后面还要加上 .in-addr.arpa，完整写法如下。

1.168.192.in-addr.arpa

目前为止，基本的 named.conf 配置已经完成。接下来配置资源记录文件，即区域文件，在 /var/named/chroot/var/named/ 目录下，有很多这样的文件，如 named.ca，该文件内容如下。

```
1   [root@www.ntp_server.com ~]# cat /var/named/chroot/var/named/named.ca
2   ;       This file holds the information on root name servers needed to
3   ;       initialize cache of Internet domain name servers
4   ;       (e.g. reference this file in the "cache  .  <file>"
5   ;       configuration file of bIND domain name servers).
6   ;
7   ;       This file is made available by InterNIC
8   ;       under anonymous FTP as
9   ;           file                /domain/named.cache
10  ;           on server           FTP.INTERNIC.NET
11  ;           -OR-                RS.INTERNIC.NET
12  ;
13  ;       last update:    April 11, 2017
14  ;       related version of root zone:   2017041101
15  ;
16  ; formerly NS.INTERNIC.NET
17  ;
18  .                        3600000      NS    A.ROOT-SERVERS.NET.
19  A.ROOT-SERVERS.NET.      3600000      A     198.41.0.4
20  A.ROOT-SERVERS.NET.      3600000      AAAA  2001:503:ba3e::2:30
21  ;
22  ; FORMERLY NS1.ISI.EDU
23  ;
24  .                        3600000      NS    B.ROOT-SERVERS.NET.
25  B.ROOT-SERVERS.NET.      3600000      A     192.228.79.201
26  B.ROOT-SERVERS.NET.      3600000      AAAA  2001:500:84::b
27  ;
28  ; FORMERLY C.PSI.NET
29  ;
30  .                        3600000      NS    C.ROOT-SERVERS.NET.
31  C.ROOT-SERVERS.NET.      3600000      A     192.33.4.12
32  C.ROOT-SERVERS.NET.      3600000      AAAA  2001:500:2::c
33  ;
34  ; FORMERLY TERP.UMD.EDU
35  ;
36  .                        3600000      NS    D.ROOT-SERVERS.NET.
37  D.ROOT-SERVERS.NET.      3600000      A     199.7.91.13
38  D.ROOT-SERVERS.NET.      3600000      AAAA  2001:500:2d::d
39  ;
40  ; FORMERLY NS.NASA.GOV
41  ;
```

```
42  .                           3600000   NS     E.ROOT-SERVERS.NET.
43  E.ROOT-SERVERS.NET.         3600000   A      192.203.230.10
44  E.ROOT-SERVERS.NET.         3600000   AAAA   2001:500:a8::e
45  ;
46  ; FORMERLY NS.ISC.ORG
47  ;
48  .                           3600000   NS     F.ROOT-SERVERS.NET.
49  F.ROOT-SERVERS.NET.         3600000   A      192.5.5.241
50  F.ROOT-SERVERS.NET.         3600000   AAAA   2001:500:2f::f
51  ;
52  ; FORMERLY NS.NIC.DDN.MIL
53  ;
54  .                           3600000   NS     G.ROOT-SERVERS.NET.
55  G.ROOT-SERVERS.NET.         3600000   A      192.112.36.4
56  G.ROOT-SERVERS.NET.         3600000   AAAA   2001:500:12::d0d
57  ;
58  ; FORMERLY AOS.ARL.ARMY.MIL
59  ;
60  .                           3600000   NS     H.ROOT-SERVERS.NET.
61  H.ROOT-SERVERS.NET.         3600000   A      198.97.190.53
62  H.ROOT-SERVERS.NET.         3600000   AAAA   2001:500:1::53
63  ;
64  ; FORMERLY NIC.NORDU.NET
65  ;
66  .                           3600000   NS     I.ROOT-SERVERS.NET.
67  I.ROOT-SERVERS.NET.         3600000   A      192.36.148.17
68  I.ROOT-SERVERS.NET.         3600000   AAAA   2001:7fe::53
69  ;
70  ; OPERATED BY VERISIGN, INC.
71  ;
72  .                           3600000   NS     J.ROOT-SERVERS.NET.
73  J.ROOT-SERVERS.NET.         3600000   A      192.58.128.30
74  J.ROOT-SERVERS.NET.         3600000   AAAA   2001:503:c27::2:30
75  ;
76  ; OPERATED BY RIPE NCC
77  ;
78  .                           3600000   NS     K.ROOT-SERVERS.NET.
79  K.ROOT-SERVERS.NET.         3600000   A      193.0.14.129
80  K.ROOT-SERVERS.NET.         3600000   AAAA   2001:7fd::1
81  ;
82  ; OPERATED BY ICANN
83  ;
84  .                           3600000   NS     L.ROOT-SERVERS.NET.
85  L.ROOT-SERVERS.NET.         3600000   A      199.7.83.42
86  L.ROOT-SERVERS.NET.         3600000   AAAA   2001:500:9f::42
87  ;
88  ; OPERATED BY WIDE
89  ;
90  .                           3600000   NS     M.ROOT-SERVERS.NET.
91  M.ROOT-SERVERS.NET.         3600000   A      202.12.27.33
92  M.ROOT-SERVERS.NET.         3600000   AAAA   2001:dc3::35
93  ; End of file
```

上述代码中第 2~93 行为根节点相关内容。

named.ca 是根的资源记录文件，默认记录 13 台根域服务器的 NS 记录和 A 记录，DNS 服务器必须要有 named.ca 文件。

5. 根域服务器

named.ca 文件中很多以 ";" 开头的行代表注释，文件格式如下。

- 第一列使用 ";" 注释，把 ";" 开头的内容去掉，一共有 13 行以 "." 开头的，为根域服务器。
- 第二列都是 3600000，代表 TTL 值，即 DNS 缓存的有效期，单位是秒。

- 第三列是 NS，代表资源记录类型，即 NS 记录。
- 最后一列是根域服务器的名称，名称为 A.ROOT-SERVERS.NET.，从 A~M，全球一共有 13 台根域服务器。

每台 DNS 服务器是通过 named.ca 文件中对根域服务器的定义"知道"13 台根域服务器。

6. 记录信息详解

named.ca 文件中记录的是根域服务器的信息，named.empty 是资源记录文件的模板，要设定的域名解析，通过此模板进行修改，具体代码如下。

```
[root@www.ntp_server.com ~]# nl /var/named/chroot/var/named/named.empty
     1  $TTL 3H
     2  @       IN SOA  @ rname.invalid. (
     3                                  0       ; serial
     4                                  1D      ; refresh
     5                                  1H      ; retry
     6                                  1W      ; expire
     7                                  3H )    ; minimum
     8          NS      @
     9          A       127.0.0.1
    10          AAAA    ::1
```

（1）设置 TTL 值。

TTL 值即 DNS 缓存的有效期，此处是预设的 DNS 缓存默认值，在配置每个资源记录时单独指定 TTL 值。其中 H 代表小时，D 代表天，W 代表周，3H 表示 3 小时。

（2）设置 SOA 记录。

@代表资源记录的结尾，即在 named.conf 中定义的区域后面的区域文件配置信息。

（3）IN。

IN 表示当前资源类型是 Internet class。

（4）SOA 记录。

SOA 记录表示目前区域的起始授权记录开始，每个区域文件只能有一个 SOA，不能重复，而且必须是所负责的区域中的第一记录。

（5）管理者邮箱。

紧接着定义该区域的授权主机和管理者信箱。注意：SOA 记录必须能够在 DNS 服务器中找到一条 A 记录（以后会详解），此处配置的分别是"@"和"rname.invalid."，建议使用的邮箱格式是类似 user@host.com 的格式。由于"@"在 DNS 中，是个特殊的字符，因此这里不能再使用"@"，而是使用"."来代替"@"，所以此处的信箱地址原本为 rname@invalid。

（6）SOA 配置参数。

SOA 配置参数是在"()"之中的 5 组数字，主要是主 DNS 服务器和从 DNS 服务器同步 DNS 数据时使用的，serial 配置格式为数字，建议读者使用时间日期加修改次数。当从 DNS 服务器进行数据同步时，会与主 DNS 服务器进行数字比较。若主 DNS 服务器数字大，则同步，否则忽略刷新（refresh）。

- 主、从 DNS 服务器是否同步需要对比主、从 DNS 服务器 serial 值的大小。
- retry 告知从 DNS 服务器在同步失败后，隔多长时间再次同步。
- expire 记录逾期时间，即告诉从 DNS 服务器如果一直不能同步成功，到这里就放弃同步。
- minimum 记录最小的 TTL 预设值。如果没有使用$TTL 来显式指定，则使用此处的配置值。

> 注意：SOA 记录中，带 "(" 的内容，只能写在一行，不能按 "Enter" 键换行。有时候我们在书上，或者网络上看到断开两行多，那可能是因为版面的关系。而 SOA 的 ")"，也不能写到 ";" 的右边，因为 ";" 代表的是注释，写到右边就和没有 ")" 是一样的。复制资源记录文件，代码如下。
>
> ```
> [root@www.DNS_master_33.com etc]#cd /var/named/chroot/var/named
> [root@www.DNS_master_33.com named]#pwd
> /var/named/chroot/var/named
> [root@www.DNS_master_33.com named]#cp -av named.empty booxin*.com.zone
> 'named.empty' -> 'booxin*.com.zone'
> [root@www.DNS_master_33.com named]#cp -av named.empty booxin*.com.rev
> 'named.empty' -> 'booxin*.com.rev'
> ```

（7）创建区域配置文件。

下面介绍使用 cat 指令在/var/named 目录下创建 booxin*.com.zone 和 booxin*.com.rev 两个文件，分别用于正向查询和反向查询。

```
1    cd /var/named/chroot/var/named
2    cat >booxin*.com.zone<<q
3    \$TTL 3H
4    @         IN   SOA   booxin*.com.   312874370*.qq.com. (
5                                                0       ; serial
6                                                1D      ; refresh
7                                                1H      ; retry
8                                                1W      ; expire
9                                                3H )    ; minimum
10             IN   NS    ns1.booxin*.com.
11             IN   NS    ns2.booxin*.com.
12             IN   MX  5 mail.booxin*.com.
13
14   ns1       IN   A     192.168.2.33
15   ns2       IN   A     192.168.2.34
16   www       IN   A     192.168.2.33
17   www       IN   A     192.168.2.34
18   mail      IN   A     192.168.2.33
19   ftp       IN   A     192.168.2.33
20   ftp       IN   A     192.168.2.34
21   q
```

上述代码第 2~21 行，使用 cat 指令（结合重定向）创建 booxin*.com.zone 正向解析文件。

```
1    cd /var/named/chroot/var/named
2    cat >booxin*.com.rev<<q
3    \$TTL 1D
4    @         IN SOA booxin*.com. 312874370*.qq.com. (
5                                                0       ; serial
6                                                1D      ; refresh
7                                                1H      ; retry
8                                                1W      ; expire
9                                                3H )    ; minimum
10             IN   NS    ns1.booxin*.com.
11             IN   NS    ns2.booxin*.com.
12
13   133       IN   PTR   ns1.booxin*.com.
14   134       IN   PTR   ns2.booxin*.com.
15   133       IN   PTR   www.booxin*.com.
16   134       IN   PTR   www.booxin*.com.
17   133       IN   PTR   mail.booxin*.com.
18   133       IN   PTR   ftp.booxin*.com.
19   134       IN   PTR   ftp.booxin*.com.
20   q
```

上述代码第 2~20 行，使 cat 指令创建反向解析文件（booxin*.com.rev）。

正向解析和反向解析完整代码输出如下。

```
[root@www.DNS_master_33.com etc]#cd /var/named/chroot/var/named
[root@www.DNS_master_33.com named]#cat >booxin*.com.zone<<q
> \$TTL 3H
> @          IN  SOA  booxin*.com.  312874370*.qq.com. (
>                                    0        ; serial
>                                    1D       ; refresh
>                                    1H       ; retry
>                                    1W       ; expire
>                                    3H )     ; minimum
>            IN   NS    ns1.booxin*.com.
>            IN   NS    ns2.booxin*.com.
>            IN   MX  5 mail.booxin*.com.
>
> ns1        IN   A     192.168.2.33
> ns2        IN   A     192.168.2.34
> www        IN   A     192.168.2.33
> www        IN   A     192.168.2.34
> mail       IN   A     192.168.2.33
> ftp        IN   A     192.168.2.33
> ftp        IN   A     192.168.2.34
> q
[root@www.DNS_master_33.com named]#
[root@www.DNS_master_33.com named]#cd /var/named/chroot/var/named
[root@www.DNS_master_33.com named]#cat >booxin*.com.rev<<q
> \$TTL 1D
> @          IN SOA booxin*.com. 312874370*.qq.com. (
>                                    0        ; serial
>                                    1D       ; refresh
>                                    1H       ; retry
>                                    1W       ; expire
>                                    3H )     ; minimum
>            IN   NS    ns1.booxin*.com.
>            IN   NS    ns2.booxin*.com.
>
> 133        IN   PTR   ns1.booxin*.com.
> 134        IN   PTR   ns2.booxin*.com.
> 133        IN   PTR   www.booxin*.com.
> 134        IN   PTR   www.booxin*.com.
> 133        IN   PTR   mail.booxin*.com.
> 133        IN   PTR   ftp.booxin*.com.
> 134        IN   PTR   ftp.booxin*.com.
> q
```

查看并修改文件的权限，完成后我们把 named.conf 的属主设置为 named，并启动 DNS 服务器，BIND 的服务名叫作 named，代码如下。

```
[root@www.DNS_master_33.com named]#pwd
/var/named/chroot/var/named
[root@www.DNS_master_33.com named]#ll
total 40
-rw-r----- 1 root  named  704 Nov  3 15:24 booxin*.com.rev
-rw-r----- 1 root  named  732 Nov  3 15:24 booxin*.com.zone
drwxr-x--- 7 root  named 4096 Nov  2 07:02 chroot
drwxrwx--- 2 named named 4096 Nov  2 07:19 data
drwxrwx--- 2 named named 4096 Nov  3 12:10 dynamic
-rw-r----- 1 root  named 3289 Apr 11  2017 named.ca
-rw-r----- 1 root  named  152 Dec 15  2009 named.empty
-rw-r----- 1 root  named  152 Jun 21  2007 named.localhost
-rw-r----- 1 root  named  168 Dec 15  2009 named.loopback
drwxrwx--- 2 named named 4096 Aug 27 23:39 slaves
[root@www.DNS_master_33.com named]#chown -R root:named booxin*.com.*
[root@www.DNS_master_33.com named]#ll booxin*.com.*
-rw-r----- 1 root named  704 Nov  3 15:24 booxin*.com.rev
-rw-r----- 1 root named  732 Nov  3 15:24 booxin*.com.zone
```

重启 DNS 服务器加载配置文件，代码如下。

```
[root@www.DNS_master_33.com ~]#/etc/init.d/named restart
Stopping named:                                          [  OK  ]
Starting named:                                          [  OK  ]
```

查看 DNS 服务器端口和进程信息，代码如下。

```
[root@www.DNS_master_33.com ~]#netstat -ntpl
Active Internet connections (only servers)
Proto Recv-Q Send-Q Local Address      Foreign Address    State       PID/Program name
tcp        0      0 192.168.2.33:53    0.0.0.0:*          LISTEN      2724/named
tcp        0      0 127.0.0.1:53       0.0.0.0:*          LISTEN      2724/named
tcp        0      0 0.0.0.0:22         0.0.0.0:*          LISTEN      1210/sshd
tcp        0      0 127.0.0.1:953      0.0.0.0:*          LISTEN      2724/named
tcp        0      0 127.0.0.1:25       0.0.0.0:*          LISTEN      1290/master
tcp        0      0 ::1:53             :::*               LISTEN      2724/named
tcp        0      0 :::22              :::*               LISTEN      1210/sshd
tcp        0      0 ::1:953            :::*               LISTEN      2724/named
tcp        0      0 ::1:25             :::*               LISTEN      1290/master
[root@www.DNS_master_33.com ~]#ps -ef |grep named
named     2724     1  0 15:32 ?        00:00:00 /usr/sbin/named -u named -t
/var/named/chroot
root      2774  1910  0 16:27 pts/0    00:00:00 grep named
```

7．设置 DNS 服务器解析记录

设置 DNS 服务器解析记录，代码如下。

```
[root@www.DNS_master_33.com ~]#cat /etc/resolv.conf
nameserver 192.168.2.33
nameserver 192.168.2.34
```

如上述代码所示，解析记录必须在主 DNS 服务器中设置。

1.4.6 测试主 DNS 服务器

1．客户端设置主 DNS 服务器地址

在任意一台 Linux 主机或 Windows 主机上设置主 DNS 服务器的地址，代码如下。

```
[root@www.DNS_slave_34.com ~]#cat /etc/resolv.conf
nameserver 192.168.2.33
nameserver 192.168.2.34
```

从上述代码可以看到，操作系统客户端已经设置了主 DNS 服务器的地址。

2．ping 指令测试 DNS 服务器设置是否生效

使用 ping 指令测试 DNS 服务器设置是否生效，代码如下。

```
[root@www.DNS_slave_34.com ~]#ping www.booxin*.com -c 3
PING www.booxin*.com (192.168.2.34) 56(84) bytes of data.
64 bytes from 192.168.2.34: icmp_seq=1 ttl=64 time=0.009 ms
64 bytes from 192.168.2.34: icmp_seq=2 ttl=64 time=0.027 ms
64 bytes from 192.168.2.34: icmp_seq=3 ttl=64 time=0.040 ms

--- www.booxin*.com ping statistics ---
3 packets transmitted, 3 received, 0% packet loss, time 10035ms
rtt min/avg/max/mdev = 0.009/0.025/0.040/0.013 ms
[root@www.DNS_slave_34.com ~]#ping www.booxin*.com -c 3
PING www.booxin*.com (192.168.2.33) 56(84) bytes of data.
64 bytes from 192.168.2.33: icmp_seq=1 ttl=64 time=0.061 ms
64 bytes from 192.168.2.33: icmp_seq=2 ttl=64 time=0.308 ms
64 bytes from 192.168.2.33: icmp_seq=3 ttl=64 time=0.448 ms

--- www.booxin*.com ping statistics ---
3 packets transmitted, 3 received, 0% packet loss, time 2001ms
rtt min/avg/max/mdev = 0.061/0.272/0.448/0.160 ms
[root@www.DNS_slave_34.com ~]#ping www.booxin*.com -c 3
PING www.booxin*.com (192.168.2.34) 56(84) bytes of data.
64 bytes from 192.168.2.34: icmp_seq=1 ttl=64 time=0.008 ms
64 bytes from 192.168.2.34: icmp_seq=2 ttl=64 time=0.038 ms
```

```
64 bytes from 192.168.2.34: icmp_seq=3 ttl=64 time=0.038 ms

--- www.booxin*.com ping statistics ---
3 packets transmitted, 3 received, 0% packet loss, time 2000ms
rtt min/avg/max/mdev = 0.008/0.028/0.038/0.014 ms
[root@www.DNS_slave_34.com ~]#
[root@www.DNS_slave_34.com ~]#ping www.booxin*.com -c 3
PING www.booxin*.com (192.168.2.33) 56(84) bytes of data.
64 bytes from 192.168.2.33: icmp_seq=1 ttl=64 time=0.111 ms
64 bytes from 192.168.2.33: icmp_seq=2 ttl=64 time=0.253 ms
64 bytes from 192.168.2.33: icmp_seq=3 ttl=64 time=0.247 ms

--- www.booxin*.com ping statistics ---
3 packets transmitted, 3 received, 0% packet loss, time 2000ms
rtt min/avg/max/mdev = 0.111/0.203/0.253/0.067 ms
[root@www.DNS_slave_34.com ~]#ping www.booxin*.com -c 3
PING www.booxin*.com (192.168.2.34) 56(84) bytes of data.
64 bytes from 192.168.2.34: icmp_seq=1 ttl=64 time=0.008 ms
64 bytes from 192.168.2.34: icmp_seq=2 ttl=64 time=0.027 ms
64 bytes from 192.168.2.34: icmp_seq=3 ttl=64 time=0.038 ms

--- www.booxin*.com ping statistics ---
3 packets transmitted, 3 received, 0% packet loss, time 2000ms
rtt min/avg/max/mdev = 0.008/0.024/0.038/0.013 ms
```

上述代码返回的 DNS 解析后的 IP 地址不是主 DNS 服务器解析的 IP 地址,使用 ping 指令测试 www.booxin*.com 域名,测试结果及代码如下。

```
[root@www.DNS_slave_34.com ~]#ping -c 3 www.booxin*.com
PING www.booxin*.com (192.168.2.33) 56(84) bytes of data.
64 bytes from 192.168.2.33: icmp_seq=1 ttl=64 time=0.061 ms
64 bytes from 192.168.2.33: icmp_seq=2 ttl=64 time=0.449 ms
64 bytes from 192.168.2.33: icmp_seq=3 ttl=64 time=0.654 ms

--- www.booxin*.com ping statistics ---
3 packets transmitted, 3 received, 0% packet loss, time 2002ms
rtt min/avg/max/mdev = 0.061/0.388/0.654/0.245 ms
```

使用 ping 指令测试主 DNS 服务器,代码如下。

```
[root@www.DNS_slave_34.com ~]#ping -c 3 www.booxin*.com
PING www.booxin*.com (192.168.2.34) 56(84) bytes of data.
64 bytes from 192.168.2.34: icmp_seq=1 ttl=64 time=0.007 ms
64 bytes from 192.168.2.34: icmp_seq=2 ttl=64 time=0.045 ms
64 bytes from 192.168.2.34: icmp_seq=3 ttl=64 time=0.018 ms

--- www.booxin*.com ping statistics ---
3 packets transmitted, 3 received, 0% packet loss, time 1999ms
rtt min/avg/max/mdev = 0.007/0.023/0.045/0.016 ms
[root@www.DNS_slave_34.com ~]#
[root@www.DNS_slave_34.com ~]#ping -c 3 www.booxin*.com
PING www.booxin*.com (192.168.2.33) 56(84) bytes of data.
64 bytes from 192.168.2.33: icmp_seq=1 ttl=64 time=0.128 ms
64 bytes from 192.168.2.33: icmp_seq=2 ttl=64 time=0.546 ms
64 bytes from 192.168.2.33: icmp_seq=3 ttl=64 time=0.433 ms

--- www.booxin.com ping statistics ---
3 packets transmitted, 3 received, 0% packet loss, time 2003ms
rtt min/avg/max/mdev = 0.128/0.369/0.546/0.176 ms
[root@www.DNS_slave_34.com ~]#ping -c 3 www.booxin*.com
PING www.booxin*.com (192.168.2.34) 56(84) bytes of data.
64 bytes from 192.168.2.34: icmp_seq=1 ttl=64 time=0.007 ms
64 bytes from 192.168.2.34: icmp_seq=2 ttl=64 time=0.020 ms
64 bytes from 192.168.2.34: icmp_seq=3 ttl=64 time=0.020 ms

--- www.booxin*.com ping statistics ---
3 packets transmitted, 3 received, 0% packet loss, time 2047ms
```

```
rtt min/avg/max/mdev = 0.007/0.015/0.020/0.007 ms
[root@www.DNS_slave_34.com ~]#ping -c 3 www.booxin*.com
PING www.booxin*.com (192.168.2.33) 56(84) bytes of data.
64 bytes from 192.168.2.33: icmp_seq=1 ttl=64 time=0.126 ms
64 bytes from 192.168.2.33: icmp_seq=2 ttl=64 time=0.425 ms
64 bytes from 192.168.2.33: icmp_seq=3 ttl=64 time=0.415 ms

--- www.booxin*.com ping statistics ---
3 packets transmitted, 3 received, 0% packet loss, time 2003ms
rtt min/avg/max/mdev = 0.126/0.322/0.425/0.138 ms
```

第一次 ping 延迟很大，第二次执行 ping 测试之后会发现返回结果的速度很快，是因为第二次查询的时候使用了缓存，可以看到 www.booxin*.com 域名还有负载均衡的功能，说明这台服务器既是 DNS 服务器又是 Web 服务器。

3. nslookup 测试工具

（1）修改 DNS 服务器解析记录。

使用 nslookup 修改 DNS 服务器解析记录代码如下。

```
cat>/etc/resolv.conf<<q
nameserver 192.168.2.33
nameserver 192.168.2.34
q
[root@www.DNS_master_33.com ~]# cat>/etc/resolv.conf<<q
> nameserver 192.168.2.33
> nameserver 192.168.2.34
> q
[root@www.DNS_master_33.com ~]# cat /etc/resolv.conf
nameserver 192.168.2.33
nameserver 192.168.2.34
[root@www.DNS_slave_34.com ~]# cat>/etc/resolv.conf<<q
nameserver 192.168.2.33
> nameserver 192.168.2.33
> nameserver 192.168.2.34
> q
[root@www.DNS_slave_34.com ~]# cat /etc/resolv.conf
nameserver 192.168.2.33
nameserver 192.168.2.34
[root@www.ntp_server.com ~]# cat /etc/resolv.conf
nameserver 192.168.2.33
nameserver 192.168.2.34
```

（2）主 DNS 服务器解析测试。

使用 nslookup 指令测试主 DNS 服务器解析 booxin*.com 域是否正确，执行过程和结果代码如下。

```
[root@www.DNS_master_33.com ~]# nslookup
> www.booxin*.com
Server:         192.168.2.33
Address:        192.168.2.33#53

Name:   www.booxin*.com
Address: 192.168.2.33
Name:   www.booxin*.com
Address: 192.168.2.34
> ftp.booxin*.com
Server:         192.168.2.33
Address:        192.168.2.33#53

Name:   ftp.booxin*.com
Address: 192.168.2.34
Name:   ftp.booxin*.com
Address: 192.168.2.33
> ftp.booxin*.com
```

```
Server:         192.168.2.33
Address:        192.168.2.33#53

Name:   ftp.booxin*.com
Address: 192.168.2.33
Name:   ftp.booxin*.com
Address: 192.168.2.34
> exit
```

上述代码使用 nslookup 指令交互式解析相关域名。

（3）从 DNS 服务器解析测试。

使用 nslookup 指令测试从 DNS 服务器，执行过程和结果代码如下。

```
[root@www.DNS_slave_34.com ~]# nslookup
> www.booxin*.com
Server:         192.168.2.33
Address:        192.168.2.33#53

Name:   www.booxin*.com
Address: 192.168.2.34
Name:   www.booxin*.com
Address: 192.168.2.33
> www.booxin*.com
Server:         192.168.2.33
Address:        192.168.2.33#53

Name:   www.booxin*.com
Address: 192.168.2.33
Name:   www.booxin*.com
Address: 192.168.2.34
> ftp.booxin*.com
Server:         192.168.2.33
Address:        192.168.2.33#53

Name:   ftp.booxin*.com
Address: 192.168.2.34
Name:   ftp.booxin*.com
Address: 192.168.2.33
> ftp.booxin*.com
Server:         192.168.2.33
Address:        192.168.2.33#53

Name:   ftp.booxin*.com
Address: 192.168.2.33
Name:   ftp.booxin*.com
Address: 192.168.2.34
> exit

[root@www.DNS_slave_34.com ~]#nslookup www.booxin*.com
Server:         192.168.2.33
Address:        192.168.2.33#53

Name:   www.booxin*.com
Address: 192.168.2.34
Name:   www.booxin*.com
Address: 192.168.2.33

[root@www.DNS_slave_34.com ~]#
```

（4）使用 NTP 时间同步服务器测试域名解析。

网络时间协议（Network Time Protocol，NTP）时间同步服务器测试域名解析，代码如下。

```
[root@www.ntp_server.com ~]# nslookup
> www.booxin*.com
Server:         192.168.2.33
Address:        192.168.2.33#53
```

```
        Name:   www.booxin*.com
        Address: 192.168.2.34
        Name:   www.booxin*.com
        Address: 192.168.2.33
        > www.booxin*.com
        Server:         192.168.2.33
        Address:        192.168.2.33#53

        Name:   www.booxin*.com
        Address: 192.168.2.33
        Name:   www.booxin*.com
        Address: 192.168.2.34
        > ftp.booxin*.com
        Server:         192.168.2.33
        Address:        192.168.2.33#53

        Name:   ftp.booxin*.com
        Address: 192.168.2.34
        Name:   ftp.booxin*.com
        Address: 192.168.2.33
        > ftp.booxin*.com
        Server:         192.168.2.33
        Address:        192.168.2.33#53

        Name:   ftp.booxin*.com
        Address: 192.168.2.33
        Name:   ftp.booxin*.com
        Address: 192.168.2.34
        > exit
```

上述代码中，添加的 booxin*.com 域名，正向、反向都可以解析出来，说明配置没有问题。解析没配置过的域名会怎么样呢？挑选几个进行测试，看会发生什么，代码如下。

```
[root@www.DNS_slave_34.com ~]#nslookup www.booxiiiin*.com
Server:         192.168.2.33
Address:        192.168.2.33#53

** server can't find www.booxiiiin*.com: NXDOMAIN

[root@www.DNS_slave_34.com ~]#nslookup duoduo*.com
Server:         192.168.2.33
Address:        192.168.2.33#53

Non-authoritative answer:
*** Can't find duoduo*.com: No answer

[root@www.DNS_slave_34.com ~]#
```

上述代码中，第一次解析 www.booxiiiin*.com 域名时，由于该域名在操作系统中没有缓存，花的时间会长一点儿。

（5）配置总结。

- 多区域文件解析配置时，需要在 named.conf 主配置文件中定义多个区域文件。
- 配置相关资源记录文件。
- 配置 DNS 服务器地址。
- 测试解析是否正确。

4．dig 测试工具

（1）查看域名，代码如下。

```
dig booxin*.com
[root@www.ntp_server.com ~]# dig booxin*.com
```

```
; <<>> DiG 9.8.2rc1-RedHat-9.8.2-0.62.rc1.el6_9.5 <<>> booxin*.com
;; global options: +cmd
;; Got answer:
;; ->>HEADER<<- opcode: QUERY, status: NOERROR, id: 30192
;; flags: qr aa rd ra; QUERY: 1, ANSWER: 0, AUTHORITY: 1, ADDITIONAL: 0

;; QUESTION SECTION:
;booxin.com.                    IN      A

;; AUTHORITY SECTION:
booxin*.com.           10800    IN      SOA     booxin*.com. 312874370*.qq.com. 0 86400 3600 604800 10800

;; Query time: 20 msec
;; SERVER: 192.168.2.33#53(192.168.2.33)
;; WHEN: Sun Mar  4 21:19:51 2018
;; MSG SIZE  rcvd: 78
[root@www.DNS_slave_34.com ~]#dig www.booxin*.com

; <<>> DiG 9.8.2rc1-RedHat-9.8.2-0.68.rc1.el6_10.3 <<>> www.booxin*.com
;; global options: +cmd
;; Got answer:
;; ->>HEADER<<- opcode: QUERY, status: NOERROR, id: 44399
;; flags: qr aa rd ra; QUERY: 1, ANSWER: 2, AUTHORITY: 2, ADDITIONAL: 2

;; QUESTION SECTION:
;www.booxin*.com.               IN      A

;; ANSWER SECTION:
www.booxin*.com.       10800    IN      A       192.168.2.34
www.booxin*.com.       10800    IN      A       192.168.2.33

;; AUTHORITY SECTION:
booxin*.com.           10800    IN      NS      ns1.booxin*.com.
booxin*.com.           10800    IN      NS      ns2.booxin*.com.

;; ADDITIONAL SECTION:
ns1.booxin*.com.       10800    IN      A       192.168.2.33
ns2.booxin*.com.       10800    IN      A       192.168.2.34

;; Query time: 0 msec
;; SERVER: 192.168.2.33#53(192.168.2.33)
;; WHEN: Mon Jul  1 22:43:18 2019
;; MSG SIZE  rcvd: 132

[root@www.DNS_slave_34.com ~]#
```

上述代码中，DNS 服务器对应的地址为 192.168.2.33 和 192.168.2.34。

(2) 查找域名对应的 CNAME 记录，代码如下。

```
[root@www.ntp_server.com ~]# dig booxin*.com -t CNAME

; <<>> DiG 9.8.2rc1-RedHat-9.8.2-0.62.rc1.el6_9.5 <<>> booxin*.com -t CNAME
;; global options: +cmd
;; Got answer:
;; ->>HEADER<<- opcode: QUERY, status: NOERROR, id: 5381
;; flags: qr aa rd ra; QUERY: 1, ANSWER: 0, AUTHORITY: 1, ADDITIONAL: 0

;; QUESTION SECTION:
;booxin*.com.                   IN      CNAME

;; AUTHORITY SECTION:
booxin*.com.           10800    IN      SOA     booxin*.com. 312874370*.qq.com. 0 86400 3600 604800 10800

;; Query time: 15 msec
```

```
;; SERVER: 192.168.2.33#53(192.168.2.33)
;; WHEN: Sun Mar  4 21:24:13 2018
;; MSG SIZE  rcvd: 78
```

(3) 查询域名对应的 A 记录,代码如下。

```
[root@www.ntp_server.com ~]# dig -t -A www.booxin*.com
;; Warning, ignoring invalid type -A

; <<>> DiG 9.8.2rc1-RedHat-9.8.2-0.62.rc1.el6_9.5 <<>> -t -A www.booxin*.com
;; global options: +cmd
;; Got answer:
;; ->>HEADER<<- opcode: QUERY, status: NOERROR, id: 51033
;; flags: qr aa rd ra; QUERY: 1, ANSWER: 2, AUTHORITY: 2, ADDITIONAL: 2

;; QUESTION SECTION:
;www.booxin*.com.                IN      A

;; ANSWER SECTION:
www.booxin*.com.        10800   IN      A       192.168.2.33
www.booxin*.com.        10800   IN      A       192.168.2.34

;; AUTHORITY SECTION:
booxin*.com.            10800   IN      NS      ns1.booxin*.com.
booxin*.com.            10800   IN      NS      ns2.booxin*.com.

;; ADDITIONAL SECTION:
ns1.booxin*.com.        10800   IN      A       192.168.2.33
ns2.booxin*.com.        10800   IN      A       192.168.2.34

;; Query time: 7 msec
;; SERVER: 192.168.2.33#53(192.168.2.33)
;; WHEN: Sun Mar  4 21:09:16 2018
;; MSG SIZE  rcvd: 132
```

上述代码中设置了 A 记录。

5. 其他测试

(1) 查询 A 记录,代码如下。

```
[root@www.ntp_server.com ~]# nslookup -qt=a www.booxin*.com
*** Invalid option: qt=a
Server:         192.168.2.33
Address:        192.168.2.33#53

Name:   www.booxin*.com
Address: 192.168.2.34
Name:   www.booxin*.com
Address: 192.168.2.33
```

上述代码中,使用 nslookup 指令查询 www.booxin*.com 地址,返回结果准确无误。

(2) 查询 MX 记录,代码如下。

```
[root@www.ntp_server.com ~]# host -t mx booxin*.com
booxin*.com mail is handled by 5 mail.booxin*.com.
```

上述代码中,使用 host 指令查询 booxin*.com 地址,返回结果准确无误。

(3) 查询 MX 记录,代码如下。

```
[root@www.ntp_server.com ~]#  nslookup
> set type=MX
> booxin*.com
Server:         192.168.2.33
Address:        192.168.2.33#53

booxin*.com     mail exchanger = 5 mail.booxin*.com.
> exit
```

上述代码中,使用 nslookup 指令交互式查询 booxin*.com 地址,返回结果准确无误。

(4) 测试 A 记录，代码如下。

```
[root@www.ntp_server.com ~]# ping  -c 2 www.booxin*.com
PING www.booxin*.com (192.168.2.33) 56(84) bytes of data.
64 bytes from 192.168.2.33: icmp_seq=1 ttl=64 time=2.63 ms
64 bytes from 192.168.2.33: icmp_seq=2 ttl=64 time=1.24 ms

--- www.booxin*.com ping statistics ---
2 packets transmitted, 2 received, 0% packet loss, time 1005ms
rtt min/avg/max/mdev = 1.242/1.939/2.637/0.698 ms
[root@www.ntp_server.com ~]# ping  -c 2 www.booxin*.com
PING www.booxin*.com (192.168.2.34) 56(84) bytes of data.
64 bytes from 192.168.2.34: icmp_seq=1 ttl=64 time=1.75 ms
64 bytes from 192.168.2.34: icmp_seq=2 ttl=64 time=3.74 ms

--- www.booxin*.com ping statistics ---
2 packets transmitted, 2 received, 0% packet loss, time 1012ms
rtt min/avg/max/mdev = 1.753/2.750/3.748/0.998 ms
```

DNS 服务器配置邮件服务时，对于不能发邮件的问题，做了端口映射。为了使其他计算机能够访问邮件服务器，需要设置防火墙或者路由，以便允许以下必要的端口连接服务器，端口对应信息如下。

- IMAP 对应端口 143。
- IMAPS 对应端口 i993。
- POP3 对应端口 110。
- POP3S 对应端口 995。

不能收邮件的问题可能出在 DNS 解析上。首先，需要一条 MX 记录，如 mail.domain.com。注意/etc/postfix/ main.cf 配置文件中的格式，邮件解析记录需要和 MX 记录中的值一致。代码如下。

```
myhostname = mail.domain.com
```

其次，需要一条 A 记录，将 mail.domain.com 指向一个固定的 IP 地址。

6. Windows 10 客户端测试

Windows 10 操作系统客户端 IP 地址设置如图 1-21 所示。

进入 Windows 操作系统的命令行模式测试域名解析，代码如下。

图 1-21 Windows 10 操作系统客户端 IP 地址设置

```
C:\>ping www.booxin*.com -n 2

Pinging www.booxin*.com [192.168.2.34] with 32  bytes of data:

Reply from 192.168.2.34: bytes=32 time<1ms TTL=64
Reply from 192.168.2.34: bytes=32 time<1ms TTL=64

Ping statistics for 192.168.2.34:
    Packets: Sent = 2, Received = 2, Lost = 0 (0% loss),
Approximate round trip times in milli-seconds:
    Minimum = 0ms, Maximum = 0ms, Average = 0ms

C:\>ipconfig /flushDNS

Windows IP Configuration
```

```
Successfully flushed the DNS Resolver Cache.

C:\>ping www.booxin*.com -n 2

Pinging www.booxin*.com [192.168.2.33] with 32 bytes of data:

Reply from 192.168.2.33: bytes=32 time<1ms TTL=64
Reply from 192.168.2.33: bytes=32 time<1ms TTL=64

Ping statistics for 192.168.2.33:
    Packets: Sent = 2, Received = 2, Lost = 0 (0% loss),
Approximate round trip times in milli-seconds:
    Minimum = 0ms, Maximum = 0ms, Average = 0ms
```

如上代码所示,第一次执行 ping www.booxin*.com 时返回的是 192.168.2.34,第二次需要刷新 DNS 缓存,然后再次执行 ping www.booxin*.com 时返回的 IP 地址是 192.168.2.33。

1.4.7　保存并备份配置文件

下载 BIND 服务器配置文件并保存,以便下次使用。

1. 保存 BIND 主配置文件

保存 BIND 主配置文件,代码如下。

```
[root@www.DNS_master_33.com ~]# sz -bey /var/named/chroot/etc/named.confrz
zmodem trl+C

  100%       1 KB     1 KB/s 00:00:01          0 Errors
```

保存每一次的实验文件,以便后期查询或备份恢复使用。

2. 保存 BIND 区域配置文件

保存 BIND 区域配置文件,代码如下。

```
[root@www.DNS_master_33.com ~]# sz -bey /var/named/chroot/var/named/booxin*.com.*
rz
  zmodem trl+C

  100%     704 bytes  704 bytes/s 00:00:01      0 Errors
  100%     811 bytes  811 bytes/s 00:00:01      0 Errors
```

保存每一次的实验文件,留作备份,以便后期查询和优化配置。

1.4.8　named.conf 配置文件详解

BIND 服务核心配置文件是 named.conf,除此之外还有其他配置文件,BIND 配置文件说明如表 1-7 所示。

表 1-7　　　　　　　　　　　BIND 配置文件说明

配置文件	说明
/etc/named.conf	BIND 主配置文件
/etc/named.rfc1912.zones	定义区域的文件
/etc/rc.d/init.d/named	BIND 脚本文件
/etc/rndc.conf	rndc 配置文件
/usr/sbin/named-checkconf	检测/etc/named.conf 文件语法
/usr/sbin/named-checkzone	检测区域和对应区域文件的语法
/usr/sbin/rndc	远程 DNS 管理工具
/usr/sbin/rndc-confgen	生成 rndc 密钥

续表

配置文件	说明
/var/named/named.ca	根解析库
/var/named/named.localhost	本地主机解析库
/var/named/slaves	从 DNS 服务器文件夹

1.4.9　BIND 服务资源记录

BIND 服务资源记录格式如表 1-8 所示。

表 1-8　　　　　　　　　　BIND 服务资源记录格式

name	TTL	IN	RRtype	value
资源记录名	有效时间	IN	类型	资源记录值

（1）SOA：该记录只能有一条，而且必须第一个被设置。
- name：只能是区域名称，通常可以简写为@。
- value：主 DNS 服务器的 FQDN。

（2）NS：可以有多条。
- name：区域名称，通常可以简写为@。
- value：DNS 服务器的 FQDN（可使用相对名称）。

（3）A：只能定义在正向区域数据文件中。
- name：FQDN（可以使用相对名称）。
- value：IP 地址。

（4）MX：可以有多个。
- name：区域名称，用于标识 SMTP 服务器。
- value：包含优先级和 FQDN。

注意：优先级为 0～99，数字越小，级别越高。

（5）CNAME。
- name：FQDN。
- value：FQDN。

（6）PTR：IP 地址→FQDN。

只能定义在反向区域数据文件中，反向区域名称格式为逆向网络地址加.in-addr.arpa 后缀。
- name: IP 地址，逆向主机地址。
- value: FQDN。

1.4.10　BIND 配置 DNS 服务器总结

使用 BIND 软件搭建并配置 DNS 服务器，以 duoduo*.com 域名配置为例。

1. 简化 BIND 主配置文件

named.conf 配置文件的代码如下。

```
options {
    directory "/var/named";
};
zone "duoduo*.com" IN {
    type master;
```

```
        file "duoduo*.com.zone";
};
```

2．区域数据配置文件

duoduo*.com 解析的区域数据配置文件代码如下。

```
cd /var/named
touch duoduo*.com.zone
$TTL 7200
duoduo*.com.            IN  SOA duoduo*.com. admin.duoduo*.com. ( 222 1H 10M 1W 1D )
duoduo*.com.            IN  NS  DNS1.duoduo*.com.
duoduo*.com.            IN  NS  DNS2.duoduo*.com.
DNS1.duoduo*.com.  IN   A   192.168.2.33
DNS2.duoduo*.com.  IN   A   192.168.2.34
www.duoduo*.com.        IN  A   192.168.2.100
```

3．检测配置文件语法

检测 BIND 服务的配置文件语法设置是否正确，检测语法代码如下。

```
[root@www.DNS_master_33.com named]#/etc/init.d/named configtest
zone duoduo*.com/IN: loaded serial 222
zone booxin*.com/IN: loaded serial 2
zone 2.168.192.\032in-addr.arpa/IN: loaded serial 2
zone localhost.localdomain/IN: loaded serial 0
zone localhost/IN: loaded serial 0
zone 1.0.0.0.0.0.0.0.0.0.0.0.0.0.0.0.0.0.0.0.0.0.0.0.0.0.0.0.0.0.0.0.ip6.arpa/IN: loaded serial 0
zone 1.0.0.127.in-addr.arpa/IN: loaded serial 0
zone 0.in-addr.arpa/IN: loaded serial 0
```

4．测试解析效果

测试 duoduo*.com 域名的解析设置是否正确，代码如下。

```
[root@www.DNS_master_33.com named]#dig @192.168.2.33 www.duoduo*.com

; <<>> DiG 9.8.2rc1-RedHat-9.8.2-0.68.rc1.el6_10.3 <<>> @192.168.2.33 www.duoduo*.com
; (1 server found)
;; global options: +cmd
;; Got answer:
;; ->>HEADER<<- opcode: QUERY, status: NOERROR, id: 24778
;; flags: qr aa rd ra; QUERY: 1, ANSWER: 1, AUTHORITY: 2, ADDITIONAL: 2

;; QUESTION SECTION:
;www.duoduo*.com.               IN      A

;; ANSWER SECTION:
www.duoduo*.com.        7200IN   A       192.168.2.100

;; AUTHORITY SECTION:
duoduo*.com.            7200IN   NS      DNS2.duoduo*.com.
duoduo*.com.            7200IN   NS      DNS1.duoduo*.com.

;; ADDITIONAL SECTION:
DNS1.duoduo*.com.7200IN    A       192.168.2.33
DNS2.duoduo*.com.7200IN    A       192.168.2.34

;; Query time: 0 msec
;; SERVER: 192.168.2.33#53(192.168.2.33)
;; WHEN: Sun Aug 11 19:51:25 2019
;; MSG SIZE  rcvd: 118
```

上述输出过程总结。

第 1 步．duoduo*.com.二级域名被解析，即 DNS1.duoduo*.com.和 DNS2.duoduo*.com.。

第 2 步．DNS1.duoduo*.com.对应的 IP 地址为 192.168.2.33，DNS2.duoduo*.com.对应的 IP 地址为 192.168.2.34，即 192.168.2.33 和 192.168.2.34 为权威 DNS 服务器的地址。

第 3 步：三级域名 www.duoduo*.com.被解析到了 192.168.2.100 这台服务器，其中，区域配置文件中的条目可以采用缩写模式，代码如下。

```
touch duoduo*.com.zone
$TTL 7200
duoduo*.com.            IN   SOA duoduo*.com. admin.duoduo*.com. ( 222 1H 10M 1W 1D )
duoduo*.com.            IN   NS   DNS1.duoduo*.com.
duoduo*.com.            IN   NS   DNS2.duoduo*.com.
DNS1            IN    A    192.168.2.33
DNS2            IN    A    192.168.2.34
www             IN    A    192.168.2.100
```

重启 DNS 服务器即可，代码如下。

```
DNS1.duoduo*.com.IN    A    192.168.2.33
DNS2.duoduo*.com.IN    A    192.168.2.34
www.duoduo*.com.       IN   A    192.168.2.100
```

上述解析记录可以被缩写，代码如下。

```
DNS1            IN    A    192.168.2.33
DNS2            IN    A    192.168.2.34
www             IN    A    192.168.2.100
```

补充知识："@" 在 BIND 语法中有特殊作用，对比以下两段代码。

第一段代码如下。

```
$TTL 7200
duoduo*.com.            IN   SOA duoduo*.com. admin.duoduo*.com. ( 222 1H 10M 1W 1D )
duoduo*.com.            IN   NS   DNS1.duoduo*.com.
duoduo*.com.            IN   NS   DNS2.duoduo*.com.
DNS1.duoduo*.com.IN    A    192.168.2.33
DNS2.duoduo*.com.IN    A    192.168.2.34
www.duoduo*.com.       IN   A    192.168.2.100
```

第二段代码如下。

```
$TTL 7200
@                       IN   SOA duoduo*.com. admin.duoduo*.com. ( 222 1H 10M 1W 1D )
duoduo*.com.            IN   NS   DNS1.duoduo*.com.
duoduo*.com.            IN   NS   DNS2.duoduo*.com.fair
DNS1.duoduo*.com.IN    A    192.168.2.33
DNS2.duoduo*.com.IN    A    192.168.2.34
www.duoduo*.com.       IN   A    192.168.2.100
```

5. BIND 配置总结

（1）严格注意语法书写，其格式非常严格。

BIND 的配置文件格式非常严格，named.conf 主配置文件中，每一行都必须以 ";" 结尾，并使用 "{}" 闭合。

（2）@是保留字，表示当前域名，书写联系人邮箱时，可使用 "." 替代 "@" 符号。

（3）记录不允许换行书写。

（4）单行记录开头不允许以空格开头。

6. CNAME 记录案例实战

为了后期方便管理和维护 DNS 解析记录，CNAME 记录适用于需要将多个域名指向同一个 IP 地址时的场景。

（1）主配置文件如下。

```
zone "mingze.com" IN {
      type master;
      file "mingze*.com.zone";
};
```

（2）区域数据文件配置，代码如下。

```
[root@www.DNS_master_33.com named]#cat duoduo*.com.zone >mingze*.com.zone
```

```
[root@www.DNS_master_33.com named]#vim mingze*.com.zone
[root@www.DNS_master_33.com named]#chown -R root:named mingze*.com.zone
[root@www.DNS_master_33.com named]#cat mingze*.com.zone
$TTL 7200
mingze*.com.          IN   SOA   mingze*.com. admin.mingze*.com. ( 222 1H 10M 1W 1D )
mingze*.com.          IN   NS    DNS1.mingze*.com.
mingze*.com.          IN   NS    DNS2.mingze*.com.
DNS1.mingze*.com.     IN   A     192.168.2.33
DNS2.mingze*.com.     IN   A     192.168.2.34
www.mingze*.com.      IN   CNAME www.duoduo*.com.
```

上述代码中 www.mingze*.com.与 www.duoduo*.com.解析记录会指向同一个 IP 地址。

(3) 检测语法并重启服务。

重新检测配置语法并重启服务,代码如下。

```
[root@www.DNS_master_33.com named]#/etc/init.d/named configtest
zone duoduo*.com/IN: loaded serial 222
zone mingze*.com/IN: loaded serial 222
zone booxin*.com/IN: loaded serial 2
zone 2.168.192.\032in-addr.arpa/IN: loaded serial 2
zone localhost.localdomain/IN: loaded serial 0
zone localhost/IN: loaded serial 0
zone 1.0.0.0.0.0.0.0.0.0.0.0.0.0.0.0.0.0.0.0.0.0.0.0.0.0.0.0.0.0.0.0.ip6.arpa/IN:
loaded serial 0
zone 1.0.0.127.in-addr.arpa/IN: loaded serial 0
zone 0.in-addr.arpa/IN: loaded serial 0
[root@www.DNS_master_33.com named]#/etc/init.d/named restart
停止 named: umount: /var/named/chroot/var/named: device is busy.
        (In some cases useful info about processes that use
         the device is found by lsof(8) or fuser(1))
                                                              [确定]
启动 named:                                                    [确定]
```

(4) 测试解析结果。

测试解析结果是否正确,代码如下。

```
[root@www.DNS_master_33.com named]#dig @192.168.2.33 www.mingze*.com

; <<>> DiG 9.8.2rc1-RedHat-9.8.2-0.68.rc1.el6_10.3 <<>> @192.168.2.33 www.mingze*.com
; (1 server found)
;; global options: +cmd
;; Got answer:
;; ->>HEADER<<- opcode: QUERY, status: NOERROR, id: 30863
;; flags: qr aa rd ra; QUERY: 1, ANSWER: 2, AUTHORITY: 2, ADDITIONAL: 2

;; QUESTION SECTION:
;www.mingze*.com.              IN    A

;; ANSWER SECTION:
www.mingze*.com.        7200IN  CNAME   www.duoduo*.com.
www.duoduo*.com.        7200IN  A       192.168.2.100

;; AUTHORITY SECTION:
duoduo*.com.            7200IN  NS      DNS2.duoduo*.com.
duoduo*.com.            7200IN  NS      DNS1.duoduo*.com.

;; ADDITIONAL SECTION:
DNS1.duoduo*.com.7200IN A       192.168.2.33
DNS2.duoduo*.com.7200IN A       192.168.2.34

;; Query time: 0 msec
;; SERVER: 192.168.2.33#53(192.168.2.33)
;; WHEN: Sun Aug 11 20:30:01 2019
;; MSG SIZE  rcvd: 143
```

上述代码将 www.mingze*.com.域名解析到 www.duoduo*.com.域名,访问 www.mingze*.com.域名和访问 www.duoduo*.com.域名时,解析到的内容是完全一致的。

第 2 章
DNS 服务进阶

2.1 BIND 实现网站负载均衡实战

2.1.1 主流负载均衡器介绍

负载均衡是指把客户端的请求均匀地分摊到多个服务器上处理，一般常见的负载均衡架构有如下几种。

（1）客户端与反向代理服务器之间的 DNS 负载均衡。

（2）反向代理服务器与应用服务器之间的负载均衡。

（3）反向代理与中间件服务器之间的负载均衡。

（4）反向代理与数据库（关系型、非关系型）之间的负载均衡。

负载均衡的关键在于如何把请求均匀地分摊到各个后端的服务器上，这取决于所选的负载均衡策略及相关算法。本节重点讨论如何使用 DNS 服务实现 Web 负载均衡，并对常见的负载均衡解决方案及其软件做简单解析。

1. Nginx 负载均衡

Nginx 是目前主流的反向代理和负载均衡服务器，可以实现基于 4 层和 7 层协议的反向代理。为了使客户端请求均匀分发到后端服务器，Nginx 内置了 5 种负载均衡策略。

（1）轮询。

Nginx 按次序把客户端请求分发给后端服务器。后端各服务器之间需要配置 Session 同步策略。

（2）加权轮询。

在轮询算法的基础上为后端服务器配置一定的权重比例，配置代码如下。

```
upstream cluster {
        server a weight=1;
        server b weight=2;
        server c weight=4;
                }
```

按照上述配置，Nginx 每收到 7 个客户端的请求，就会把其中的 1 个转发给后端服务器 a，把其中的 2 个转发给后端服务器 b，把其中的 4 个转发给后端服务器 c。

加权轮询算法的目的，就是要生成一个服务器序列。每当有请求到来时，就依次从该序列中取出一个服务器用于处理该请求。针对上面的例子，加权轮询算法会生成序列{c, c, b, c, a, b, c}。这样，Nginx 每收到 7 个客户端的请求，就会按序列顺序转发给后端服务器 a、b、c，若收到第 8 个请求，则重新从该序列的头部开始轮询。

加权轮询的优点是可以根据情况进行调整，可控。但仍然需要进行 Session 同步。

2. LVS 负载均衡

LVS 的全称是 Linux Virtual Server，即 Linux 虚拟服务器。在 Linux 2.6 中，它已经成为内核的一部分，在此之前的内核版本则需要重新编译内核。

LVS 主要用于多服务器的负载均衡，工作在网络层，可以实现高性能、高可用的服务器集群技术，其特点如下。

（1）廉价。

可把许多低性能的服务器组合在一起形成一个超级服务器。

（2）易用。

配置非常简单，且有多种负载均衡的方法。

（3）稳定可靠。

即使集群中的某台服务器无法正常工作，也不影响整体效果。

（4）可扩展性好。

可以无缝添加多台服务器到负载均衡列表中，易于扩展。

（5）一些不足。

由于 LVS 自身的工作特性（工作在网络层），所以 LVS 本身不支持正则表达式处理，不能做动静分离。而现在许多网站在这方面都有较强的需求，这是 Nginx 或 HAProxy+Keepalived 的优势所在。

3．健康检查

负载均衡算法，一般要伴随健康检查算法一起使用。

健康检查算法的作用是对所有的服务器进行存活和健康检测。如果后端某一台服务器的服务出现问题，健康检查算法会将此服务器从服务列表中"摘掉"，让负载均衡算法看不到这台服务器的存在。

2.1.2　BIND 实现轮询基础知识

本节基于 DNS 实现负载均衡。实现目标：使用 BIND 软件，配置基于 DNS 的负载均衡，即一个域名对应后端多个服务器（基于 IP 地址）。

1．基础知识

DNS 服务器除了能解析域名，还具有**负载均衡**的功能，DNS 服务器负载均衡的工作原理如图 2-1 所示。

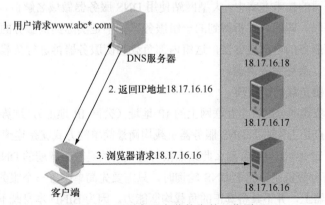

图 2-1　DNS 服务器负载均衡的工作原理

由图 2-1 可以看出，在 DNS 服务器中配置了 3 条 A 记录，代码如下。

```
www.abc*.com IN A 18.17.16.18;
www.abc*.com IN A 18.17.16.17;
www.abc*.com IN A 18.17.16.16;
```

每次域名解析请求都会根据对应的负载均衡算法计算出一个不同的 IP 地址并返回给客户端，这样 A 记录中配置多个服务器就可以构成一个集群，并可以实现负载均衡。如图 2-1 所示，用户向 www.abc*.com 发起请求后，DNS 服务器根据 A 记录和负载均衡算法计算得到一个 IP 地址 18.17.16.16，并返回给客户端，客户端根据该 IP 地址，访问真实的物理服务器 18.17.16.16。这些操作对用户来说都是不透明的，用户只需知道 www.abc*.com 域名即可。

2. DNS 服务器负载均衡优点

DNS 服务器负载均衡的优点如下。

- 将负载均衡的工作交给 DNS 服务器，避免了网站管理、维护负载均衡服务器的麻烦。
- 技术实现比较灵活、方便，简单易行，成本低，适用于大多数 TCP/IP 应用。
- 对部署在服务器上的应用来说，不需要进行任何的代码修改，即可实现不同服务器上的应用访问。
- 服务器可以位于互联网的任意位置。
- 许多 DNS 还支持基于地理位置的域名解析，会将域名解析成距离用户地理位置最近的一个服务器地址，这样就可以加速用户访问，改善性能。

3. DNS 服务器负载均衡缺点

DNS 服务器负载均衡也存在如下缺点。

- 目前的 DNS 服务器是多级解析的，每一级 DNS 服务器都可能缓存 A 记录。某台服务器下线之后，即使修改了 A 记录，要使 DNS 服务器生效也需要较长的时间。这段时间内，DNS 服务器仍然会将域名解析到已下线的服务器上，最终导致用户访问失败。
- 不能按服务器的处理能力来分配负载。因为 DNS 服务器负载均衡采用的是简单的轮询算法，不能区分服务器之间的差异，不能反映服务器当前运行状态，所以其负载均衡效果并不是太好。
- 可能会造成额外的网络问题。为了使 DNS 服务器和其他 DNS 服务器及时交互，保证 DNS 服务器数据及时更新和地址能随机分配，一般都要将 DNS 服务器的刷新时间值设置得较小，但刷新时间值太小可能会使 DNS 服务器流量大增而造成额外的网络问题。

目前互联网公司的技术方案中，大型网站使用 DNS 服务器做域名解析，利用域名解析作为第一级负载均衡手段。即域名解析得到的一组服务器并不是实际提供服务的物理服务器，而是同样提供负载均衡服务的内部服务器，这组内部负载均衡服务器再进行负载均衡，将请求发送到真实的服务器上，最终完成请求。

4. DNS 服务器负载均衡实现

如果给每台服务器都分配一个互联网上的 IP 地址（公网 IP 地址），这势必会占用过多的 IP 地址。判断一个站点是否采用了 DNS 服务器负载均衡最简单的方式就是连续 ping 这个域名。如果多次解析返回的 IP 地址不相同，那么此站点很可能采用的是较为普遍的 DNS 服务器负载均衡。

DNS 轮询没有分级机制。配置 DNS 轮询时，只需要先简单配置一个服务器列表，然后将请求转发到每个服务器上，并不具备真正的负载均衡能力。因为 BIND 本身既不关注服务器负载，也不会对后端服务器做健康检查，所以如果一个服务器宕机，请求仍然会发送到宕机的服务器上。

DNS 服务器负载均衡的优点是简单易用。如果读者维护 Web 服务器集群，想通过一个简单的方法在它们之间实现负载均衡方案，那么 DNS 轮询策略可以满足前期需求。使用 BIND 软件配置 DNS 服务器负载均衡代码如下。

```
fileserv.duoduo*.com.   IN  A   192.169.2.87
fileserv.duoduo*.com.   IN  A   192.168.2.88
fileserv.duoduo*.com.   IN  A   192.168.2.89
```

上述代码在域名解析层面实现了 BIND 负载均衡。

2.1.3　BIND 实现 Web 服务器负载均衡

随着公司规模越来越大，Linux 系统管理员会发现客户端的请求连接数增加时，服务器的响应延时也会随之增加。虽然使用在服务器内部增加 RAM、升级处理器，安装更快的驱动器和总线纵向扩展等方法，短期内会有一定的帮助，但最终会发现一台服务器无法完成预期的任务。因此使用多台服务器配置负载均衡是一个不错的想法，在服务器资源池中增加一定数量的服务器可提升服务器的吞吐能力。

后端服务器资源池中有多台服务器。当一台服务器宕机后，是否会有其他服务器可以无缝接替它的工作而不至于造成业务中断？

上述问题可以通过 DNS 服务器的负载均衡配置进行解决。通过使用循环域名服务（Round-Robin Domain Name System，RR-DNS）可以实现均衡负载的功能，即向一个域名发出的入站请求可以被转发到多个后端服务器的 IP 地址上。BIND 实现 Web 服务器负载均衡架构需要 4 台虚拟机，实验环境部署如表 2-1 所示。

表 2-1　实验环境部署

角色	操作系统	IP 地址	主机名和域名
DNS 服务器	CentOS 6.9 x86_64	**192.168.2.33**	DNS.booxin*.com
Web 服务器	CentOS 6.9 x86_64	192.168.2.35	www.booxin*.com
Web 服务器	CentOS 6.9 x86_64	192.168.2.36	www.booxin*.com
Windows 客户端	Windows 10	192.168.2.37	test_client

1. 配置 DNS 服务器

在 BIND 配置文件中添加两条 A 记录到 www.booxin*.com 区域配置文件中，代码如下，区域文件具体内容如图 2-2 所示。

```
www            IN    A      192.168.2.35
www            IN    A      192.168.2.36
```

系统管理员可以根据需要加入更多解析记录，解析 www.booxin*.com 时将有 50%的概率解析到 192.168.2.35 服务器上，也有 50%的概率解析到 192.168.2.36 服务器上。使用 RR-DNS 方法实现负载均衡也会带来一些问题。

图 2-2　区域文件具体内容

（1）DNS 服务器是一个分布式系统，是按照一定的层次结构组织的。

当用户将域名解析请求提交给本地 DNS 服务器时，本地 DNS 服务器不能直接解析该域名，而是转发请求至上一级 DNS 服务器，上一级 DNS 服务器再依次向上提交，直到 RR-DNS 服务器把这个域名解析到其中一台服务器的 IP 地址。从用户到 RR-DNS 服务器间存在多台服务器，而它们都会缓存已解析的域名到 IP 地址的映射，这会导致该 DNS 服务器组下所有用户都会访问同一 Web 服务器，出现不同 Web 服务器间的负载不均衡。

为了保证 DNS 服务器中域名到 IP 地址的映射不被长久缓存，RR-DNS 在域名到 IP 地址的映射上设置了一个 TTL 值。一段时间后，DNS 服务器将这个映射从缓冲中淘汰，当用户请求时，它会重新向上一级 DNS 服务器提交请求并进行映射。

上述问题涉及如何设置 TTL 值。若该值设置过大，处于 TTL 期间，很多请求会被映射到同一台 Web 服务器上，同样会导致负载不均衡。若该值设置过小，例如 0，会导致本地域名服务器频繁地向 RR-DNS 提交请求，增加了域名解析的网络流量，同样会使 RR-DNS 成为系统中一个新的瓶颈。

（2）客户端会缓存从域名到 IP 地址的映射，而不受 TTL 值的影响，用户的访问请求会被送到同一台 Web 服务器上。

由于用户访问请求的突发性和访问方式不同，例如有的用户访问一下就离开了，而有的用户访问可长达几个小时，因此各台服务器的负载仍存在不均衡。

（3）系统的可靠性和可维护性不好。

若一台服务器失效，会导致正在解析域名的用户服务中断，即使用户单击"Reload"按钮也无济于事。

系统管理员也不可能随时随地对一台服务器进行维护。

进行操作系统和应用软件升级时，需要修改 RR-DNS 服务器中的 IP 地址列表，把该服务器的 IP 地址从中剔除，然后等待一段时间，等所有域名服务器将该域名到这台服务器的 IP 地址映射淘汰，和所有映射到这台服务器的客户端不再使用该站点为止。

RR-DNS 方法只是一个简单的负载均衡方案，如果有更高需求，可以研究 LVS 集群（IPVS 和 KTCPVS、TCPHA），实现基于 IP 地址的负载均衡和基于内容的负载均衡。

2. Web 服务器配置

（1）192.168.2.35 服务器验证 IP 地址和 DNS 服务器地址设置。

```
[root@www.booxin*.com-35 ~]#ip ro
192.168.1.0/24 dev eth0  proto kernel  scope link  src 192.168.2.35
169.254.0.0/16 dev eth0  scope link  metric 1002
default via 192.168.1.1 dev eth0
[root@www.booxin*.com-35 ~]#grep -i DNS /etc/sysconfig/network-scripts/ifcfg-eth0
DNS1=192.168.2.33
[root@www.booxin*.com-35 ~]#grep -i nameserver /etc/resolv.conf
nameserver 192.168.2.33
nameserver 202.106.195.68
```

上述代码验证 DNS 服务器地址和 IP 地址配置是否正确，返回结果为 192.168.2.33，说明配置正确。

（2）192.168.2.36 服务器验证 IP 地址和 DNS 服务器地址设置。

```
[root@www.booxin*.com-36~]#ip ro
192.168.1.0/24 dev eth0  proto kernel  scope link  src 192.168.2.36
169.254.0.0/16 dev eth0  scope link  metric 1002
default via 192.168.1.1 dev eth0
[root@www.booxin*.com-36~]#grep -i DNS /etc/sysconfig/network-scripts/ifcfg-eth0
DNS1=192.168.2.33
[root@www.booxin*.com-36~]#grep -i nameserver /etc/resolv.conf
nameserver 192.168.2.33
nameserver 202.106.195.60
```

上述代码验证 DNS 服务器地址和 IP 地址配置是否正确，从输出结果中可以看到二者保持了一致，说明配置正确。

（3）192.168.2.35 服务器通过 ping 指令验证负载均衡配置是否正确。

```
1   [root@www.booxin*.com-35 ~]#ping www.booxin*.com -c 2
2   PING www.booxin*.com (192.168.2.36) 56(84) bytes of data.
3   64 bytes from 192.168.2.36: icmp_seq=1 ttl=64 time=0.155 ms
4   64 bytes from 192.168.2.36: icmp_seq=2 ttl=64 time=0.179 ms
5
6   --- www.booxin*.com ping statistics ---
7   2 packets transmitted, 2 received, 0% packet loss, time 1000ms
8   rtt min/avg/max/mdev = 0.155/0.167/0.179/0.012 ms
9   [root@www.booxin*.com-35 ~]#ping www.booxin*.com -c 2
10  PING www.booxin*.com (192.168.2.35) 56(84) bytes of data.
11  64 bytes from 192.168.2.35: icmp_seq=1 ttl=64 time=0.006 ms
12  64 bytes from 192.168.2.35: icmp_seq=2 ttl=64 time=0.015 ms
13
14  --- www.booxin*.com ping statistics ---
15  2 packets transmitted, 2 received, 0% packet loss, time 1000ms
16  rtt min/avg/max/mdev = 0.006/0.010/0.015/0.005 ms
```

上述代码验证负载均衡服务器的配置是否正确，其中第 2～4 行将 www.booxin*.com 域名解析为 192.168.2.36，第 10～12 行将 www.booxin*.com 解析为 192.168.2.35。

（4）192.168.2.36 服务器通过 ping 指令验证负载均衡配置是否正确。

```
1   [root@www.booxin*.com-36~]#ping www.booxin*.com -c 2
2   PING www.booxin*.com (192.168.2.36) 56(84) bytes of data.
3   64 bytes from 192.168.2.36: icmp_seq=1 ttl=64 time=0.008 ms
4   64 bytes from 192.168.2.36: icmp_seq=2 ttl=64 time=0.015 ms
5
6   --- www.booxin*.com ping statistics ---
7   2 packets transmitted, 2 received, 0% packet loss, time 999ms
8   rtt min/avg/max/mdev = 0.008/0.011/0.015/0.004 ms
9   [root@www.booxin*.com-36~]#ping www.booxin*.com -c 2
10  PING www.booxin*.com (192.168.2.35) 56(84) bytes of data.
11  64 bytes from 192.168.2.35: icmp_seq=1 ttl=64 time=0.129 ms
12  64 bytes from 192.168.2.35: icmp_seq=2 ttl=64 time=0.151 ms
13
14  --- www.booxin*.com ping statistics ---
15  2 packets transmitted, 2 received, 0% packet loss, time 1000ms
16  rtt min/avg/max/mdev = 0.129/0.140/0.151/0.011 ms
```

上述代码验证负载均衡服务器的配置是否正确，使用 ping 指令测试 www.booxin*.com 的返回结果，其中第 2～4 行返回的 IP 地址为 192.168.2.36，第 10～12 行返回的 IP 地址为 192.168.2.35，可见对 www.booxin*.com 负载均衡配置准确无误。

（5）192.168.2.35 服务器设置 Web 服务。

使用 yum 指令在线安装 Nginx 软件，代码如下。

```
[root@www.booxin*.com-35 ~]#yum -y install nginx >/dev/null  2>&1
[root@www.booxin*.com-35 ~]#echo $?
0
[root@www.booxin*.com-35 ~]#rpm -qa |grep nginx
nginx-mod-http-image-filter-1.10.2-1.el6.x86_64
nginx-mod-http-xslt-filter-1.10.2-1.el6.x86_64
nginx-1.10.2-1.el6.x86_64
nginx-all-modules-1.10.2-1.el6.noarch
nginx-filesystem-1.10.2-1.el6.noarch
nginx-mod-mail-1.10.2-1.el6.x86_64
nginx-mod-stream-1.10.2-1.el6.x86_64
nginx-mod-http-perl-1.10.2-1.el6.x86_64
nginx-mod-http-geoip-1.10.2-1.el6.x86_64
```

上述代码的作用是安装 Nginx，并使用 rpm 指令查看安装的软件包组件。

```
[root@www.booxin*.com-35 ~]#cd /etc/nginx/conf.d/
[root@www.booxin*.com-35 conf.d]#cp -av default.conf www.booxin*.conf
'default.conf' -> 'www.booxin*.conf'
```

```
[root@www.booxin*.com-35 conf.d]#vim www.booxin*.conf
[root@www.booxin*.com-35 conf.d]#cat www.booxin*.conf
server {
   listen      80 ;
   server_name www.booxin*.com;
   root        /usr/share/nginx/html;
   include /etc/nginx/default.d/*.conf;

   location / {
   }

   error_page 404 /404.html;
       location = /40x.html {
   }

   error_page 500 502 503 504 /50x.html;
       location = /50x.html {
   }
}
[root@www.booxin*.com-35 conf.d]#cd /usr/share/nginx/html/
[root@www.booxin*.com-35 html]#rz -bey
-bash: rz: command not found
[root@www.booxin*.com-35 html]#yum -y install lrzsz >/dev/null 2>&1
[root@www.booxin*.com-35 html]#rz -bey
[root@www.booxin*.com-35 html]#/etc/init.d/nginx restart
Stopping nginx:                                           [FAILED]
Starting nginx:                                           [ OK ]
[root@www.booxin*.com-35 html]#ip ro
192.168.1.0/24 dev eth0  proto kernel  scope link  src 192.168.2.35
169.254.0.0/16 dev eth0  scope link  metric 1002
default via 192.168.1.1 dev eth0
[root@www.booxin*.com-35 html]#ll
total 24
-rw-r--r-- 1 root root  307 Feb 16 22:10 35.html
-rw-r--r-- 1 root root 3652 Oct 31  2016 404.html
-rw-r--r-- 1 root root 3695 Oct 31  2016 50x.html
-rw-r--r-- 1 root root 3698 Oct 31  2016 index.html
-rw-r--r-- 1 root root  368 Oct 31  2016 nginx-logo.png
-rw-r--r-- 1 root root 2811 Oct 31  2016 poweredby.png
[root@www.booxin*.com-35 html]#rm -fv index.html
removed 'index.html'
[root@www.booxin*.com-35 html]#ln -sv 35.html index.html
'index.html' -> '35.html'
[root@www.booxin*.com-35 html]#ll
total 20
-rw-r--r-- 1 root root  307 Feb 16 22:10 35.html
-rw-r--r-- 1 root root 3652 Oct 31  2016 404.html
-rw-r--r-- 1 root root 3695 Oct 31  2016 50x.html
lrwxrwxrwx 1 root root    8 Feb 16 22:18 index.html -> 35.html
-rw-r--r-- 1 root root  368 Oct 31  2016 nginx-logo.png
-rw-r--r-- 1 root root 2811 Oct 31  2016 poweredby.png
[root@www.booxin*.com-35 html]#vim index.html
<!DOCTYPE html>
<html lang="en">
  <head>
    <meta charset="utf-8">
    <title>35-test-page</title>
    <style type="text/css">
      span {
        color:pink;
        font-size:80px;
      }
    </style>
  </head>
```

```
    <body>
        <span>This is a 192.168.2.35 test page</span>
    </body>
</html>
```

上述代码创建 Nginx.conf 配置文件,先设置服务器的 ServerName,然后编写一个简单的 HTML 静态页面,浏览器访问效果如图 2-3 所示。

(6) 192.168.2.36 服务器设置 Web 服务。

图 2-3 浏览器访问效果 1

安装 Nginx 软件,提供 Web 服务,代码如下。

```
[root@[root@www.booxin*.com-36~]#yum -y install nginx >/dev/null  2>&1
[root@[root@www.booxin*.com-36~]#echo $?
0
[root@[root@www.booxin*.com-36~]#rpm -qa |grep nginx
nginx-mod-http-image-filter-1.10.2-1.el6.x86_64
nginx-mod-http-xslt-filter-1.10.2-1.el6.x86_64
nginx-1.10.2-1.el6.x86_64
nginx-all-modules-1.10.2-1.el6.noarch
nginx-filesystem-1.10.2-1.el6.noarch
nginx-mod-mail-1.10.2-1.el6.x86_64
nginx-mod-stream-1.10.2-1.el6.x86_64
nginx-mod-http-perl-1.10.2-1.el6.x86_64
nginx-mod-http-geoip-1.10.2-1.el6.x86_64
```

设置 Nginx.conf 配置文件,设置 ServerName,并编写简单的 HTML 静态页面,代码如下。

```
[root@[root@www.booxin*.com-36html]#ll
total 20
-rw-r--r-- 1 root root  305 Feb 16 22:25 36.html
-rw-r--r-- 1 root root 3652 Oct 31  2016 404.html
-rw-r--r-- 1 root root 3695 Oct 31  2016 50x.html
lrwxrwxrwx 1 root root    8 Feb 16 22:30 index.html -> 36.html
-rw-r--r-- 1 root root  368 Oct 31  2016 nginx-logo.png
-rw-r--r-- 1 root root 2811 Oct 31  2016 poweredby.png
[root@[root@www.booxin*.com-36html]#cat index.html
<!DOCTYPE html>
<html lang="en">
  <head>
    <meta charset="utf-8">
    <title>127-test-page</title>
    <style type="text/css">
      span {
        color:blue;
        font-size:80px;
      }
     </style>
  </head>

  <body>
    <span>This is a 192.168.2.36 test page</span>
  </body>
</html>
[root@[root@www.booxin*.com-36html]#cat /etc/nginx/conf.d/www.booxin*.com.conf
server {
    listen       80 ;
    server_name  www.booxin*.com;
    root         /usr/share/nginx/html;

    # Load configuration files for the default server block.
    include /etc/nginx/default.d/*.conf;

    location / {
    }
```

```
    error_page 404 /404.html;
        location = /40x.html {
    }

    error_page 500 502 503 504 /50x.html;
        location = /50x.html {
    }
}
```

浏览器访问效果如图 2-4 所示。

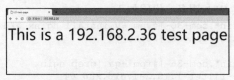

图 2-4　浏览器访问效果 2

3．Windows 10 客户端配置

Windows 10 客户端配置如图 2-5 所示。

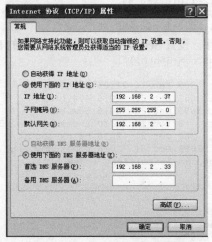

图 2-5　Windows 10 客户端配置

4．Windows 10 客户端测试

测试结果如图 2-6 和图 2-7 所示。

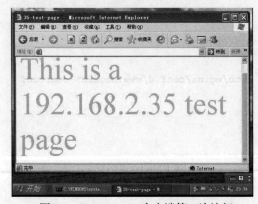

图 2-6　Windows 10 客户端第一次访问

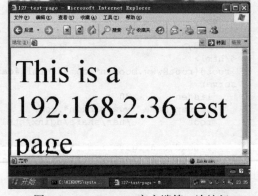

图 2-7　Windows 10 客户端第二次访问

注意：由于第一次通过浏览器访问后会在本地留下缓存，因此需要使用如下指令清除缓存，具体代码如下。

```
C:\>ipconfig /flushDNS

Windows IP Configuration

Successfully flushed the DNS Resolver Cache.

C:\>ping www.booxin*.com

Pinging www.booxin*.com [192.168.2.36] with 32 bytes of data:

Reply from 192.168.2.36: bytes=32 time<1ms TTL=64
Reply from 192.168.2.36: bytes=32 time<1ms TTL=64

Ping statistics for 192.168.2.36:
    Packets: Sent = 2, Received = 2, Lost = 0 (0% loss),
Approximate round trip times in milli-seconds:
    Minimum = 0ms, Maximum = 0ms, Average = 0ms
Control-C
^C
C:\>^Q^Q
```

上述代码清除 DNS 缓存后，可以看到再次执行 ping www.booxin*.com，发现其已经是 DNS 服务器设置的第二台服务器了。

2.1.4　BIND 实现 Web 服务器负载均衡总结

使用 BIND 实现 Web 服务器负载均衡配置，总结如下。
- 设置 DNS 服务器负载均衡。
- 检测 DNS 服务语法、启动 DNS 服务。
- 设置客户端（Linux）。

Linux 操作系统客户端设置地址 1，代码如下。

```
[root@www.bj-Apache.com conf.d]#cat /etc/resolv.conf
nameserver 192.168.2.33
nameserver 192.168.2.34
```

Linux 操作系统客户端设置地址 2，代码如下。

```
[root@www.bj-nginx.com ~]#cat /etc/resolv.conf
nameserver 192.168.2.33
nameserver 192.168.2.34
```

Windows 操作系统 DNS 客户端设置地址如图 2-8 所示。

Linux 客户端通过传参的方式进行访问，以便监听 www.booxin*.com 的服务器，区分是来自哪个客户端的哪次访问，代码如下。

```
[root@www.DNS_slave_34.com ~]#curl -G -d "name=1" http://www.booxin*.com
<!DOCTYPE html>
<html lang="en">
  <head>
    <meta charset="utf-8">
    <title>35-test-page</title>
    <style type="text/css">
      span {
        color:pink;
        font-size:80px;
      }
```

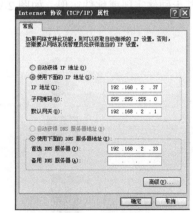

图 2-8　Windows 操作系统 DNS 客户端设置地址

```
        </style>
    </head>
    <body>
        <span>This is a 192.168.2.35 test page</span>
    </body>
</html>
```

2.1.5 BIND 实现 DNS 轮询探讨

DNS 轮询会受到多方面的影响。
- A 记录的 TTL 值大小的影响。
- 其他 DNS 服务器缓存的影响，Windows 客户端也有一个 DNS 缓存。

上述因素都会影响 DNS 轮询的效果，因此 DNS 轮询并不能作为一个负载均衡的解决方案，只能作为一个负载分配方案。可以使用参数调整轮询的效果，在 named.conf 配置文件中可以设置 BIND 的 round-robin 以调整结果的顺序，代码如下。

```
options {
rrset-order { order random; };
};
```

rrset-order 支持 3 个参数：fixed、random、cyclic。
- **fixed**：会将多个 A 记录按配置文件的顺序固定列出。
- **random**：会随机给出 A 记录。
- **cyclic**：会循环给出 A 记录。

> 注意：BIND 9 默认不支持 cyclic 算法，使用该算法需重新编译 BIND，且加入支持该算法的编译选项。

2.1.6 BIND 实现网站负载均衡深入探讨

当网站的访问量较大时，会考虑负载均衡技术，该技术也是每一个架构师的基本功。尤其是在后期架构扩展或架构重构的时候，架构师对全局的掌控能力和对架构细节的把握是非常重要的。

负载均衡对架构的可扩展性、健壮性等，不但具有指导性的意义，更关键的是可以提升业务在服务器宕机或不可用时继续为用户提供服务的能力。传统的负载均衡实现思路是单点的，基于硬件实现或软件实现，如图 2-9 所示。

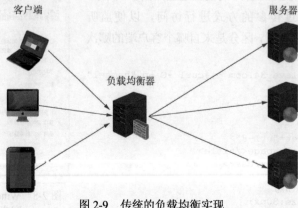

图 2-9　传统的负载均衡实现

图 2-9 所示为使用 BIND 软件实现负载均衡，该架构基本可以解决一般的需求问题，且维护起来相对简单。

1．传统思路的局限性

如图 2-9 所示，传统思路存在非常明显的局限性，网站的响应速度很大程度上依赖于负载均衡节点的能力，而且一旦负载均衡节点本身宕机不可用后，整个网站就会完全瘫痪。

2．CDN 实现负载均衡

作为互联网上承载大部分流量的一大基础设施，内容分发网络（Content Delivery Network，CDN）对负载分流的解决思路很具有启发性，如图 2-10 所示。

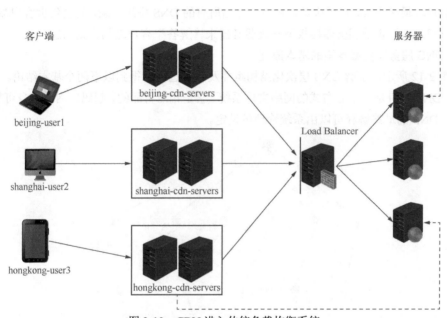

图 2-10　CDN 进入传统负载均衡系统

如图 2-10 所示，用户的访问被第一层负载均衡器分流了，所有的请求不再被聚集到一个节点上，而是被分摊在了各个负载均衡器的各个节点上。即使存在单点故障，也只会影响到一部分用户，而且可以使用其他技术实现故障转移。同样的思路也可以用于传统的 B/S 架构中，系统管理员可以把用户的请求直接分流到不同的服务器上，而不必经过一个统一的节点中转，这个分流可以通过 DNS 技术结合负载均衡器协同实现。

3．DNS 的基本工作流程

（1）Windows 10 操作系统配置 DNS 服务器地址。

访问互联网需要查询 DNS 服务器，Windows 10 操作系统 DNS 服务器设置如图 2-11 所示。

（2）DNS 的基本工作流程（以浏览器访问 Web 服务器为例）。

浏览器访问 Web 服务器的基本流程如下。

图 2-11　Windows 10 操作系统 DNS 服务器设置

- 浏览器通过访问 DNS 服务器，查询域名所对应的 IP 地址。
- DNS 服务器将对应的 IP 地址返回给浏览器。
- 浏览器通过 IP 地址向 Web 服务器发送资源请求，如文字、图片、音频、视频。
- Web 服务器根据收到的资源请求，向计算机获取资源。
- 计算机将资源返回给 Web 服务器。
- Web 服务器将资源返回给浏览器。

4．授权机制（SOA）

根域服务器拥有一切域名的起始解释权。对于客户端的一般查询，根域服务器一般不会直接告诉客户端最终答案，根域服务器也不会存储所有的 DNS 资源记录，它会告诉客户端应该去哪里查询记录，即**根域服务器授权下一级服务器**来"解答"客户端的查询问题。

5．DNS 服务器负载均衡的基本原理

如图 2-12 所示，了解 DNS 层次化结构组件及基本过程，得到以下两个基本结论。

（1）DNS 本身是一个分布式的网络文件系统，它是相对可靠的，起码比网站本身可靠得多。

（2）DNS 的最终解释可以由系统管理员设定。

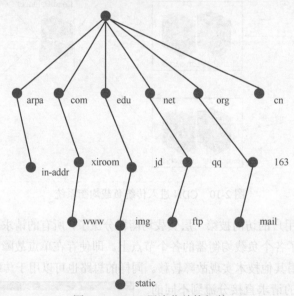

图 2-12　DNS 层次化结构组件

根据上述结论，系统管理员可以将所有服务器地址按照需求制定频次，返回给用户。以 github 网站（它的网址后缀是.com）为例，首先获取它的 SOA 服务器（因为 DNS 缓存查询服务器会缓存结果，直接查询域名，会返回相同的结果），.com 的 DNS 服务器也是 13 台，它们是[a-m].gtld-servers.net，可以根据需要选择一台查询 github 网站的 SOA 起始授权记录。

```
    [root@www.DNS_master_33.com ~]#dig @a.gtld-servers.net github网站

; <<>> DiG 9.8.2rc1-RedHat-9.8.2-0.68.rc1.el6_10.1 <<>> @a.gtld-servers.net
; (2 servers found)
;; global options: +cmd
;; Got answer:
;; ->>HEADER<<- opcode: QUERY, status: NOERROR, id: 47267
;; flags: qr rd; QUERY: 1, ANSWER: 0, AUTHORITY: 8, ADDITIONAL: 1
;; WARNING: recursion requested but not available
```

```
;; QUESTION SECTION:
;; AUTHORITY SECTION:
github.com.          172800   IN   NS   ns1.p16.dynect.net.
github.com.          172800   IN   NS   ns3.p16.dynect.net.
github.com.          172800   IN   NS   ns2.p16.dynect.net.
github.com.          172800   IN   NS   ns4.p16.dynect.net.
github.com.          172800   IN   NS   ns-520.awsDNS-01.net.
github.com.          172800   IN   NS   ns-421.awsDNS-52.com.
github.com.          172800   IN   NS   ns-1707.awsDNS-21.co.uk.
github.com.          172800   IN   NS   ns-1283.awsDNS-32.org.

;; ADDITIONAL SECTION:
ns-421.awsDNS-52.com.    172800   IN   A    205.251.193.165

;; Query time: 313 msec
;; SERVER: 192.5.6.30#53(192.5.6.30)
;; WHEN: Sun Feb 17 20:53:03 2019
;; MSG SIZE  rcvd: 264
```

上述代码中，总共获取了 8 个 SOA 服务器的 NS 记录，再选择一个 DNS 服务器查询 github.com 对应的解析记录，多尝试几次看看最终的 IP 地址会不会变化，代码及其输出如下。

```
[root@www.DNS_master_33.com ~]#dig @ns1.p16.dynect.net github.com

; <<>> DiG 9.8.2rc1-RedHat-9.8.2-0.68.rc1.el6_10.1 <<>> @ns1.p16.dynect.net github.com
; (2 servers found)
;; global options: +cmd
;; Got answer:
;; ->>HEADER<<- opcode: QUERY, status: NOERROR, id: 60845
;; flags: qr aa rd; QUERY: 1, ANSWER: 3, AUTHORITY: 8, ADDITIONAL: 0
;; WARNING: recursion requested but not available

;; QUESTION SECTION:
;github.com.               IN   A

;; ANSWER SECTION:
github.com.          60    IN   A    13.229.188.59
github.com.          60    IN   A    13.250.177.223
github.com.          60    IN   A    52.74.223.119

;; AUTHORITY SECTION:
github.com.          900   IN   NS   ns3.p16.dynect.net.
github.com.          900   IN   NS   ns-520.awsDNS-01.net.
github.com.          900   IN   NS   ns1.p16.dynect.net.
github.com.          900   IN   NS   ns-421.awsDNS-52.com.
github.com.          900   IN   NS   ns-1707.awsDNS-21.co.uk.
github.com.          900   IN   NS   ns-1283.awsDNS-32.org.
github.com.          900   IN   NS   ns4.p16.dynect.net.
github.com.          900   IN   NS   ns2.p16.dynect.net.

;; Query time: 267 msec
;; SERVER: 208.78.70.16#53(208.78.70.16)
;; WHEN: Sun Feb 17 20:54:42 2019
;; MSG SIZE  rcvd: 296
```

上述代码查询了 3 次记录，注意 ANSWER SECTION 返回了 3 个结果，分别是 13.229.188.59、13.250.177.223、52.74.223.119。从查询结果中可以看到，github.com 网站利用 DNS 实现了负载均衡，用户访问域名时最终会通过负载均衡器转发请求到达不同的后端服务器，从而返回不同的 IP 地址。

6. 智能 DNS 服务器供应商

DNS 服务器负载均衡是一种比较高级的服务，一般域名供应商的 DNS 服务器不支持。

推荐使用 DNSPod 或者阿里云 DNS 供应商，而且现在的 DNS 供应商一般都支持域名的健康检查服务，尤其是对于跨机房或者分布在多个区域的机房，该功能会非常有用。

7. 总结

其实 DNS 服务器可以实现的功能远不止这些，还可以进行**故障转移、按地区解析**等。例如阿里云的 GTM 可以实现全局仲裁 DNS，该场景在双机房中应用较多。

域名从互联网诞生之初就存在了，但是对它的研究，以及衍生出来的使用方法才刚刚被发掘。随着互联网的广泛应用，此类技术会越来越多，如阿里云的 GTM 等。

2.2 DNS 服务器部署实战

2.2.1 从 DNS 服务器应用场景

1. 为什么需要配置从 DNS 服务器

从 DNS 服务器也叫辅助 DNS 服务器。如果网络上某个节点只有一台 DNS 服务器。首先，服务器的抗压能力是有限的，当压力达到一定的程度时，服务器就会宕机罢工；其次，如果这台服务器出现了硬件故障，那么服务器管理的区域的域名将无法访问。

为了解决这些问题，最好的办法就是使用多台 DNS 服务器同时工作，并实现数据的同步，这样多台服务器就都可以实现域名解析操作。

2. 实现故障转移，增强系统健壮性

当只有一台 DNS 服务器时，所有的域名解析过程都由这台 DNS 服务器负责，不仅压力极大，而且极不安全。因为这台 DNS 服务器宕机后，所有域名的解析服务都将停止，整个网站也就无法被访问了，所以无论是出于负载均衡的考虑，还是服务可用性的考虑，至少都应该配置 2 台或 2 台以上的 DNS 服务器。

其中一台必须是主 DNS 服务器，其余的是从 DNS 服务器，主 DNS 服务器和从 DNS 服务器都可以配置成向外提供解析服务，但从 DNS 服务器上的区域数据从何而来？不是从 DNS 服务器自己写区域数据文件，而是从主 DNS 服务器上复制而来的。主从架构，其实是从主 DNS 服务器复制区域数据文件到从 DNS 服务器的过程，其对应的 DNS 专业术语为"区域传送"。

3. 配置主、从 DNS 服务器的基本条件

主 DNS 服务器架设好后，从 DNS 服务器的架设就相对简单多了。架设主、从 DNS 服务器需要如下两个基本条件。

- 两台主机不一定处在同一网段，但是两台主机之间必须要实现网络通信。
- 从 DNS 服务器必须要有主 DNS 服务器的授权，才可以正常操作。

2.2.2 DNS 主从同步原理

DNS 主从同步原理如图 2-13 所示，DNS 主从同步原理具体描述如下。

2.2 DNS 服务器部署实战 | 73

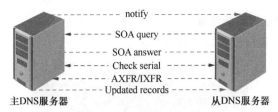

图 2-13 DNS 主从同步原理

- 主 DNS 服务器每次修改完成并重启服务后，将传送 notify 给所有的从 DNS 服务器。
- 从 DNS 服务器将查询主 DNS 服务器的 SOA 记录。（SOA query。）
- 主 DNS 服务器收到请求后，将 SOA 记录发送给从 DNS 服务器，从 DNS 服务器收到后对比查询结果中的 serial 值。（SOA answer, Check serial。）
- 如果 serial 值小于或等于本机设定值，则将结束数据同步过程；但如果 serial 值大于本机设定值，从 DNS 服务器将发送 zone transfer 请求要求 AXFR/IXFR，主 DNS 服务器响应 zone transfer 请求并传送数据，直到整个从 DNS 服务器更新完成。（AXFR/IXFR，Updated records。）

2.2.3 DNS 主从同步架构选型

架构选型不要局限于原生的 DNS 主从同步架构，实际工作中有很多方案可以实现基于 DNS 服务器的区域数据文件和配置文件同步，如 rsync+inotofy、lsyncd+inotify、svn、git，或者使用 NFS 共享等，都可以实现高质量的 DNS 主从同步。

2.2.4 DNS 主从实验环境介绍

配置 192.168.2.33 为主 DNS 服务器，192.168.2.34 为从 DNS 服务器，DNS 服务器实验环境部署如表 2-2 所示。

表 2-2　　　　　　　　　　　DNS 服务器实验环境部署

角色	操作系统	IP 地址	主机名和域名
主 DNS 服务器	CentOS 6.9 x86_64	192.168.2.33	dns.m.booxin*.com
从 DNS 服务器	CentOS 6.9 x86_64	192.168.2.34	dns.s.booxin*.com
Windows 10 客户端	Windows 10	192.168.2.37	Hanyw

2.2.5 主 DNS 服务器设置

1. 主 DNS 服务器正向解析设置

主 DNS 服务器正向解析设置如图 2-14 所示。

```
30 // hanyanwei-2019-07-07 start
31 zone "booxin.com" IN {
32     type master;
33     file "booxin.com.zone";
34     allow-transfer { 192.168.2.34; };
35     notify yes;
36 };
37 // hanyanwei-2019-07-07 end
```

图 2-14 主 DNS 服务器正向解析设置

2. 主 DNS 服务器反向解析设置

主 DNS 服务器反向解析设置如图 2-15 所示。

```
42 // hanyanwei-2019-07-07   start
43 zone "2.168.192. in-addr.arpa" IN {
44         type master;
45         file "booxin.com.rev";
46         allow-update { none;  };
47         allow-transfer { 192.168.2.3 };
48         notify yes;
49
50 };
51 // hanyanwei-2019-07-07   end
```

图 2-15 主 DNS 服务器反向解析设置

3. 检测配置文件语法

主 DNS 服务器检测配置文件语法是否有误，代码如下。

```
[root@www.DNS_master_33.com named]#named-checkconf -z
zone booxin*.com/IN: loaded serial 0
zone 1.168.192.\032in-addr.arpa/IN: loaded serial 0
zone localhost.localdomain/IN: loaded serial 0
zone localhost/IN: loaded serial 0
zone 1.0.0.0.0.0.0.0.0.0.0.0.0.0.0.0.0.0.0.0.0.0.0.0.0.0.0.0.0.0.0.0.ip6.arpa/
IN: loaded serial 0
zone 1.0.0.127.in-addr.arpa/IN: loaded serial 0
zone 0.in-addr.arpa/IN: loaded serial 0
```

上述代码检测配置文件语法准确无误。

4. 查看 BIND 主配置文件

查看 BIND 主配置文件，代码如下。

```
[root@www.DNS_master_33.com ~]#nl /etc/named.conf
     1  options {
     2      listen-on port 53 { 127.0.0.1; 192.168.2.33; };
     3      directory    "/var/named";
     4      dump-file    "/var/named/data/cache_dump.db";
     5          statistics-file "/var/named/data/named_stats.txt";
     6          memstatistics-file "/var/named/data/named_mem_stats.txt";
     7      allow-query     { 192.168.2.0/24; };
     8      recursion yes;
     9      DNSsec-enable yes;
    10      DNSsec-validation yes;
    11      BINDkeys-file "/etc/named.iscdlv.key";
    12      managed-keys-directory "/var/named/dynamic";
    13  };
    14  logging {
    15          channel default_debug {
    16                  file "data/named.run";
    17                  severity dynamic;
    18          };
    19  };
    20  zone "." IN {
    21      type hint;
    22      file "named.ca";
    23  };
    24  zone "booxin*.com" IN {
    25      type master;
    26      file "booxin*.com.zone";
    27  };
    28  zone "2.168.192. in-addr.arpa" IN {
    29      type master;
    30      file "booxin*.com.rev";
    31  };
    32  include "/etc/named.rfc1912.zones";
    33  include "/etc/named.root.key";
```

上述代码中，第 1~33 行为 BIND 主配置文件内容，可根据实际需求进行调整。

5．检测 BIND 配置文件语法

检测 BIND 配置文件语法，代码如下。

```
[root@www.DNS_master_33.com ~]#/etc/init.d/named configtest
zone booxin*.com/IN: loaded serial 0
zone 2.168.192.\032in-addr.arpa/IN: loaded serial 0
zone localhost.localdomain/IN: loaded serial 0
zone localhost/IN: loaded serial 0
zone 1.0.0.0.0.0.0.0.0.0.0.0.0.0.0.0.0.0.0.0.0.0.0.0.0.0.0.0.0.0.0.0.ip6.arpa/
IN: loaded serial 0
zone 1.0.0.127.in-addr.arpa/IN: loaded serial 0
zone 0.in-addr.arpa/IN: loaded serial 0
```

上述代码检测配置文件语法准确无误。

6．serial 加 1 并重启服务

设置 serial，如图 2-16 所示。

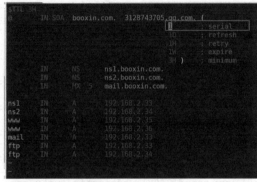

图 2-16　设置 serial

```
[root@www.DNS_master_33.com named]#pwd
/var/named/chroot/var/named
[root@www.DNS_master_33.com named]#ll
总用量 40
-rw-r--r-- 1 root   named   704 7月  16 22:20 booxin*.com.rev
-rw-r--r-- 1 root   named   725 7月  17 21:22 booxin*.com.zone
drwxr-x--- 7 root   named  4096 6月  30 22:43 chroot
drwxrwx--- 2 named  named  4096 7月  17 21:25 data
drwxrwx--- 2 named  named  4096 7月  23 22:22 dynamic
-rw-r----- 1 root   named  3289 4月  11 2017 named.ca
-rw-r----- 1 root   named   152 12月 15 2009 named.empty
-rw-r----- 1 root   named   152 6月  21 2007 named.localhost
-rw-r----- 1 root   named   168 12月 15 2009 named.loopback
drwxrwx--- 2 named  named  4096 6月  19 00:19 slaves
[root@www.DNS_master_33.com named]#vim booxin*.com.zone
[root@www.DNS_master_33.com named]#vim booxin*.com.rev
```

上述代码需要修改域名区域配置文件（正向解析、反向解析均需设置）。

7．重启 BIND 服务

重启 BIND 服务，代码如下。

```
[root@www.DNS_master_33.com named]#/etc/init.d/named restart
停止 named: umount: /var/named/chroot/var/named: device is busy.
        (In some cases useful info about processes that use
         the device is found by lsof(8) or fuser(1))
                                                        [确定]
启动 named:                                             [确定]
[root@www.DNS_master_33.com named]#cd
```

```
[root@www.DNS_master_33.com ~]#/etc/init.d/named restart
停止 named:                                                    [确定]
启动 named:                                                    [确定]
```

上述代码重启 BIND 服务。

8. 查看主 DNS 服务器日志

主 DNS 服务器日志如下。

```
14510:Jul  7 09:54:01 www named[2408]: listening on IPv4 interface lo,
127.0.0.1#53
14511:Jul  7 09:54:01 www named[2408]: listening on IPv4 interface eth0,
192.168.2.33#53
14512:Jul  7 09:54:01 www named[2408]: listening on IPv6 interface lo, ::1#53
14513:Jul  7 09:54:01 www named[2408]: generating Session key for dynamic DNS
14514:Jul  7 09:54:01 www named[2408]: sizing zone task pool based on 8 zones
14515:Jul  7 09:54:01 www named[2408]: set up managed keys zone for view
default, file '/var/named/dynamic/managed-keys.BIND'
14516:Jul  7 09:54:01 www named[2408]: Warning: 'empty-zones-enable/disable-
empty-zone' not set: disabling RFC 1918 empty zones
14517:Jul  7 09:54:01 www named[2408]: automatic empty zone: 127.IN-ADDR.ARPA
14518:Jul  7 09:54:01 www named[2408]: automatic empty zone: 254.169.IN-ADDR.ARPA
14519:Jul  7 09:54:01 www named[2408]: automatic empty zone: 2.0.192.IN-ADDR.ARPA
14520:Jul  7 09:54:01 www named[2408]: automatic empty zone: 100.51.198.IN-ADDR.
ARPA
14521:Jul  7 09:54:01 www named[2408]: automatic empty zone: 113.0.203.IN-ADDR.
ARPA
14522:Jul  7 09:54:01 www named[2408]: automatic empty zone: 255.255.255.255.
IN-ADDR.ARPA
14523:Jul  7 09:54:01 www named[2408]: automatic empty zone: 0.0.0.0.0.0.0.0.
0.0.0.0.0.0.0.0.0.0.0.0.0.0.0.0.0.0.0.0.0.0.0.0.IP6.ARPA
14524:Jul  7 09:54:01 www named[2408]: automatic empty zone: D.F.IP6.ARPA
14525:Jul  7 09:54:01 www named[2408]: automatic empty zone: 8.E.F.IP6.ARPA
14526:Jul  7 09:54:01 www named[2408]: automatic empty zone: 9.E.F.IP6.ARPA
14527:Jul  7 09:54:01 www named[2408]: automatic empty zone: A.E.F.IP6.ARPA
14528:Jul  7 09:54:01 www named[2408]: automatic empty zone: B.E.F.IP6.ARPA
14529:Jul  7 09:54:01 www named[2408]: automatic empty zone: 8.B.D.0.1.0.0.2.
IP6.ARPA
14530:Jul  7 09:54:01 www named[2408]: command channel listening on 127.0.0.1#953
14531:Jul  7 09:54:01 www named[2408]: command channel listening on ::1#953
14532:Jul  7 09:54:01 www named[2408]: zone 2.168.192.\032in-addr.arpa/IN: loaded
serial 0
14533:Jul  7 09:54:01 www named[2408]: zone 0.in-addr.arpa/IN: loaded serial 0
14534:Jul  7 09:54:01 www named[2408]: zone 1.0.0.127.in-addr.arpa/IN: loaded
serial 0
14535:Jul  7 09:54:01 www named[2408]: zone 1.0.0.0.0.0.0.0.0.0.0.0.0.0.0.0.
0.0.0.0.0.0.0.0.0.0.0.0.0.0.0.0.ip6.arpa/IN: loaded serial 0
14536:Jul  7 09:54:01 www named[2408]: zone booxin*.com/IN: loaded serial 1
14537:Jul  7 09:54:01 www named[2408]: zone localhost.localdomain/IN: loaded
serial 0
14538:Jul  7 09:54:01 www named[2408]: zone localhost/IN: loaded serial 0
14539:Jul  7 09:54:01 www named[2408]: managed-keys-zone ./IN: loaded serial 5
14540:Jul  7 09:54:01 www named[2408]: running
14541:Jul  7 09:54:01 www named[2408]: zone 2.168.192.\032in-addr.arpa/IN:
sending notifies (serial 0)
14542:Jul  7 09:54:01 www named[2408]: zone booxin*.com/IN: sending notifies
(serial 1)
```

从上述代码可以看出，主 DNS 服务器已经正常提供服务。

9. 使用 dig 指令测试 DNS

测试 DNS 的代码如下。

```
[root@www.DNS_master_33.com ~]#dig -t A www.booxin*.com @192.168.2.33

; <<>> DiG 9.8.2rc1-RedHat-9.8.2-0.68.rc1.el6_10.3 <<>> -t A www.booxin*.com
@192.168.2.33
```

```
;; global options: +cmd
;; Got answer:
;; ->>HEADER<<- opcode: QUERY, status: NOERROR, id: 92
;; flags: qr aa rd ra; QUERY: 1, ANSWER: 2, AUTHORITY: 2, ADDITIONAL: 2

;; QUESTION SECTION:
;www.booxin*.com.          IN   A

;; ANSWER SECTION:
www.booxin*.com.     10800    IN   A    192.168.2.36
www.booxin*.com.     10800    IN   A    192.168.2.35

;; AUTHORITY SECTION:
booxin*.com.     10800    IN   NS   ns1.booxin*.com.
booxin*.com.     10800    IN   NS   ns2.booxin*.com.

;; ADDITIONAL SECTION:
ns1.booxin*.com.     10800    IN   A    192.168.2.33
ns2.booxin*.com.     10800    IN   A    192.168.2.34

;; Query time: 0 msec
;; SERVER: 192.168.2.33#53(192.168.2.33)
;; WHEN: Tue Jul 23 23:09:21 2019
;; MSG SIZE  rcvd: 132
```

2.2.6 从 DNS 服务器设置

1. 从 DNS 服务器安装 BIND

从 DNS 服务器安装 BIND 软件，代码如下。

```
yum install BIND BIND-utils BIND-chroot -y
```

2. 配置文件

从 DNS 服务器的安装很简单，只需配置 named.conf 文件即可，区域文件无须手动建立，如图 2-17 所示。

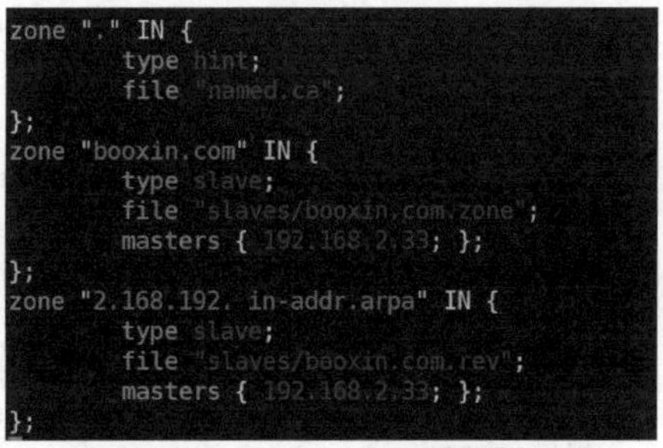

```
zone "." IN {
        type hint;
        file "named.ca";
};
zone "booxin.com" IN {
        type slave;
        file "slaves/booxin.com.zone";
        masters { 192.168.2.33; };
};
zone "2.168.192. in-addr.arpa" IN {
        type slave;
        file "slaves/booxin.com.rev";
        masters { 192.168.2.33; };
};
```

图 2-17　从 DNS 服务器区域文件

从 DNS 服务器主配置文件 named.conf 代码如下。

```
[root@www.DNS_slave_34.com ~]#nl /etc/named
.conf
     1  options {
     2         listen-on port 53 { 127.0.0.1;
192.168.2.34; };
     3         rrset-order { order cyclic; };
     4         directory    "/var/named";
     5         dump-file    "/var/named/data/
cache_dump.db";
     6             statistics-file "/var/named/
data/named_stats.txt";
     7             memstatistics-file "/var/named/data/named_mem_stats.txt";
     8         allow-query      { 192.168.2.0/24; };
     9         recursion yes;

    10         BINDkeys-file "/etc/named.iscdlv.key";

    11         managed-keys-directory "/var/named/dynamic";
    12  };

    13  logging {
    14          channel default_debug {
    15                  file "data/named.run";
    16                  severity dynamic;
    17          };

    18      channel query_info {
```

```
19                file "/var/named/log/query.log" versions 1 size 100m;
20                severity info;
21                print-category yes;
22                print-severity yes;
23                print-time yes;
24            };
25        };
26        zone "." IN {
27            type hint;
28            file "named.ca";
29        };
30        zone "booxin*.com" IN {
31            type slave;
32            file "slaves/booxin*.com.zone";
33            masters { 192.168.2.33; };
34        };
35        zone "2.168.192. in-addr.arpa" IN {
36            type slave;
37            file "slaves/booxin*.com.rev";
38            masters { 192.168.2.33; };
39        };
40        include "/etc/named.rfc1912.zones";
41        include "/etc/named.root.key";
```

上述代码为从 DNS 服务器主配置文件 named.conf，可根据实际需求进行调整和修改。

> 注意：第 31 行和 36 行的 type 必须设置为 slave 类型，第 33 行和 38 行的 masters 选项必须指向主 DNS 服务器的 IP 地址，第 32 行和 37 行的 slaves 目录所属主和所属组用户均为 named。

3. 检测 BIND 配置文件语法

检测 BIND 配置文件语法，代码如下。

```
[root@www.DNS_slave_34.com named]#named-checkconf -z
zone localhost.localdomain/IN: loaded serial 0
zone localhost/IN: loaded serial 0
zone 1.0.0.0.0.0.0.0.0.0.0.0.0.0.0.0.0.0.0.0.0.0.0.0.0.0.0.0.0.0.0.0.ip6.arpa/
IN: loaded serial 0
zone 1.0.0.127.in-addr.arpa/IN: loaded serial 0
zone 0.in-addr.arpa/IN: loaded serial 0
```

上述代码检测配置文件语法准确无误。

4. 重启 BIND 服务并查看日志

重启 BIND 服务并查看日志，代码如下。

```
[root@www.DNS_slave_34.com ~]#/etc/init.d/named restart
停止 named:                                              [确定]
启动 named:                                              [确定]

[root@www.DNS_slave_34.com named]#cat /var/log/messages |grep named
Feb 17 21:49:01 www named[1581]: starting BIND 9.8.2rc1-RedHat-9.8.2-0.68.rc1.
el6_10.1 -u named -t /var/named/chroot
Feb 17 21:49:01 www named[1581]: built with '--build=x86_64-redhat-linux-gnu'
'--host=x86_64-redhat-linux-gnu' '--target=x86_64-redhat-linux-gnu' '--program-
prefix=' '--prefix=/usr' '--exec-prefix=/usr' '--BINDir=/usr/bin' '--sBINDir=
/usr/sbin' '--sysconfdir=/etc' '--datadir=/usr/share' '--includedir=/usr/include'
'--libdir=/usr/lib64' '--libexecdir=/usr/libexec' '--sharedstatedir=/var/lib'
'--mandir=/usr/share/man' '--infodir=/usr/share/info' '--with-libtool'
'--localstatedir=/var' '--enable-threads' '--enable-ipv6' '--enable-filter-aaaa'
'--with-pic' '--disable-static' '--disable-openssl-version-check' '--enable-
rpz-nsip' '--enable-rpz-nsdname' '--with-dlopen=yes' '--with-dlz-ldap=yes'
'--with-dlz-postgres=yes' '--with-dlz-mysql=yes' '--with-dlz-filesystem=yes'
'--with-gssapi=yes' '--disable-isc-spnego' '--with-docbook-xsl=/usr/share/sgml/
docbook/xsl-stylesheets' '--enable-fixed-rrset' 'build_alias=x86_64-redhat-
linux-gnu' 'host_alias=x86_64-redhat-linux-gnu' 'target_alias=x86_64-redhat-
```

```
linux-gnu' 'CFLAGS= -O2 -g -pipe -Wall -Wp,-D_FORTIFY_SOURCE=2 -fexceptions
-fstack-protector --param=ssp-buffer-size=4 -m64 -mtune=generic' 'CPPFLAGS=
-DDIG_SIGCHASE'
Feb 17 21:49:01 www named[1581]: ---------------------------------------------------
Feb 17 21:49:01 www named[1581]: BIND 9 is maintained by Internet Systems
Consortium,
Feb 17 21:49:01 www named[1581]: Inc. (ISC), a non-profit 501(c)(3) public-benefit
Feb 17 21:49:01 www named[1581]: corporation.  Support and training for BIND 9 are
Feb 17 21:49:01 www named[1581]: available
Feb 17 21:49:01 www named[1581]: ---------------------------------------------------
Feb 17 21:49:01 www named[1581]: adjusted limit on open files from 4096 to 1048576
Feb 17 21:49:01 www named[1581]: found 1 CPU, using 1 worker thread
Feb 17 21:49:01 www named[1581]: using up to 4096 sockets
Feb 17 21:49:01 www named[1581]: loading configuration from '/etc/named.conf'
Feb 17 21:49:01 www named[1581]: reading built-in trusted keys from file
'/etc/named.iscdlv.key'
Feb 17 21:49:01 www named[1581]: using default UDP/IPv4 port range: [1024, 65535]
Feb 17 21:49:01 www named[1581]: using default UDP/IPv6 port range: [1024, 65535]
Feb 17 21:49:01 www named[1581]: listening on IPv4 interface lo, 127.0.0.1#53
Feb 17 21:49:01 www named[1581]: listening on IPv4 interface eth0, 192.168.2.34#53
Feb 17 21:49:01 www named[1581]: listening on IPv6 interface lo, ::1#53
Feb 17 21:49:01 www named[1581]: generating Session key for dynamic DNS
Feb 17 21:49:01 www named[1581]: sizing zone task pool based on 8 zones
Feb 17 21:49:01 www named[1581]: set up managed keys zone for view _default, file
'/var/named/dynamic/managed-keys.BIND'
Feb 17 21:49:01 www named[1581]: Warning: 'empty-zones-enable/disable-empty-zone'
not set: disabling RFC 1918 empty zones
Feb 17 21:49:01 www named[1581]: automatic empty zone: 127.IN-ADDR.ARPA
Feb 17 21:49:01 www named[1581]: automatic empty zone: 254.169.IN-ADDR.ARPA
Feb 17 21:49:01 www named[1581]: automatic empty zone: 2.0.192.IN-ADDR.ARPA
Feb 17 21:49:01 www named[1581]: automatic empty zone: 100.51.198.IN-ADDR.ARPA
Feb 17 21:49:01 www named[1581]: automatic empty zone: 113.0.203.IN-ADDR.ARPA
Feb 17 21:49:01 www named[1581]: automatic empty zone: 255.255.255.255.IN-ADDR.ARPA
Feb 17 21:49:01 www named[1581]: automatic empty zone: 0.0.0.0.0.0.0.0.0.0.
0.0.0.0.0.0.0.0.0.0.0.0.0.0.0.0.0.0.0.0.IP6.ARPA
Feb 17 21:49:01 www named[1581]: automatic empty zone: D.F.IP6.ARPA
Feb 17 21:49:01 www named[1581]: automatic empty zone: 8.E.F.IP6.ARPA
Feb 17 21:49:01 www named[1581]: automatic empty zone: 9.E.F.IP6.ARPA
Feb 17 21:49:01 www named[1581]: automatic empty zone: A.E.F.IP6.ARPA
Feb 17 21:49:01 www named[1581]: automatic empty zone: B.E.F.IP6.ARPA
Feb 17 21:49:01 www named[1581]: automatic empty zone: 8.B.D.0.1.0.0.2.IP6.ARPA
Feb 17 21:49:01 www named[1581]: command channel listening on 127.0.0.1#953
Feb 17 21:49:01 www named[1581]: command channel listening on ::1#953
Feb 17 21:49:01 www named[1581]: zone 0.in-addr.arpa/IN: loaded serial 0
Feb 17 21:49:01 www named[1581]: zone 1.0.0.127.in-addr.arpa/IN: loaded serial 0
Feb 17 21:49:01 www named[1581]: zone 1.0.0.0.0.0.0.0.0.0.0.0.0.0.0.0.0.0.0.0.
0.0.0.0.0.0.0.0.0.0.0.ip6.arpa/IN: loaded serial 0
Feb 17 21:49:01 www named[1581]: zone localhost.localdomain/IN: loaded serial 0
Feb 17 21:49:01 www named[1581]: zone localhost/IN: loaded serial 0
Feb 17 21:49:01 www named[1581]: managed-keys-zone ./IN: loaded serial 0
Feb 17 21:49:01 www named[1581]: running
Feb 17 21:49:01 www named[1581]: error (network unreachable) resolving './DNSKEY
/IN': 2001:503:ba3e::2:30#53
Feb 17 21:49:01 www named[1581]: error (network unreachable) resolving './NS/IN':
2001:503:ba3e::2:30#53
Feb 17 21:49:02 www named[1581]: zone booxin*.com/IN: refresh: unexpected rcode
(SERVFAIL) from master 192.168.2.34#53 (source 0.0.0.0#0)
Feb 17 21:49:02 www named[1581]: zone 1.168.192.\032in-addr.arpa/IN: refresh:
unexpected rcode (SERVFAIL) from master 192.168.2.34#53 (source 0.0.0.0#0)
Feb 17 21:49:11 www named[1581]: client 192.168.2.33#26630: received notify
forzone '1.168.192.\032in-addr.arpa'
Feb 17 21:49:11 www named[1581]: zone 1.168.192.\032in-addr.arpa/IN: refused
notify from non-master: 192.168.2.33#26630
Feb 17 21:49:11 www named[1581]: client 192.168.2.33#26630: received notify
forzone 'booxin*.com'
```

```
Feb 17 21:49:11 www named[1581]: zone booxin*.com/IN: refused notify from non-
master: 192.168.2.33#26630
Feb 17 21:50:00 www named[1581]: zone booxin*.com/IN: refresh: unexpected rcode
(SERVFAIL) from master 192.168.2.34#53 (source 0.0.0.0#0)
Feb 17 21:50:00 www named[1581]: zone 1.168.192.\032in-addr.arpa/IN: refresh:
unexpected rcode (SERVFAIL) from master 192.168.2.34#53 (source 0.0.0.0#0)
Feb 17 21:51:55 www named[1581]: zone booxin*.com/IN: refresh: unexpected rcode
(SERVFAIL) from master 192.168.2.34#53 (source 0.0.0.0#0)
Feb 17 21:51:59 www named[1581]: zone 1.168.192.\032in-addr.arpa/IN: refresh:
unexpected rcode (SERVFAIL) from master 192.168.2.34#53 (source 0.0.0.0#0)
Feb 17 21:53:19 www named[1581]: client 192.168.2.33#25946: received notify for
zone '1.168.192.\032in-addr.arpa'
Feb 17 21:53:19 www named[1581]: zone 1.168.192.\032in-addr.arpa/IN: refused
notify from non-master: 192.168.2.33#25946
Feb 17 21:53:19 www named[1581]: client 192.168.2.33#1192: received notify for
zone 'booxin*.com'
Feb 17 21:53:19 www named[1581]: zone booxin*.com/IN: refused notify from non-
master: 192.168.2.33#1192
Feb 17 21:55:41 www named[1581]: zone booxin*.com/IN: refresh: unexpected rcode
(SERVFAIL) from master 192.168.2.34#53 (source 0.0.0.0#0)
Feb 17 21:55:59 www named[1581]: zone 1.168.192.\032in-addr.arpa/IN: refresh:
unexpected rcode (SERVFAIL) from master 192.168.2.34#53 (source 0.0.0.0#0)
Feb 17 21:57:10 www named[1581]: received control channel command 'stop'
Feb 17 21:57:10 www named[1581]: shutting down: flushing changes
Feb 17 21:57:10 www named[1581]: stopping command channel on 127.0.0.1#953
Feb 17 21:57:10 www named[1581]: stopping command channel on ::1#953
Feb 17 21:57:10 www named[1581]: no longer listening on 127.0.0.1#53
Feb 17 21:57:10 www named[1581]: no longer listening on 192.168.2.34#53
Feb 17 21:57:10 www named[1581]: no longer listening on ::1#53
Feb 17 21:57:10 www named[1581]: exiting
Feb 17 21:57:11 www named[1703]: starting BIND 9.8.2rc1-RedHat-9.8.2-0.68.rc1.
el6_10.1 -u named -t /var/named/chroot
Feb 17 21:57:11 www named[1703]: built with '--build=x86_64-redhat-linux-gnu'
'--host=x86_64-redhat-linux-gnu' '--target=x86_64-redhat-linux-gnu' '--program-
prefix=' '--prefix=/usr' '--exec-prefix=/usr' '--BINDir=/usr/bin' '--sBINDir=
/usr/sbin' '--sysconfdir=/etc' '--datadir=/usr/share' '--includedir=/usr/include'
'--libdir=/usr/lib64' '--libexecdir=/usr/libexec' '--sharedstatedir=/var/lib'
'--mandir=/usr/share/man' '--infodir=/usr/share/info' '--with-libtool'
'--localstatedir=/var' '--enable-threads' '--enable-ipv6' '--enable-filter-aaaa'
'--with-pic' '--disable-static' '--disable-openssl-version-check' '--enable-
rpz-nsip' '--enable-rpz-nsdname' '--with-dlopen=yes' '--with-dlz-ldap=yes'
'--with-dlz-postgres=yes' '--with-dlz-mysql=yes' '--with-dlz-filesystem=yes'
'--with-gssapi=yes' '--disable-isc-spnego' '--with-docbook-xsl=/usr/share/sgml/
docbook/xsl-stylesheets' '--enable-fixed-rrset' 'build_alias=x86_64-redhat-
linux-gnu' 'host_alias=x86_64-redhat-linux-gnu' 'target_alias=x86_64-redhat-
linux-gnu' 'CFLAGS= -O2 -g -pipe -Wall -Wp,-D_FORTIFY_SOURCE=2 -fexceptions
-fstack-protector --param=ssp-buffer-size=4 -m64 -mtune=generic' 'CPPFLAGS=
-DDIG_SIGCHASE'
Feb 17 21:57:11 www named[1703]: ----------------------------------------------
Feb 17 21:57:11 www named[1703]: BIND 9 is maintained by Internet Systems
Consortium,
Feb 17 21:57:11 www named[1703]: Inc. (ISC), a non-profit 501(c)(3) public-benefit
Feb 17 21:57:11 www named[1703]: corporation.  Support and training for BIND 9 are
Feb 17 21:57:11 www named[1703]: available
Feb 17 21:57:11 www named[1703]: ----------------------------------------------
Feb 17 21:57:11 www named[1703]: adjusted limit on open files from 4096 to 1048576
Feb 17 21:57:11 www named[1703]: found 1 CPU, using 1 worker thread
Feb 17 21:57:11 www named[1703]: using up to 4096 sockets
Feb 17 21:57:11 www named[1703]: loading configuration from '/etc/named.conf'
Feb 17 21:57:11 www named[1703]: reading built-in trusted keys from file '/etc/
named.iscdlv.key'
Feb 17 21:57:11 www named[1703]: using default UDP/IPv4 port range: [1024, 65535]
Feb 17 21:57:11 www named[1703]: using default UDP/IPv6 port range: [1024, 65535]
Feb 17 21:57:11 www named[1703]: listening on IPv4 interface lo, 127.0.0.1#53
Feb 17 21:57:11 www named[1703]: listening on IPv4 interface eth0, 192.168.2.34#53
```

```
Feb 17 21:57:11 www named[1703]: listening on IPv6 interface lo, ::1#53
Feb 17 21:57:11 www named[1703]: generating Session key for dynamic DNS
Feb 17 21:57:11 www named[1703]: sizing zone task pool based on 8 zones
Feb 17 21:57:11 www named[1703]: set up managed keys zone for view _default,
file '/var/named/dynamic/managed-keys.BIND'
Feb 17 21:57:11 www named[1703]: Warning: 'empty-zones-enable/disable-empty-zone'
not set: disabling RFC 1918 empty zones
Feb 17 21:57:11 www named[1703]: automatic empty zone: 127.IN-ADDR.ARPA
Feb 17 21:57:11 www named[1703]: automatic empty zone: 254.169.IN-ADDR.ARPA
Feb 17 21:57:11 www named[1703]: automatic empty zone: 2.0.192.IN-ADDR.ARPA
Feb 17 21:57:11 www named[1703]: automatic empty zone: 100.51.198.IN-ADDR.ARPA
Feb 17 21:57:11 www named[1703]: automatic empty zone: 113.0.203.IN-ADDR.ARPA
Feb 17 21:57:11 www named[1703]: automatic empty zone: 255.255.255.255.IN-ADDR.
ARPA
Feb 17 21:57:11 www named[1703]: automatic empty zone: 0.0.0.0.0.0.0.0.0.0.0.
0.0.0.0.0.0.0.0.0.0.0.0.0.0.0.0.0.IP6.ARPA
Feb 17 21:57:11 www named[1703]: automatic empty zone: D.F.IP6.ARPA
Feb 17 21:57:11 www named[1703]: automatic empty zone: 8.E.F.IP6.ARPA
Feb 17 21:57:11 www named[1703]: automatic empty zone: 9.E.F.IP6.ARPA
Feb 17 21:57:11 www named[1703]: automatic empty zone: A.E.F.IP6.ARPA
Feb 17 21:57:11 www named[1703]: automatic empty zone: B.E.F.IP6.ARPA
Feb 17 21:57:11 www named[1703]: automatic empty zone: 8.B.D.0.1.0.0.2.IP6.ARPA
Feb 17 21:57:11 www named[1703]: command channel listening on 127.0.0.1#953
Feb 17 21:57:11 www named[1703]: command channel listening on ::1#953
Feb 17 21:57:11 www named[1703]: zone 0.in-addr.arpa/IN: loaded serial 0
Feb 17 21:57:11 www named[1703]: zone 1.0.0.127.in-addr.arpa/IN: loaded serial 0
Feb 17 21:57:11 www named[1703]: zone 1.0.0.0.0.0.0.0.0.0.0.0.0.0.0.0.0.0.0.
0.0.0.0.0.0.0.0.0.0.0.0.ip6.arpa/IN: loaded serial 0
Feb 17 21:57:11 www named[1703]: zone localhost.localdomain/IN: loaded serial 0
Feb 17 21:57:11 www named[1703]: zone localhost/IN: loaded serial 0
Feb 17 21:57:11 www named[1703]: managed-keys-zone ./IN: loaded serial 2
Feb 17 21:57:11 www named[1703]: running
Feb 17 21:57:11 www named[1703]: zone 1.168.192.\032in-addr.arpa/IN: Transfer
started.
Feb 17 21:57:11 www named[1703]: transfer of '1.168.192.\032in-addr.arpa/IN'
from 192.168.2.33#53: connected using 192.168.2.34#46348
Feb 17 21:57:11 www named[1703]: zone 1.168.192.\032in-addr.arpa/IN: transferred
serial 0
Feb 17 21:57:11 www named[1703]: transfer of '1.168.192.\032in-addr.arpa/IN'
from 192.168.2.33#53: Transfer completed: 1 messages, 11 records, 300 bytes,
0.001 secs (300000 bytes/sec)
Feb 17 21:57:11 www named[1703]: zone 1.168.192.\032in-addr.arpa/IN: sending
notifies (serial 0)
Feb 17 21:57:11 www named[1703]: dumping master file: tmp-jyYIUGCwwv: open:
permission denied
Feb 17 21:57:11 www named[1703]: zone booxin*.com/IN: Transfer started.
Feb 17 21:57:11 www named[1703]: transfer of 'booxin*.com/IN' from 192.168.2.33
#53: connected using 192.168.2.34#54671
Feb 17 21:57:11 www named[1703]: zone booxin*.com/IN: transferred serial 1
Feb 17 21:57:11 www named[1703]: transfer of 'booxin*.com/IN' from 192.168.2.33
#53: Transfer completed: 1 messages, 12 records, 291 bytes, 0.002 secs (145500
bytes/sec)
Feb 17 21:57:11 www named[1703]: zone booxin*.com/IN: sending notifies (serial 1)
Feb 17 21:57:11 www named[1703]: dumping master file: tmp-ySBlagqzFz: open:
permission denied
```

可通过上述代码查看启动、同步等信息是否有误。

5. 查看 BIND 同步文件

查看 BIND 同步文件，代码如下。

```
[root@www.DNS_slave_34.com ~]#ls -lShrt --full-time  /var/named/chroot/var
/named/slaves/
总用量 8.0K
-rw-r--r-- 1 named named 507 2019-07-23 22:48:46.321758063 +0800 booxin*.com.rev
-rw-r--r-- 1 named named 477 2019-07-23 22:48:46.822758619 +0800 booxin*.com.zone
```

```
[root@www.DNS_slave_34.com ~]#cat /var/named/chroot/var/named/slaves
/booxin*.com.zone
$ORIGIN .
$TTL 10800      ; 3 hours
booxin*.com             IN SOA  booxin*.com. 3128743705.qq.com. (
                                2          ; serial
                                86400      ; refresh (1 day)
                                3600       ; retry (1 hour)
                                604800     ; expire (1 week)
                                10800      ; minimum (3 hours)
                                )
                        NS      ns1.booxin*.com.
                        NS      ns2.booxin*.com.
                        MX      5 mail.booxin*.com.
$ORIGIN booxin*.com.
ftp             A       192.168.2.33
                A       192.168.2.34
mail            A       192.168.2.33
ns1             A       192.168.2.33
ns2             A       192.168.2.34
www             A       192.168.2.35
                A       192.168.2.36
```

上述代码中,从 DNS 服务器与主 DNS 服务器同步后,查看文件内容,可以看到该文件内容和主 DNS 服务器上的文件内容是一样的。

6. 测试从 DNS 服务器

测试从 DNS 服务器,代码如下。

```
[root@www.DNS_slave_34.com ~]#dig -t A www.booxin*.com @192.168.2.34

; <<>> DiG 9.8.2rc1-RedHat-9.8.2-0.68.rc1.el6_10.3 <<>> -t A www.booxin*.com
@192.168.2.34
;; global options: +cmd
;; Got answer:
;; ->>HEADER<<- opcode: QUERY, status: NOERROR, id: 40718
;; flags: qr aa rd ra; QUERY: 1, ANSWER: 2, AUTHORITY: 2, ADDITIONAL: 2

;; QUESTION SECTION:
;www.booxin*.com.               IN      A

;; ANSWER SECTION:
www.booxin*.com.        10800   IN      A       192.168.2.35
www.booxin*.com.        10800   IN      A       192.168.2.36

;; AUTHORITY SECTION:
booxin*.com.            10800   IN      NS      ns2.booxin*.com.
booxin*.com.            10800   IN      NS      ns1.booxin*.com.

;; ADDITIONAL SECTION:
ns1.booxin*.com.        10800   IN      A       192.168.2.33
ns2.booxin*.com.        10800   IN      A       192.168.2.34

;; Query time: 0 msec
;; SERVER: 192.168.2.34#53(192.168.2.34)
;; WHEN: Tue Jul 23 23:12:30 2019
;; MSG SIZE  rcvd: 132

[root@www.DNS_slave_34.com ~]#dig  -x  192.168.2.34 @192.168.2.34

; <<>> DiG 9.8.2rc1-RedHat-9.8.2-0.68.rc1.el6_10.3 <<>> -x 192.168.2.34 @192.
168.2.34
;; global options: +cmd
;; Got answer:
;; ->>HEADER<<- opcode: QUERY, status: SERVFAIL, id: 9929
```

```
;; flags: qr rd ra; QUERY: 1, ANSWER: 0, AUTHORITY: 0, ADDITIONAL: 0

;; QUESTION SECTION:
;34.2.168.192.in-addr.arpa.    IN    PTR

;; Query time: 1217 msec
;; SERVER: 192.168.2.34#53(192.168.2.34)
;; WHEN: Tue Jul 23 23:13:14 2019
;; MSG SIZE  rcvd: 43

[root@www.DNS_slave_34.com ~]#dig  -x  192.168.2.33 @192.168.2.34

; <<>> DiG 9.8.2rc1-RedHat-9.8.2-0.68.rc1.el6_10.3 <<>> -x 192.168.2.33 @192.168.2.34
;; global options: +cmd
;; Got answer:
;; ->>HEADER<<- opcode: QUERY, status: NXDOMAIN, id: 11468
;; flags: qr rd ra; QUERY: 1, ANSWER: 0, AUTHORITY: 1, ADDITIONAL: 0

;; QUESTION SECTION:
;33.2.168.192.in-addr.arpa.    IN    PTR

;; AUTHORITY SECTION:
168.192.in-addr.arpa.    10800    IN    SOA    prisoner.iana.org. hostmaster.root-servers.org. 1 604800 60 604800 604800

;; Query time: 2966 msec
;; SERVER: 192.168.2.34#53(192.168.2.34)
;; WHEN: Tue Jul 23 23:14:36 2019
;; MSG SIZE  rcvd: 120
```

上述返回结果中，从 DNS 服务器上正向解析和反向解析都能测试成功。

2.2.7 主从同步数据的安全性

DNS 服务器的数据同步默认是没有限定主机的，网络上只要有一台 DNS 服务器向你的 DNS 服务器请求数据，就能实现数据同步，相当不安全。

建议读者使用参数 allow-transfer，指定可以同步数据的主机 IP 地址。主 DNS 服务器的数据可以给其他主 DNS 服务器同步，相应地，从 DNS 服务器的数据也可以给其他从 DNS 服务器同步，建议为所有的主、从 DNS 服务器设置该参数。

1. 指定可以从主 DNS 服务器上同步数据的主机

```
[root@www.DNS_master_33.com ~]#grep allow /etc/named.conf
    allow-query    { 192.168.2.0/24; };
```

在每块区域上添加参数 allow-transfer，"{}"内填写可以同步的主机 IP 地址，一般填写从 DNS 服务器的 IP 地址开始。可以使用 dig 指令测试区域同步，格式如下。

```
dig -t axfr ZONE_NAME @DNS_SERVCER_IP
```

配置文件中授权指定 IP 地址可以同步数据，如图 2-18 所示。

2. 指定可以从从 DNS 服务器上同步数据的主机

配置/etc/named.conf 文件，显式指定同步从 DNS 服务器数据的主机，代码如下。

```
zone "booxin*.com" IN {
    type slave;
    file "slaves/booxin*.com.zone";
    masters { 192.168.2.33; };
};
zone "2.168.192. in-addr.arpa" IN {
    type slave;
    file "slaves/booxin*.com.rev";
```

```
            masters { 192.168.2.33; };
};
```

本实验只有一台从 DNS 服务器,所以根本不会有主机从这台机器同步数据,我们设置成不允许任何人同步,如图 2-19 所示。

图 2-18 DNS 同步数据　　　　　　　　　　图 2-19 从 DNS 服务器安全配置

2.2.8 DNS 主从配置优化

DNS 主从配置优化大致可以分为如下几个方面。

- 禁止递归查询。
- 客户端安装缓存软件 NSCD。
- chroot 安全加固。
- 负载均衡。
- DNS 视图技术(智能 DNS)。

2.2.9 DNS 主从搭建总结

DNS 主从搭建总结如下。

1. 主 DNS 服务器核心配置

主 DNS 服务器核心配置代码如下。

```
zone "booxin*.com" IN {
        type master;
        file "booxin*.com.zone";
        also-notify { 192.168.2.34; };
        allow-transfer { 192.168.2.34; };
};

zone "2.168.192. in-addr.arpa" IN {
        type master;
        file "booxin*.com.rev";
        also-notify { 192.168.2.34; };
        allow-transfer { 192.168.2.34; };
        };
```

上述代码中标记为加粗的,是主 DNS 服务器必须要设置的主从同步重要参数。

2. 从 DNS 服务器核心配置

从 DNS 服务器与主 DNS 服务器的不同之处,如标记的加粗代码所示。

```
zone "booxin*.com" IN {
        type slave;
        file "slaves/booxin*.com.zone";
        masters { 192.168.2.33; };
        allow-transfer { none; };
};
zone "2.168.192. in-addr.arpa" IN {
```

```
            type slave;
            file "slaves/booxin*.com.rev";
            masters { 192.168.2.33; };
            allow-transfer {  none; };
            };
```

3．从 DNS 服务器测试

查看主从同步是否成功，代码如下。

```
[root@www.DNS_slave_34.com named]#ls -lhrt --full-time slaves/
总用量 8.0K
-rw-r--r-- 1 named named 507 2019-09-09 21:34:33.459651606 +0800 booxin*.com.rev
-rw-r--r-- 1 named named 709 2019-09-09 21:34:33.958652353 +0800 booxin*.com.zone
```

上述代码表示，DNS 主从同步已经成功。

4．配置应用端指向 DNS 服务器

主 DNS 服务器设置服务器地址，将 DNS 服务器指向自身，代码如下。

```
[root@www.DNS_master_33.com named]#cat /etc/resolv.conf
nameserver 192.168.2.33
nameserver 8.8.8.8
```

从 DNS 服务器设置服务器地址，代码如下。

```
[root@www.DNS_slave_34.com named]#cat /etc/resolv.conf
nameserver 192.168.2.33
nameserver 192.168.2.34
```

5．其他注意事项

新建配置文件的权限、所属用户和组等，遇到问题查看相关程序日志。

2.3 DNS 服务常用分析指令

2.3.1 DNS 服务查询基础指令

测试域名解析的指令除了 dig，还有两个指令，即 host 和 nslookup。

1．host 指令入门

host 指令格式如下。

```
# host [-t type] {name} [server]
```

使用 host 指令解析 www.jd*.com 网站的 A 记录（见图 2-20），输出如下。

```
[root@VM_0_12_centos ~]# host -t A www.jd*.com
www.jd*.com is an alias for www.jd*.com.gslb.qianxun.com.
www.jd*.com.gslb.qianxun.com is an alias for www.jdcdn*.com.
www.jdcdn*.com has address 150.138.120.1
```

2．nslookup 指令入门

nslookup 指令是以前的 DNS 解析指令，在 Windows 操作系统中的 DOS 指令行里面也可以使用，nslookup 指令格式有两种：交互式和非交互式。代码如下。

```
nslookup >
 server DNS_SERVER_IP
    set q=TYPE
{name}
```

使用 nslookup 指令解析 jd*.com（见图 2-21），输出如下。

```
[root@VM_0_12_centos ~]# nslookup  jd*.com
Server:         183.60.83.19
Address:        183.60.83.19#53

Non-authoritative answer:
```

```
      Name:    jd*.com
      Address: 120.52.148.118
```

```
[root@VM_0_12_centos ~]# host -t A www.jd*.com
www.jd*.com is an alias for www.jd*.com.gslb.qianxun.com
www.jd*.com.gslb.qianxun.com is an alias for www.jdcdn*.com.
www.jdcdn*.com has address 150.138.120.1
[root@VM_0_12_centos ~]#
```

```
[root@VM_0_12_centos ~]# nslookup jd*.com
Server:         183.60.83.19
Address:        183.60.83.19#53

Non-authoritative answer:
Name:   jd*.com
Address: 120.52.148.118

[root@VM_0_12_centos ~]#
```

图 2-20　host 解析网站 A 记录 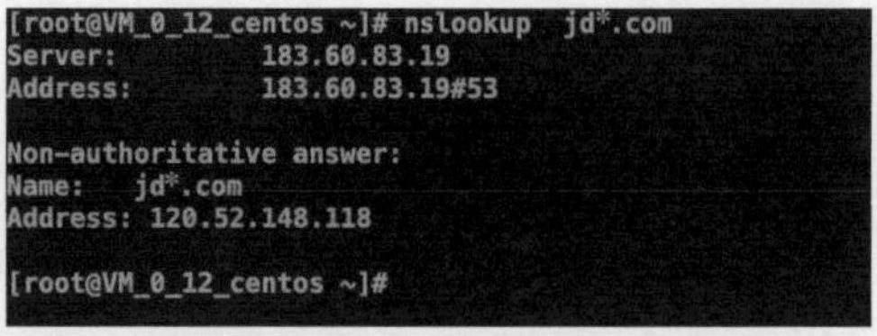　　　　　图 2-21　nslookup 指令解析网站

2.3.2　DNS 高级查询指令之 dig

输入 dig jd*.com 指令，返回代码如下。

```
[root@VM_16_14_centos ~]# dig jd*.com

; <<>> DiG 9.9.4-RedHat-9.9.4-50.el7 <<>> jd*.com
;; global options: +cmd
;; Got answer:
;; ->>HEADER<<- opcode: QUERY, status: NOERROR, id: 31263
;; flags: qr rd ra; QUERY: 1, ANSWER: 1, AUTHORITY: 8, ADDITIONAL: 5

;; OPT PSEUDOSECTION:
; EDNS: version: 0, flags:; udp: 4096
;; QUESTION SECTION:
;jd*.com.               IN      A

;; ANSWER SECTION:
jd*.com.        120     IN      A       120.52.148.118

;; AUTHORITY SECTION:
jd*.com.        120     IN      NS      ns4.jd*.com.
jd*.com.        120     IN      NS      ns4.jdcache*.com.
jd*.com.        120     IN      NS      ns1.jdcache*.com.
jd*.com.        120     IN      NS      ns1.jd*.com.
jd*.com.        120     IN      NS      ns2.jd*.com.
jd*.com.        120     IN      NS      ns3.jdcache*.com.
jd*.com.        120     IN      NS      ns3.jd*.com.
jd*.com.        120     IN      NS      ns2.jdcache*.com.

;; ADDITIONAL SECTION:
ns1.jd*.com.    81      IN      A       111.13.28.10
ns2.jd*.com.    81      IN      A       111.206.226.10
ns3.jd*.com.    81      IN      A       120.52.149.254
ns4.jd*.com.    81      IN      A       106.39.177.32

;; Query time: 83 msec
;; SERVER: 183.60.83.19#53(183.60.83.19)
;; WHEN: 一 7月 15 14:39:41 CST 2019
;; MSG SIZE  rcvd: 267
```

上述代码中一些返回参数说明如下。

（1）status：NOERROR 表示查询没有什么错误，Query time 表示查询完成时间。

（2）SERVER：183.60.83.19#53(183.60.83.19)表示本地 DNS 服务器地址和端口号。

（3）QUESTION SECTION：表示需要查询的内容，这里需要查询域名的 A 记录。

（4）ANSWER SECTION：表示查询结果，返回 A 记录的 IP 地址。120 表示本次查询缓存时间，在 120s 内本地 DNS 服务器可以直接从缓存返回结果。

(5) AUTHORITY SECTION：表示从哪台 DNS 服务器获取到具体的 A 记录信息。注意，本地 DNS 服务器只是查询，而 AUTHORITY SECTION 返回的服务器是权威 DNS 服务器，由它来维护 jd*.com 的域名信息，代码如下。

```
ns4.jdcache*.com.
ns1.jdcache*.com.
ns1.jd*.com.
ns2.jd*.com.
ns3.jdcache*.com.
ns3.jd*.com.
ns2.jdcache*.com.
```

(6) ADDITIONAL SECTION：表示 DNS 服务器对应的 IP 地址，这些 IP 地址对应的服务器安装了 BIND 软件，或者具备提供解析域名的能力。

dig 指令常用的 3 个选项如下。

- +short：显示简短的信息。
- -t：指定查询的记录类型，可以是 CNAME、A、MX、NS，分别表示 CNAME 记录、A 记录、MX 记录、DNS 服务器，默认值是 A。
- -x：表示反向查找，也就是根据 IP 地址查找域名。

2.3.3 查询 DNS 服务器记录类型

使用 dig 指令解析 DNS 服务器记录类型，默认 dig 指令解析后返回的结果是 A 记录类型，其他类型还包括 MX、NS、SOA 等，使用 dig -t a/mx/soa/mx jd*.com 指令进行自定义查询类型。

1. dig -t a www.weibo.com +noall +answer

北京节点探测结果如下。

```
[root@instance-disbrfer ~]# dig -t a www.weibo.com +noall +answer

; <<>> DiG 9.9.4-RedHat-9.9.4-61.el7 <<>> -t a www.weibo.com +noall +answer
;; global options: +cmd
www.weibo.com.          41      IN      A       180.149.138.56
www.weibo.com.          41      IN      A       180.149.134.142
www.weibo.com.          41      IN      A       180.149.134.141
```

上述结果中返回了 www.weibo.com 域名的 3 条 A 记录。

香港节点探测结果如下。

```
[root@VM_16_14_centos ~]# dig -t a www.weibo.com +noall +answer

; <<>> DiG 9.9.4-RedHat-9.9.4-50.el7 <<>> -t a www.weibo.com +noall +answer
;; global options: +cmd
www.weibo.com.                  44      IN      CNAME   weibo.com.edgekey.net.
weibo.com.edgekey.net.          3597    IN      CNAME   e4141.dscb.akamaiedge.net.
e4141.dscb.akamaiedge.net.      4       IN      A       104.74.20.156
```

上述结果中，+noall +answer 选项表示返回简短信息，这里表示查询 A 记录。

2. dig -t ns weibo.com

假如用户输入 dig -t ns www.weibo.com 是查询不出任何 NS 记录的，原因在于只有**一级域名**（或者**顶级域名**）才有 NS 记录，所以通过 FQDN 是查询不出 NS 记录的，要输入 dig –t ns weibo.com 才行。

北京节点探测结果如下。

```
[root@instance-disbrfer ~]# dig -t ns weibo.com

; <<>> DiG 9.9.4-RedHat-9.9.4-61.el7 <<>> -t ns weibo.com
;; global options: +cmd
```

```
;; Got answer:
;; ->>HEADER<<- opcode: QUERY, status: NOERROR, id: 31692
;; flags: qr rd ra; QUERY: 1, ANSWER: 6, AUTHORITY: 0, ADDITIONAL: 7

;; OPT PSEUDOSECTION:
; EDNS: version: 0, flags:; udp: 4096
;; QUESTION SECTION:
;weibo.com.                     IN      NS

;; ANSWER SECTION:
weibo.com.              21616   IN      NS      ns1.sina.com.cn.
weibo.com.              21616   IN      NS      ns3.sina.com.cn.
weibo.com.              21616   IN      NS      ns2.sina.com.cn.
weibo.com.              21616   IN      NS      ns3.sina.com.
weibo.com.              21616   IN      NS      ns4.sina.com.
weibo.com.              21616   IN      NS      ns4.sina.com.

;; ADDITIONAL SECTION:
ns1.sina.com.cn.        15538   IN      A       36.51.252.8
ns2.sina.com.cn.        20614   IN      A       180.149.138.199
ns3.sina.com.           8485    IN      A       180.149.138.199
ns3.sina.com.cn.        36416   IN      A       123.125.29.99
ns4.sina.com.           8485    IN      A       123.125.29.99
ns4.sina.com.cn.        15628   IN      A       121.14.1.22

;; Query time: 2 msec
;; SERVER: 172.16.0.2#53(172.16.0.2)
;; WHEN: Tue Jul 16 16:10:56 CST 2019
;; MSG SIZE  rcvd: 258
```

香港节点探测结果如下。

```
[root@VM_16_14_centos ~]# dig -t ns weibo.com

; <<>> DiG 9.9.4-RedHat-9.9.4-50.el7 <<>> -t ns weibo.com
;; global options: +cmd
;; Got answer:
;; ->>HEADER<<- opcode: QUERY, status: NOERROR, id: 63732
;; flags: qr rd ra; QUERY: 1, ANSWER: 6, AUTHORITY: 0, ADDITIONAL: 1

;; OPT PSEUDOSECTION:
; EDNS: version: 0, flags:; udp: 4096
;; QUESTION SECTION:
;weibo.com.                     IN      NS

;; ANSWER SECTION:
weibo.com.              86400   IN      NS      ns4.sina.com.
weibo.com.              86400   IN      NS      ns4.sina.com.cn.
weibo.com.              86400   IN      NS      ns3.sina.com.cn.
weibo.com.              86400   IN      NS      ns2.sina.com.cn.
weibo.com.              86400   IN      NS      ns1.sina.com.cn.
weibo.com.              86400   IN      NS      ns3.sina.com.

;; Query time: 9 msec
;; SERVER: 183.60.83.19#53(183.60.83.19)
;; WHEN: 二 7月 16 16:11:24 CST 2019
;; MSG SIZE  rcvd: 162
```

微博的服务器是微博官方建立的，有6个域名和IP地址。这么多地址，主要是防止单点问题出现，比如某个DNS服务器连接不上，可以连接其他DNS服务器。

3. dig -t a www.baidu.com

北京节点探测结果如下。

```
[root@instance-disbrfer ~]# dig -t a www.baidu.com

; <<>> DiG 9.9.4-RedHat-9.9.4-61.el7 <<>> -t a www.baidu.com
```

```
;; global options: +cmd
;; Got answer:
;; ->>HEADER<<- opcode: QUERY, status: NOERROR, id: 18397
;; flags: qr rd ra; QUERY: 1, ANSWER: 3, AUTHORITY: 5, ADDITIONAL: 1

;; OPT PSEUDOSECTION:
; EDNS: version: 0, flags:; udp: 4096
;; QUESTION SECTION:
;www.baidu.com.                 IN      A

;; ANSWER SECTION:
www.baidu.com.          600     IN      CNAME   www.a.shifen.com.
www.a.shifen.com.       115     IN      A       220.181.38.149
www.a.shifen.com.       115     IN      A       220.181.38.150

;; AUTHORITY SECTION:
a.shifen.com.           714     IN      NS      ns4.a.shifen.com.
a.shifen.com.           714     IN      NS      ns2.a.shifen.com.
a.shifen.com.           714     IN      NS      ns3.a.shifen.com.
a.shifen.com.           714     IN      NS      ns5.a.shifen.com.
a.shifen.com.           714     IN      NS      ns1.a.shifen.com.

;; Query time: 2 msec
;; SERVER: 172.16.0.2#53(172.16.0.2)
;; WHEN: Tue Jul 16 16:12:16 CST 2019
;; MSG SIZE  rcvd: 191
```

香港节点探测结果如下。

```
[root@VM_16_14_centos ~]# dig -t a www.baidu.com

; <<>> DiG 9.9.4-RedHat-9.9.4-50.el7 <<>> -t a www.baidu.com
;; global options: +cmd
;; Got answer:
;; ->>HEADER<<- opcode: QUERY, status: NOERROR, id: 33596
;; flags: qr rd ra; QUERY: 1, ANSWER: 4, AUTHORITY: 2, ADDITIONAL: 3

;; OPT PSEUDOSECTION:
; EDNS: version: 0, flags:; udp: 4096
;; QUESTION SECTION:
;www.baidu.com.                 IN      A

;; ANSWER SECTION:
www.baidu.com.          1120    IN      CNAME   www.a.shifen.com.
www.a.shifen.com.       102     IN      CNAME   www.wshifen.com.
www.wshifen.com.        191     IN      A       119.63.197.151
www.wshifen.com.        191     IN      A       119.63.197.139

;; AUTHORITY SECTION:
wshifen.com.            523     IN      NS      ns3.wshifen.com.
wshifen.com.            523     IN      NS      ns4.wshifen.com.

;; ADDITIONAL SECTION:
ns3.wshifen.com.        523     IN      A       180.76.8.250
ns4.wshifen.com.        523     IN      A       180.76.9.250

;; Query time: 1 msec
;; SERVER: 183.60.83.19#53(183.60.83.19)
;; WHEN: 二 7月 16 16:12:11 CST 2019
;; MSG SIZE  rcvd: 195
```

上述代码中返回的 CNAME（别名解析）表示查询 www.baidu.com 的信息其实是 www.a.shifen.com 返回的 A 记录。

4. dig -t mx weibo.com

香港节点探测结果如下。

```
[root@VM_16_14_centos ~]# dig -t mx weibo.com
```

```
; <<>> DiG 9.9.4-RedHat-9.9.4-50.el7 <<>> -t mx weibo.com
;; global options: +cmd
;; Got answer:
;; ->>HEADER<<- opcode: QUERY, status: NOERROR, id: 776
;; flags: qr rd ra; QUERY: 1, ANSWER: 1, AUTHORITY: 6, ADDITIONAL: 1

;; OPT PSEUDOSECTION:
; EDNS: version: 0, flags:; udp: 4096
;; QUESTION SECTION:
;weibo.com.                     IN      MX

;; ANSWER SECTION:
weibo.com.              42      IN      MX      10 mx.sina.net.

;; AUTHORITY SECTION:
weibo.com.              86255   IN      NS      ns3.sina.com.cn.
weibo.com.              86255   IN      NS      ns2.sina.com.cn.
weibo.com.              86255   IN      NS      ns1.sina.com.cn.
weibo.com.              86255   IN      NS      ns3.sina.com.
weibo.com.              86255   IN      NS      ns4.sina.com.
weibo.com.              86255   IN      NS      ns4.sina.com.cn.

;; Query time: 1 msec
;; SERVER: 183.60.83.19#53(183.60.83.19)
;; WHEN: 二 7月 16 16:13:49 CST 2019
;; MSG SIZE  rcvd: 189
```

北京节点探测结果如下。

```
[root@instance-disbrfer ~]# dig -t mx weibo.com

; <<>> DiG 9.9.4-RedHat-9.9.4-61.el7 <<>> -t mx weibo.com
;; global options: +cmd
;; Got answer:
;; ->>HEADER<<- opcode: QUERY, status: NOERROR, id: 17692
;; flags: qr rd ra; QUERY: 1, ANSWER: 1, AUTHORITY: 6, ADDITIONAL: 7

;; OPT PSEUDOSECTION:
; EDNS: version: 0, flags:; udp: 4096
;; QUESTION SECTION:
;weibo.com.                     IN      MX

;; ANSWER SECTION:
weibo.com.              60      IN      MX      10 mx.sina.net.

;; AUTHORITY SECTION:
weibo.com.              45599   IN      NS      ns1.sina.com.cn.
weibo.com.              45599   IN      NS      ns4.sina.com.
weibo.com.              45599   IN      NS      ns4.sina.com.cn.
weibo.com.              45599   IN      NS      ns3.sina.com.
weibo.com.              45599   IN      NS      ns2.sina.com.cn.
weibo.com.              45599   IN      NS      ns3.sina.com.cn.

;; ADDITIONAL SECTION:
ns1.sina.com.cn.        46380   IN      A       36.51.252.8
ns2.sina.com.cn.        26916   IN      A       180.149.138.199
ns3.sina.com.           32097   IN      A       180.149.138.199
ns3.sina.com.cn.        83171   IN      A       123.125.29.99
ns4.sina.com.           32097   IN      A       123.125.29.99
ns4.sina.com.cn.        27144   IN      A       121.14.1.22

;; Query time: 8 msec
;; SERVER: 172.16.0.2#53(172.16.0.2)
;; WHEN: Tue Jul 16 16:14:24 CST 2019
;; MSG SIZE  rcvd:
```

> 注意：这里不能输入 dig -t mx www.jd*.com，因为 MX 记录一般配置在一级域名下。

5. 查看域名信息

使用 dig 指令查看域名信息，代码如下。

```
[root@zabbix_server ~]# dig +noall +answer qq.com
qq.com.          555 IN   A    111.161.64.48
qq.com.          555 IN   A    111.161.64.40
```

> 注意：这里输出了两行，都是 A 记录，分别将 qq.com 解析成 111.161.64.48 和 111.161.64.40 两个 IP 地址，这是一个完整的域名解析过程。

```
1   [root@VM_16_14_centos ~]# dig www.jd*.com +noall +answer
2
3   ; <<>> DiG 9.9.4-RedHat-9.9.4-50.el7 <<>> www.jd*.com +noall +answer
4   ;; global options: +cmd
5   www.jd*.com.        120 IN   CNAME   www.jd*.com.gslb.qianxun.com.
6   www.jd*.com.gslb.qianxun.com. 60 IN  CNAME   jd*-abroad.cdn20.com.
7   jd*-abroad.cdn20.com. 60  IN  A      163.171.198.117
8   jd*-abroad.cdn20.com. 60  IN  A      45.116.82.62
9
10
11  [root@zabbix_server ~]# dig www.jd*.com +noall +answer
12
13  ; <<>> DiG 9.9.4-RedHat-9.9.4-74.el7_6.1 <<>> www.jd*.com +noall +answer
14  ;; global options: +cmd
15  www.jd*.com.        32  IN   CNAME   www.jd*.com.gslb.qianxun.com.
16  www.jd*.com.gslb.qianxun.com. 39 IN  CNAME   www.jdcdn*.com.
17  www.jdcdn*.com.     633 IN   A       59.36.203.1
```

上述代码第 11 行使用 dig www.jd*.com +noall +answer 指令，第 15~16 行是 CNAME 信息，先将 www.jd*.com 解析成 www.jd*.com.gslb.qianxun.com，再将 www.jd*.com.gslb.qianxun.com 解析为 www.jdcdn*.com，第 17 行是一条 A 记录，将 www.jdcdn*.com 解析为 59.36.203.1，这是一个完整的域名解析过程。

6. 查找域名的 MX 记录

查找域名的 MX 记录，代码如下。

```
[root@zabbix_server ~]# dig qq.com -t MX +short
30 mx1.qq.com.
10 mx3.qq.com.
20 mx2.qq.com.
```

从输出可以看出，QQ 域名邮箱服务有 3 条记录。

7. 查找域名对应的 CNAME

查找域名对应的 CNAME，代码如下。

```
1   [root@laohan_httpd_server ~]# dig www.jd*.com
2
3   ; <<>> DiG 9.8.2rc1-RedHat-9.8.2-0.68.rc1.el6_10.3 <<>> www.jd*.com
4   ;; global options: +cmd
5   ;; Got answer:
6   ;; ->>HEADER<<- opcode: QUERY, status: NOERROR, id: 36825
7   ;; flags: qr rd ra; QUERY: 1, ANSWER: 3, AUTHORITY: 0, ADDITIONAL: 0
8
9   ;; QUESTION SECTION:
10  ;www.jd*.com.              IN   A
11
12  ;; ANSWER SECTION:
13  www.jd*.com.        104 IN   CNAME   www.jd*.com.gslb.qianxun.com.
14  www.jd*.com.gslb.qianxun.com. 44 IN  CNAME   www.jdcdn*.com.
15  www.jdcdn*.com.     51  IN   A       150.138.120.1
16
17  ;; Query time: 0 msec
```

```
18      ;; SERVER: 183.60.83.19#53(183.60.83.19)
19      ;; WHEN: Mon Apr  6 11:01:52 2020
20      ;; MSG SIZE  rcvd: 106
21
22      [root@laohan_httpd_server ~]# dig -t CNAME    www.jd*.com
23
24      ; <<>> DiG 9.8.2rc1-RedHat-9.8.2-0.68.rc1.el6_10.3 <<>> -t CNAME www.jd*.com
25      ;; global options: +cmd
26      ;; Got answer:
27      ;; ->>HEADER<<- opcode: QUERY, status: NOERROR, id: 35096
28      ;; flags: qr rd ra; QUERY: 1, ANSWER: 1, AUTHORITY: 0, ADDITIONAL: 0
29
30      ;; QUESTION SECTION:
31      ;www.jd*.com.                  IN    CNAME
32
33      ;; ANSWER SECTION:
34      www.jd*.com.          120 IN   CNAME    www.jd*.com.gslb.qianxun.com.
35
36      ;; Query time: 64 msec
37      ;; SERVER: 183.60.83.19#53(183.60.83.19)
38      ;; WHEN: Mon Apr  6 11:01:56 2020
39      ;; MSG SIZE  rcvd: 66
```

上述代码第 1 行查询 www.jd*.com 的 CNAME 记录,第 2～21 行为返回结果。第 22 行代码查询 www.jd*.com 的 CNAME 记录,第 23～39 行为返回结果。

8. 根据 IP 地址反向查找域名

根据 IP 地址反向查找域名,代码如下。

```
[root@VM_16_14_centos ~]# dig -x 114.114.114.114 +short
public1.114DNS.com.
[root@VM_16_14_centos ~]# dig -x 8.8.8.8 +short
DNS.google.
```

从输出可以看出,IP 地址为 114.114.114.114 的 DNS 服务器有个域名叫作 public1.114DNS.com.。

9. 查询域名的解析 DNS 服务器地址

查询域名的解析 DNS 服务器地址,代码如下。

```
[root@VM_16_14_centos ~]#   dig qq.com -t NS +short
ns4.qq.com.
ns1.qq.com.
ns2.qq.com.
ns3.qq.com.
[root@VM_16_14_centos ~]#   dig google.com -t NS +short
ns1.google.com.
ns2.google.com.
ns3.google.com.
ns4.google.com.
[root@VM_16_14_centos ~]#   dig weibo.com -t NS +short
ns4.sina.com.
ns3.sina.com.
ns4.sina.com.cn.
ns1.sina.com.cn.
ns3.sina.com.cn.
ns2.sina.com.cn.
[root@VM_16_14_centos ~]#   dig ziroom*.com -t NS +short
vip2.aliDNS.com.
vip1.aliDNS.com.
```

从上述代码输出可以看到,qq.com 的 DNS 服务器地址为 ns1.qq.com～ns4.qq.com。

2.3.4　DNS 迭代查询的具体流程

客户端(如浏览器)查询本地域名 DNS 信息的时候,是使用**递归查询**的方式,而本地 DNS 服务器为了获取某个域名的 DNS 信息,会使用迭代查询的方式。通过 dig +trace weibo.com 来

2.3 DNS 服务常用分析指令

进行查询,香港节点探测结果如下。

```
[root@VM_16_14_centos ~]# dig +trace weibo.com

; <<>> DiG 9.9.4-RedHat-9.9.4-50.el7 <<>> +trace weibo.com
;; global options: +cmd
.                       68868   IN      NS      l.root-servers.net.
.                       68868   IN      NS      m.root-servers.net.
.                       68868   IN      NS      b.root-servers.net.
.                       68868   IN      NS      c.root-servers.net.
.                       68868   IN      NS      d.root-servers.net.
.                       68868   IN      NS      e.root-servers.net.
.                       68868   IN      NS      f.root-servers.net.
.                       68868   IN      NS      g.root-servers.net.
.                       68868   IN      NS      h.root-servers.net.
.                       68868   IN      NS      a.root-servers.net.
.                       68868   IN      NS      i.root-servers.net.
.                       68868   IN      NS      j.root-servers.net.
.                       68868   IN      NS      k.root-servers.net.
;; Received 811 bytes from 183.60.83.19#53(183.60.83.19) in 420 ms

com.                    172800  IN      NS      a.gtld-servers.net.
com.                    172800  IN      NS      b.gtld-servers.net.
com.                    172800  IN      NS      c.gtld-servers.net.
com.                    172800  IN      NS      d.gtld-servers.net.
com.                    172800  IN      NS      e.gtld-servers.net.
com.                    172800  IN      NS      f.gtld-servers.net.
com.                    172800  IN      NS      g.gtld-servers.net.
com.                    172800  IN      NS      h.gtld-servers.net.
com.                    172800  IN      NS      i.gtld-servers.net.
com.                    172800  IN      NS      j.gtld-servers.net.
com.                    172800  IN      NS      k.gtld-servers.net.
com.                    172800  IN      NS      l.gtld-servers.net.
com.                    172800  IN      NS      m.gtld-servers.net.
com.                    86400   IN      DS      30909 8 2
E2D3C916F6DEEAC73294E8268FB5885044A833FC5459588F4A9184CF C41A5766
com.                    86400   IN      RRSIG   DS 8 1 86400 20190729050000
20190716040000 59944 . J0E+CyMK4b2+33ZO9Oc9QZgJNbh1k67mVP7OXvu8v0RoLQIO8B2lit0B
SqbGNwppVHxuix9dv0QoJ8GasUjshBeEpA450Xr7TUcTdpbgbfY2EBTI
D+Ut5Eg2vTI0MXxNaG2jhUYDf0Oomrh/E0IV4NFk2Z+QdgVixojxqnAg
zTdoGE9Qj3BHSbfieG4RutfvVHn8XyvBhy6Ix5HuTBA3CuNC+LMip5ug
5ka939i+siKcxBx3bD/zYLbK1THZG0IZreKNH2yyFl9CnxUlRTCwnM+k
Mgi7jUPYlQ7dzovJGXC4mj9hDevWZW19fFS1PzfX+ke/Ubd55K1SUwJk Fh7jwA==
;; Received 1169 bytes from 192.5.5.241#53(f.root-servers.net) in 18 ms

weibo.com.              172800  IN      NS      ns1.sina.com.cn.
weibo.com.              172800  IN      NS      ns2.sina.com.cn.
weibo.com.              172800  IN      NS      ns3.sina.com.cn.
weibo.com.              172800  IN      NS      ns4.sina.com.cn.
weibo.com.              172800  IN      NS      ns4.sina.com.
weibo.com.              172800  IN      NS      ns3.sina.com.
CK0POJMG874LJREF7EFN8430QVIT8BSM.com. 86400 IN NSEC3 1 1 0 -
CK0Q1GIN43N1ARRC9OSM6QPQR81H5M9A NS SOA RRSIG DNSKEY NSEC3PARAM
CK0POJMG874LJREF7EFN8430QVIT8BSM.com. 86400 IN RRSIG NSEC3 8 2 86400 20190722044529
20190715033529 3800 com. QruEtMEx3JMwz1cwK+gArQ3ZkNbCR4x+8ul54fAkec4IeN6LvHGTo61m
Dq5w/D0fzhLIhwXRtbxSpDPspvqYpEys6f69kM1U4NzGvTb8/iz7/b0T
es+n3i+Gx0S+/X7ajkvu+J6GmVpgsZPIMxpSJjGSLZXtlEB4YyPBprWZ sck=
DLN3V1VIN2MEMHLPNC16JB1R1ATDH2V4.com. 86400 IN NSEC3 1 1 0 -
DLN5PBROIQFHD0V9HVVLCN86SL75HJBK NS DS RRSIG
DLN3V1VIN2MEMHLPNC16JB1R1ATDH2V4.com. 86400 IN RRSIG NSEC3 8 2 86400 20190722052104
20190715041104 3800 com. U/86DTjPN+oMhvmKZXfkZo7IVvlvrPyxsozVHoAcvV6Xq73q9rsRLCF2
uZ5+of4rKfqp9nuP3aGvEHhqnZoDA1S6qQPcrScwzyJeB72OsjsVIJtA
iqGpFN4GXOEco4KKBO5SKTVBEgdELbe6NiWxQq8nbvimwcpkXwC5h3pH We8=
;; Received 679 bytes from 192.31.80.30#53(d.gtld-servers.net) in 65 ms

weibo.com.              60      IN      A       36.51.254.234
```

```
weibo.com.              86400     IN      NS      ns1.sina.com.cn.
weibo.com.              86400     IN      NS      ns3.sina.com.cn.
weibo.com.              86400     IN      NS      ns3.sina.com.
weibo.com.              86400     IN      NS      ns4.sina.com.
weibo.com.              86400     IN      NS      ns2.sina.com.cn.
weibo.com.              86400     IN      NS      ns4.sina.com.cn.
;; Received 274 bytes from 121.14.1.22#53(ns4.sina.com.cn) in 9 ms
```

北京节点探测结果如下。

```
[root@instance-disbrfer ~]# dig +trace weibo.com

; <<>> DiG 9.9.4-RedHat-9.9.4-61.el7 <<>> +trace weibo.com
;; global options: +cmd
.                       281417    IN      NS      f.root-servers.net.
.                       281417    IN      NS      c.root-servers.net.
.                       281417    IN      NS      j.root-servers.net.
.                       281417    IN      NS      d.root-servers.net.
.                       281417    IN      NS      l.root-servers.net.
.                       281417    IN      NS      a.root-servers.net.
.                       281417    IN      NS      b.root-servers.net.
.                       281417    IN      NS      k.root-servers.net.
.                       281417    IN      NS      i.root-servers.net.
.                       281417    IN      NS      e.root-servers.net.
.                       281417    IN      NS      g.root-servers.net.
.                       281417    IN      NS      h.root-servers.net.
.                       281417    IN      NS      m.root-servers.net.
.                       518400    IN      RRSIG   NS 8 0 518400 20190729050000
20190716040000 59944 . vWZ5F6N96z6eiYiTv8AuIbZuJkgx9ZV+WalWH3IGsHwDWdsKpeDgsZRI
0W+D69s7MSRBNovNyoZizYz5BKbJsW2jUcJATls8xPC1B2kYSHZOUMY3
jDeYCWKKh8mfog19zjITpr4dMa43Y/oQJaNJIQZQ3ZiRXsp/TNf8uDZv
xCYwqKBMqpYUhD7lLldlFLefB6PxC9+EHudMt5jcSXGcTmZF3QuN8ow8
ASYNYkpjr6M3wRCWFK+EaCfvwEzIVCcNXwklWELQvewRGVP/hqWqP/Bs
O1OWbloifmoC1fVuBH6nkDapouitEO1b5ulQYssm8+fI0QfjUSPxUJCL THajbw==
;; Received 1097 bytes from 172.16.0.2#53(172.16.0.2) in 1424 ms

com.                    172800    IN      NS      c.gtld-servers.net.
com.                    172800    IN      NS      h.gtld-servers.net.
com.                    172800    IN      NS      b.gtld-servers.net.
com.                    172800    IN      NS      i.gtld-servers.net.
com.                    172800    IN      NS      j.gtld-servers.net.
com.                    172800    IN      NS      d.gtld-servers.net.
com.                    172800    IN      NS      m.gtld-servers.net.
com.                    172800    IN      NS      f.gtld-servers.net.
com.                    172800    IN      NS      l.gtld-servers.net.
com.                    172800    IN      NS      e.gtld-servers.net.
com.                    172800    IN      NS      k.gtld-servers.net.
com.                    172800    IN      NS      a.gtld-servers.net.
com.                    172800    IN      NS      g.gtld-servers.net.
com.                    86400     IN      DS      30909 8 2
E2D3C916F6DEEAC73294E8268FB5885044A833FC5459588F4A9184CF C41A5766
com.                    86400     IN      RRSIG   DS 8 1 86400 20190728170000
20190715160000 59944 . M741IeR7eWA+yn7zTfIkDlVltDwy8gOcnuwCb5xIXiuO3R1/LU7big2e
SxZeCF1MNjzPXdfySJ0zoC4mQzM26WOqlNmlNB5xfaxH9SMIZ0iDFsvI
m0UZ+hatSSsCrZGfcoPjj2wvUx200gfMQQd9V1X1N5B9tVOhfGq7AklR
MFh5BFFcHOig0+Oynyo13WgwofpOlAap8NPIVw/13oaP9E9h4XfVe/9x
Hxq88IoXJw9QFEjgtHMm4Mn/7ehGYv2kwe8gYCVRzKiRHiqc00CXr4FJ
N+676INFXD+KvPxVgJkdjwCUr8blGnnOFHqlwZOiCH1vDgJlLCsqL4yO A60J7A==
;; Received 1169 bytes from 192.203.230.10#53(e.root-servers.net) in 2976 ms

weibo.com.              172800    IN      NS      ns1.sina.com.cn.
weibo.com.              172800    IN      NS      ns2.sina.com.cn.
weibo.com.              172800    IN      NS      ns3.sina.com.cn.
weibo.com.              172800    IN      NS      ns4.sina.com.
weibo.com.              172800    IN      NS      ns3.sina.com.
CK0POJMG874LJREF7EFN8430QVIT8BSM.com. 86400 IN NSEC3 1 1 0 -
```

```
CK0Q1GIN43N1ARRC9OSM6QPQR81H5M9A NS SOA RRSIG DNSKEY NSEC3PARAM
CK0POJMG874LJREF7EFN8430QVIT8BSM.com. 86400 IN RRSIG NSEC3 8 2 86400 20190722044529
20190715033529 3800 com. QruEtMEx3JMwz1cwK+gArQ3ZkNbCR4x+8ul54fAkec4IeN6LvHGTo61m
Dq5w/D0fzhLIhwXRtbxSpDPspvqYpEys6f69kM1U4NzeGvTb8/iz7/b0T
es+n3i+Gx0S+/X7ajkvu+J6GmVpgsZPIMxpSJjGSLZXtlEB4YyPBprWZ sck=
DLN3V1VIN2MEMHLPNC16JB1R1ATDH2V4.com. 86400 IN NSEC3 1 1 0 -
DLN5PBROIQFHD0V9HVVLCN86SL75HJBK NS DS RRSIG
DLN3V1VIN2MEMHLPNC16JB1R1ATDH2V4.com. 86400 IN RRSIG NSEC3 8 2 86400 20190722052104
20190715041104 3800 com. U/86DTjPN+oMhvmKZXfkZo7IVvlvrPyxsozVHoAcvV6Xq73q9rsRLCF2
uZ5+of4rKfqp9nuP3aGvEHhqnZoDA1S6qQPcrScwzyJeB72OsjsVIJtA
iqGpFN4GXOEco4KKBO5SKTVBEgdELbe6NiWxQq8nbvimwcpkXwC5h3pH We8=
;; Received 679 bytes from 192.5.6.30#53(a.gtld-servers.net) in 255 ms

weibo.com.              60      IN      A       123.125.104.26
weibo.com.              60      IN      A       180.149.138.56
weibo.com.              60      IN      A       123.125.104.197
weibo.com.              86400   IN      NS      ns4.sina.com.
weibo.com.              86400   IN      NS      ns2.sina.com.cn.
weibo.com.              86400   IN      NS      ns4.sina.com.cn.
weibo.com.              86400   IN      NS      ns1.sina.com.cn.
weibo.com.              86400   IN      NS      ns3.sina.com.
weibo.com.              86400   IN      NS      ns3.sina.com.cn.
;; Received 306 bytes from 123.125.29.99#53(ns3.sina.com.cn) in 5 ms
```

本地 DNS 服务器首先查询根域服务器（a-m.root-servers.net），得到 13 个根域服务器中的一个，并返回对应某个根域名的 NS 服务器，根域名 NS 服务器（193.0.14.129#53 (k.root-servers.net)）自身并不知道 weibo.com 域名的具体信息。接着转发客户端去查询.com 顶级域名服务器，并返回.com 域名服务器的 NS 记录，.com 域名服务器（192.55.83.30#53 (m.gtld-servers.net)）自身也不知道 weibo.com 的信息。但是它可以查询到供应商的地址，并返回供应商 DNS 服务器的 NS 地址，供应商 DNS 服务器（180.149.138.199#53(ns2.sina.com.cn)）返回结果，最终返回 weibo.com 的 A 记录如下。

```
weibo.com.              60      IN      A       123.125.104.26
weibo.com.              60      IN      A       123.125.104.197
```

2.3.5 DNS 查询指令之 host 进阶

host 指令是常用的分析域名查询指令，可以用来测试域名系统是否正常工作，其格式如下。

```
host(选项)(参数)
```

host 指令常用选项如表 2-3 所示。

表 2-3   host 指令常用选项

选项	说明
-a	显示详细的 DNS 信息
-d	打开调试方式
-c	指定查询类型，默认值为 "IN"
-C	查询指定主机的完整的 SOA 记录
-r	在查询域名时，不使用递归的查询方式
-t	指定查询的域名信息类型
-v	显示指令执行的详细信息
-w	如果域名服务器没有给出应答信息，则总是等待，直到域名服务器给出应答
-W	指定域名查询的最长时间，如果在指定时间内域名服务器没有给出应答信息，则退出指令
-4	使用 IPv4
-6	使用 IPv6

1. 查询 jd*.com 主机信息

查询 jd*.com 主机信息如下。

```
[root@VM_16_14_centos ~]# host jd*.com
jd*.com has address 120.52.148.118
jd*.com mail is handled by 10 mx.jd*.com.
[root@VM_16_14_centos ~]# host www.jd*.com
www.jd*.com is an alias for www.jd*.com.gslb.qianxun.com.
www.jd*.com.gslb.qianxun.com is an alias for jd-abroad*.cdn20.com.
jd-abroad*.cdn20.com has address 163.171.198.117
jd-abroad*.cdn20.com has address 45.116.82.62
```

2. 查询 www.ziroom*.com 主机信息

查询 www.ziroom*.com 主机信息如下。

```
[root@www.DNS_master_33.com ~]#host -t SOA www.ziroom*.com
www.ziroom*.com is an alias for csdc.ziroom*.com.
csdc.ziroom*.com is an alias for gtm-cn-v0h0r3hu405.gtm-a1b2.com.
[root@www.DNS_master_33.com ~]#host -t SOA ziroom*.com
ziroom*.com has SOA record vip1.aliDNS.com. hostmaster.hichina.com. 2018081616 3600 1200 86400 360
[root@www.DNS_master_33.com ~]#host -t NS ziroom*.com
ziroom*.com name server vip1.aliDNS.com.
ziroom*.com name server vip2.aliDNS.com.
[root@www.DNS_master_33.com ~]#host -t A ziroom*.com
ziroom*.com has address 119.254.76.108
ziroom*.com has address 119.254.76.106
ziroom*.com has address 119.254.83.229
ziroom*.com has address 119.254.83.228
```

2.3.6　DNS 查询指令之 nslookup 进阶

nslookup 指令用于查询 DNS 的记录，查看域名解析是否正常，在网络出现故障的时候用来诊断网络问题。

1. 直接查询

查询一个域名的 A 记录，基本格式如下。

```
nslookup domain [DNS-server]
```

如果没指定 DNS-server，用系统默认的 DNS 服务器，代码如下。

```
[root@localhost ~]# nslookup baidu.com
Server:         10.30.7.177
Address:        10.30.7.177#53

Non-authoritative answer:
Name:   baidu.com
Address: 123.125.114.144
Name:   baidu.com
Address: 111.13.101.208
Name:   baidu.com
Address: 180.149.132.47
Name:   baidu.com
Address: 220.181.57.217
```

上述代码中查询 baidu.com 域名的 A 记录。

2. 查询其他记录

直接查询返回的是 A 记录，可以指定参数，查询其他记录，如 AAAA、MX 等，格式如下。

```
nslookup -qt=type domain [DNS-server]
```

其中，type 可以是以下类型，如表 2-4 所示。

表 2-4　　　　　　　　　　　nslookup 指令查询解析类型

类型	说明
A	地址记录
AAAA	地址记录
AFSDB	Andrew 文件系统数据库服务器记录
ATMA	ATM 地址记录
CNAME	别名记录
HINFO	硬件配置记录，包括 CPU、操作系统信息
ISDN	域名对应的 ISDN 号码
MB	存放指定邮箱的服务器
MG	邮件组记录
MINFO	邮件组和邮箱的信息记录
MR	改名的邮箱记录
MX	邮件服务器记录
NS	名字服务器记录
PTR	反向记录
RP	负责人记录
RT	路由穿透记录
SRV	TCP 服务器信息记录
TXT	域名对应的文本信息
X25	域名对应的 X.25 地址记录

查询主机的 MX 记录代码如下。

```
[root@localhost ~]# nslookup -qt=mx baidu.com 8.8.8.8
*** Invalid option: qt=mx
Server:     8.8.8.8
Address:    8.8.8.8#53

Non-authoritative answer:
Name:    baidu.com
Address: 111.13.101.208
Name:    baidu.com
Address: 123.125.114.144
Name:    baidu.com
Address: 180.149.132.47
Name:    baidu.com
Address: 220.181.57.217
```

3．查询更具体的信息

nslookup 查询格式如下。

```
nslookup -d [其他参数] domain [DNS-server]
```

查询时，加上 -d 选项，即可查询域名的缓存。

2.3.7　DNS 服务类型查询指令总结

DNS 服务类型查询指令总结如表 2-5 所示。

表 2-5　　　　　　　　　　　DNS 服务类型查询指令总结

指令	通用性	易用性
dig	常用于 Linux 操作系统	专业调试指令

续表

指令	通用性	易用性
host	支持多平台，如 Windows、Linux 操作系统平台	简单、易用
nslookup	支持多平台，应用广泛	简单、易懂

2.4 用 BIND 实现子域授权和区域转发

2.4.1 实现 DNS 服务器子域授权

1．遇到的问题

假设在一个公司（域）内有两个部门（子域），这两个部门是独立运营的，并且希望实现 DNS 服务器的自我管理，则该场景可以使用子域授权。

子域即在原有的域上再划分出一个小的区域并指定新的子 DNS 服务器。在子区域中，如果有客户端请求解析，则只要找新的子 DNS 服务器即可。这样做的好处不仅可以减轻主 DNS 服务器的压力，还有利于分层管理。

> **注意**：任何的子域必须得到其父域的授权才可以解析域名。父域和子域不一定在同一个网段内，但是彼此之间必须能相互通信。

2．子域授权原理

在一个域下划分子域，并给子域指定一个新的 DNS 服务器，这种类似划分行政级别的管理方法在技术层面是可以实现的。子域授权其实就是将一个比较大的域分割成小区域（子域），每个小区域（子域）可以交由一组或多组服务器管理。这些服务器只解析其管辖范围内的域名，超出其范围的解析请求一般会转发给父域或直接转发给根域。

子域是相对而言的，对根域来说顶级域就是它的子域。以此类推，此处的子域授权是针对二级域名来说的，即三级域名授权。

正向区域的子域授权：使用胶水记录（Glue Record），在父域中添加一条 NS 记录和一条 A 记录即可。如果客户端的请求超出子域的解析范围，那么我们就需要定义转发服务器。

为了实现分层解析，减轻主 DNS 服务器的压力，可在父域中配置、设计子域授权这个功能。

3．案例——配置企业级子域授权

本案例要求为公司构建父/子 DNS 服务器，其中，父 DNS 服务器负责解析二级域 booxin*.com，而子 DNS 服务器负责解析三级域 ops.booxin*.com。现要求当客户端向父 DNS 服务器查询子域中的域名 www.ops.booxin*.com 的时候也能获得正确结果。如图 2-22 所示，需要在父 DNS 服务器上配置对子域的授权。

需要完成的配置过程如下。

- 构建父 DNS 服务器（booxin*.com）。
- 构建子 DNS 服务器（ops.booxin*.com）。
- 在父 DNS 服务器上配置子域授权。
- 测试子域授权查询。

2.4 用 BIND 实现子域授权和区域转发

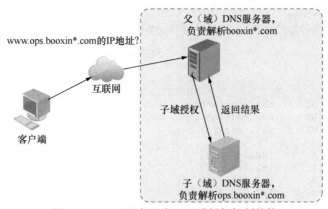

图 2-22 DNS 服务器实现子域授权机制结构

方案如下。

使用两台 CentOS 6 虚拟机,其中一台作为父 DNS 服务器(192.168.2.33),另外一台作为子 DNS 服务器(192.168.2.35),其他 CentOS 计算机、父/子 DNS 服务器中的任何一台都可以作为测试用的客户端,如图 2-23 所示。

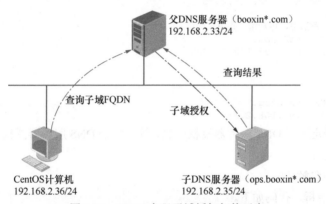

图 2-23 BIND 实现子域授权架构示意

子域授权环境部署规划如表 2-6 所示。

表 2-6 子域授权环境部署规划

角色	操作系统	IP 地址	主机名和域名
父 DNS 服务器	CentOS 6.9 x86_64	192.168.2.33	DNS.booxin*.com
子 DNS 服务器	CentOS 6.9 x86_64	192.168.2.35	DNS.ops.booxin*.com

在父 DNS 服务器上配置子 DNS 服务器授权时,在父域中配置 NS 记录和对应的 A 记录,需要修改父域的解析记录文件,添加内容格式如下。

- 子域名称.   IN   NS   子 DNS 服务器的 FQDN。
- 子 DNS 服务器的 FQDN.   IN   A   子 DNS 服务器的 IP 地址。

4. 父域配置

主配置文件配置子域名解析记录,首先架设一个父 DNS 服务器,编辑/var/named/booxin*.com.zone 文件,父 DNS 服务器的 IP 地址为 192.168.2.33,资源记录数据文件代码如下。

```
[root@www.DNS_master_33.com ~]#cat /var/named/chroot/var/named/booxin*.com.zone
$TTL 3H
```

```
$ORIGIN booxin*.com.
@       IN SOA  booxin*.com.  3128743705.qq.com. (
                              2        ; serial
                              1D       ; refresh
                              1H       ; retry
                              1W       ; expire
                              3H )     ; minimum
        IN      NS      ns1.booxin*.com.
        IN      NS      ns2.booxin*.com.
        IN      MX  5   mail.booxin*.com.

ns1     IN      A       192.168.2.33
ns2     IN      A       192.168.2.34
www     IN      A       192.168.2.37
www     IN      A       192.168.2.38
mail    IN      A       192.168.2.33
ftp     IN      A       192.168.2.33
ftp     IN      A       192.168.2.34


; 子域授权-韩艳威 2019/07/25
ops             IN      NS      ns.ops.
ns.ops.         IN      A       192.168.2.35

;ns.ops 有两个名称服务,再新增一条记录即可(主从配置)
;ops    IN      NS      ns2.ops.
;ns2.ops.       IN      A       192.168.2.37


blog    IN      NS      ns.blog.
ns.blog. IN     A       192.168.2.36
```

父 DNS 服务器定义子 DNS 服务器授权信息,等于给子 DNS 服务器授权,以上最后几行属于子 DNS 服务器配置。

5. 父域配置域名解析

父域配置域名解析,代码如下。

```
[root@www.DNS_master_33.com ~]#cat /etc/resolv.conf
nameserver 192.168.2.33
```

6. 子域配置

打开/etc/named.rfc1912.zones 文件,在最末尾定义子域的区域,代码如下。

```
[root@www.booxin*.com-35 ~]#tail /etc/named.rfc1912.zones
        allow-update { none; };
};


// 添加子域配置信息
zone "ops.booxin*.com" IN {
    type master;
    file "ops.booxin*.zone";
};
```

上述代码定义 ops.booxin*.zone 子域和区域数据文件位置。

7. 定义子 DNS 服务器的资源记录文件

查看 ops.booxin*.zone 区域配置文件,代码如下。

```
[root@www.booxin*.com-35 ~]#cat /var/named/ops.booxin*.zone
$TTL 600
$ORIGIN ops.booxin*.com.
```

```
@    IN   SOA      ops.booxin.com.    nsadmin.booxin*.com. (
                   2017052301
                   1H
                   2M
                   3D
                   1D )
@    IN   NS       ns
ns   IN   A        192.168.2.35
www  IN   A        192.168.2.38
```

上述代码定义 ns 记录和 www 记录。

8. 配置子 DNS 服务器

配置子 DNS 服务器，代码如下。

```
[root@www.booxin*.com-35 ~]#cat /etc/resolv.conf
; generated by /sbin/dhclient-script
nameserver 192.168.2.35
```

上述代码定义了 DNS 服务器的解析指向本机 IP 地址。

9. 父域、子域重启 DNS 服务器

父域重启 DNS 服务器，代码如下。

```
[root@www.DNS_master_33.com ~]#/etc/init.d/named restart
停止 named:                                              [确定]
启动 named:                                              [确定]
```

子域重启 DNS 服务器，代码如下。

```
[root@www.booxin*.com-35 ~]#/etc/init.d/named restart
Stopping named:                                         [  OK  ]
Starting named:                                         [  OK  ]
```

10. 父域解析测试

（1）测试父域解析。

```
[root@www.DNS_master_33.com named]#dig -t NS booxin*.com

; <<>> DiG 9.8.2rc1-RedHat-9.8.2-0.68.rc1.el6_10.3 <<>> -t NS booxin*.com
;; global options: +cmd
;; Got answer:
;; ->>HEADER<<- opcode: QUERY, status: NOERROR, id: 34556
;; flags: qr aa rd ra; QUERY: 1, ANSWER: 2, AUTHORITY: 0, ADDITIONAL: 2

;; QUESTION SECTION:
;booxin*.com.                 IN    NS

;; ANSWER SECTION:
booxin*.com.         10800    IN    NS    ns1.booxin*.com.
booxin*.com.         10800    IN    NS    ns2.booxin*.com.

;; ADDITIONAL SECTION:
ns1.booxin*.com.     10800    IN    A     192.168.2.33
ns2.booxin*.com.     10800    IN    A     192.168.2.34

;; Query time: 0 msec
;; SERVER: 192.168.2.33#53(192.168.2.33)
;; WHEN: Sat Jul 27 10:07:23 2019
;; MSG SIZE  rcvd: 96
```

上述代码中，父域解析 booxin*.com，返回结果正常。

（2）父域解析子域测试。

```
[root@www.DNS_master_33.com ~]#dig -t NS ops.booxin*.com

; <<>> DiG 9.8.2rc1-RedHat-9.8.2-0.68.rc1.el6_10.3 <<>> -t NS ops.booxin*.com
;; global options: +cmd
;; Got answer:
;; ->>HEADER<<- opcode: QUERY, status: SERVFAIL, id: 21847
```

```
;; flags: qr rd ra; QUERY: 1, ANSWER: 0, AUTHORITY: 0, ADDITIONAL: 0

;; QUESTION SECTION:
;ops.booxin*.com.              IN      NS

;; Query time: 0 msec
;; SERVER: 192.168.2.33#53(192.168.2.33)
;; WHEN: Sat Jul 27 09:04:43 2019
;; MSG SIZE  rcvd: 32
```

上述代码中，父域解析子域，返回结果正常。

（3）子域 A 记录测试。

父域解析子域的 A 记录，代码如下。

```
[root@www.DNS_master_33.com named]#dig -t A www.ops.booxin*.com

; <<>> DiG 9.8.2rc1-RedHat-9.8.2-0.68.rc1.el6_10.3 <<>> -t A www.ops.booxin*.com
;; global options: +cmd
;; Got answer:
;; ->>HEADER<<- opcode: QUERY, status: NOERROR, id: 24032
;; flags: qr rd ra; QUERY: 1, ANSWER: 1, AUTHORITY: 1, ADDITIONAL: 1

;; QUESTION SECTION:
;www.ops.booxin*.com.           IN      A

;; ANSWER SECTION:
www.ops.booxin*.com.    600     IN      A       192.168.2.38

;; AUTHORITY SECTION:
ops.booxin*.com.        546     IN      NS      ns.ops.booxin*.com.

;; ADDITIONAL SECTION:
ns.ops.booxin*.com.     546     IN      A       192.168.2.35

;; Query time: 0 msec
;; SERVER: 192.168.2.33#53(192.168.2.33)
;; WHEN: Sun Jul 28 10:47:20 2019
;; MSG SIZE  rcvd: 85
```

11．子域解析测试

（1）子域解析父域。

```
[root@www.booxin*.com-35 named]#dig -t NS booxin*.com

; <<>> DiG 9.8.2rc1-RedHat-9.8.2-0.68.rc1.el6_10.3 <<>> -t NS booxin*.com
;; global options: +cmd
;; Got answer:
;; ->>HEADER<<- opcode: QUERY, status: NOERROR, id: 13645
;; flags: qr rd ra; QUERY: 1, ANSWER: 2, AUTHORITY: 0, ADDITIONAL: 18

;; QUESTION SECTION:
;booxin*.com.                   IN      NS

;; ANSWER SECTION:
booxin*.com.            86400   IN      NS      DNS13.hichina.com.
booxin*.com.            86400   IN      NS      DNS14.hichina.com.

;; ADDITIONAL SECTION:
DNS14.hichina.com.      172800  IN      A       106.11.211.56
DNS14.hichina.com.      172800  IN      A       106.11.211.66
DNS14.hichina.com.      172800  IN      A       140.205.41.16
DNS14.hichina.com.      172800  IN      A       140.205.41.26
DNS14.hichina.com.      172800  IN      A       140.205.81.16
DNS14.hichina.com.      172800  IN      A       140.205.81.26
DNS14.hichina.com.      172800  IN      A       106.11.141.116
DNS14.hichina.com.      172800  IN      A       106.11.141.126
```

```
DNS14.hichina.com.      172800  IN  AAAA2400:3200:2000:33::1
DNS13.hichina.com.      172800  IN  A     106.11.141.115
DNS13.hichina.com.      172800  IN  A     106.11.141.125
DNS13.hichina.com.      172800  IN  A     106.11.211.55
DNS13.hichina.com.      172800  IN  A     106.11.211.65
DNS13.hichina.com.      172800  IN  A     140.205.41.15
DNS13.hichina.com.      172800  IN  A     140.205.41.25
DNS13.hichina.com.      172800  IN  A     140.205.81.15
DNS13.hichina.com.      172800  IN  A     140.205.81.25
DNS13.hichina.com.      172800  IN  AAAA2400:3200:2000:32::1

;; Query time: 471 msec
;; SERVER: 192.168.2.35#53(192.168.2.35)
;; WHEN: Sat Jul 27 10:11:37 2019
;; MSG SIZE  rcvd: 388
```

从上述代码的返回结果中可以看到，子域无法解析父域，子域中没有父域的相关数据记录，所以无法对父域进行解析。

（2）解析子域。

```
[root@www.booxin*.com-35 named]#dig -t NS ops.booxin*.com

; <<>> DiG 9.8.2rc1-RedHat-9.8.2-0.68.rc1.el6_10.3 <<>> -t NS ops.booxin*.com
;; global options: +cmd
;; Got answer:
;; ->>HEADER<<- opcode: QUERY, status: NOERROR, id: 7114
;; flags: qr aa rd ra; QUERY: 1, ANSWER: 1, AUTHORITY: 0, ADDITIONAL: 1

;; QUESTION SECTION:
;ops.booxin*.com.               IN      NS

;; ANSWER SECTION:
ops.booxin*.com.        600     IN      NS      ns.ops.booxin*.com.

;; ADDITIONAL SECTION:
ns.ops.booxin*.com.     600     IN      A       192.168.2.35

;; Query time: 0 msec
;; SERVER: 192.168.2.35#53(192.168.2.35)
;; WHEN: Sat Jul 27 10:03:21 2019
;; MSG SIZE  rcvd: 65
```

思考：如何让子域直接解析父域？2.4.2 小节会进行详解。

2.4.2　实现 DNS 服务器域名解析转发

2.4.1 小节中，子域解析父域中的主机名，只能通过递归到根域去解析，如何让子域不通过根域直接解析父域呢？

1. 定义转发域，设置转发器

192.168.2.35 服务器设置转发器代码如下。

```
[root@www.booxin*.com-35 named]#tail  /etc/named.rfc1912.zones
};

// 添加转发规则
zone "booxin*.com" IN {
      type forward;
      forward only;
      forwarders { 192.168.2.33; 192.168.2.34; };
};
```

2. 重启 BIND 服务

重启或重新加载 BIND 服务，使配置立即生效，实现代码如下。

```
[root@www.booxin*.com-35 named]#/etc/init.d/named restart
Stopping named:                                          [  OK  ]
Starting named:                                          [  OK  ]

[root@www.booxin*.com-35 ~]#/etc/init.d/named reload
Reloading named:                                         [  OK  ]
```

上述代码重启或重新加载 BIND 服务，使配置立即生效。

3. 重新解析父域测试

子域再次解析父域，代码如下。

```
[root@www.booxin*.com-35 named]#dig -t NS booxin*.com

; <<>> DiG 9.8.2rc1-RedHat-9.8.2-0.68.rc1.el6_10.3 <<>> -t NS booxin*.com
;; global options: +cmd
;; Got answer:
;; ->>HEADER<<- opcode: QUERY, status: NOERROR, id: 50903
;; flags: qr rd ra; QUERY: 1, ANSWER: 2, AUTHORITY: 0, ADDITIONAL: 0

;; QUESTION SECTION:
;booxin*.com.                   IN    NS

;; ANSWER SECTION:
booxin*.com.          86400    IN    NS    DNS13.hichina.com.
booxin*.com.          86400    IN    NS    DNS14.hichina.com.

;; Query time: 13 msec
;; SERVER: 202.106.195.68#53(202.106.195.68)
;; WHEN: Sat Jul 27 10:14:29 2019
;; MSG SIZE  rcvd: 76
```

子域解析父域时，第一次解析到了公网，通过公网返回解析记录。

```
[root@www.booxin*.com-35 named]#dig -t NS booxin*.com

; <<>> DiG 9.8.2rc1-RedHat-9.8.2-0.68.rc1.el6_10.3 <<>> -t NS booxin*.com
;; global options: +cmd
;; Got answer:
;; ->>HEADER<<- opcode: QUERY, status: NOERROR, id: 8966
;; flags: qr rd ra; QUERY: 1, ANSWER: 2, AUTHORITY: 0, ADDITIONAL: 2

;; QUESTION SECTION:
;booxin*.com.                   IN    NS

;; ANSWER SECTION:
booxin*.com.          10793    IN    NS    ns2.booxin*.com.
booxin*.com.          10793    IN    NS    ns1.booxin*.com.

;; ADDITIONAL SECTION:
ns1.booxin*.com.      10793    IN    A     192.168.2.33
ns2.booxin*.com.      10793    IN    A     192.168.2.34

;; Query time: 0 msec
;; SERVER: 192.168.2.35#53(192.168.2.35)
;; WHEN: Sat Jul 27 10:14:36 2019
;; MSG SIZE  rcvd: 96
```

子域解析父域时，第二次解析本地的服务器，通过本地服务器返回解析记录。

4. 子域授权和转发总结

（1）构建父 DNS 服务器。

- 安装 BIND 软件。
- 建立 named.conf 配置文件，并添加主域解析记录。

- 建立主域解析记录数据文件。
- 启动 BIND 服务，确保域名可用。

（2）构建子 DNS 服务器。
- 安装 BIND 软件。
- 建立 named.conf 配置文件，添加子域解析记录。
- 建立子域解析记录数据文件。
- 启动 BIND 服务，确保子域名解析可用。

（3）配置并测试子域授权。
- 配置子域授权。
- 配置转发记录。
- 设置各自的 DNS 客户端配置。

（4）子域授权和转发配置注意事项。

开启转发功能时，应关闭两台 DNS 服务器的 DNSsec 校验，代码如下。

```
# vim /etc/named.conf
        DNSsec-enable yes;
        DNSsec-validation no;
        DNSsec-lookaside auto;
```

2.5 用 BIND 实现域名解析

2.5.1 直接域名、泛域名及子域

1. 直接域名解析

许多用户有直接使用域名访问 Web 网站的习惯，即在浏览器中不输入 www 等主机名，而是直接使用如 baidu.com 或 csdn.net 等域名来访问网站。并不是所有的 Web 网站都支持这种访问方式，只有 DNS 服务器能直接解析域名的网站才支持，以 csdn.net 网站为例，使用 dig 指令解析结果如图 2-24 所示。

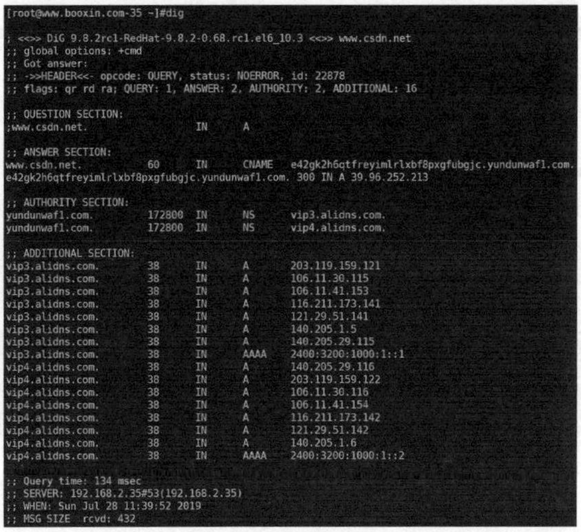

图 2-24 解析 csdn.net

域名 csdn.net 解析后 IP 地址如图 2-24 所示,与 www.csdn.net 域名的解析结果一样,如图 2-25 所示。

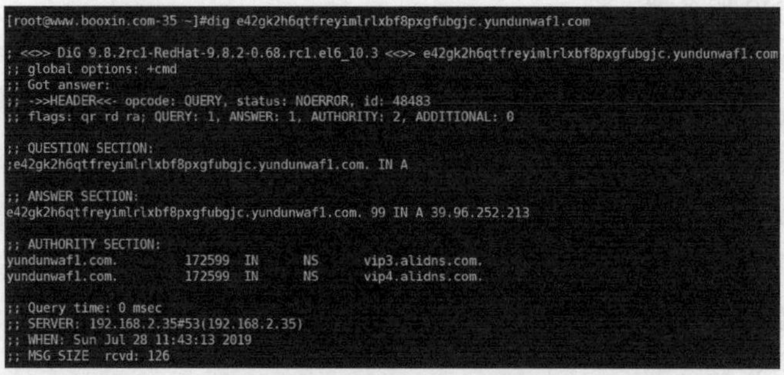

图 2-25 解析 csdn.net 域名

2. 泛域名解析

如果在 shop.taobao.com 域名中加入以下语句,还可以实现一种泛域名的效果,代码如下。

```
*.shop.taobao.com.      1800    IN    CNAME    shop.taobao.com.
```

泛域名是指一个域名下的所有主机和子域名都被解析到同一个地址上。在以上配置中,所有以 ".shop.taobao.com" 结尾的域名的地址都将解析为 shop.taobao.com。另外,默认情况下泛域名解析的优先级最高,如果区域文件中存在其他主机的资源记录,它们都将失效,泛域名解析的测试结果,如图 2-26 和图 2-27 所示。

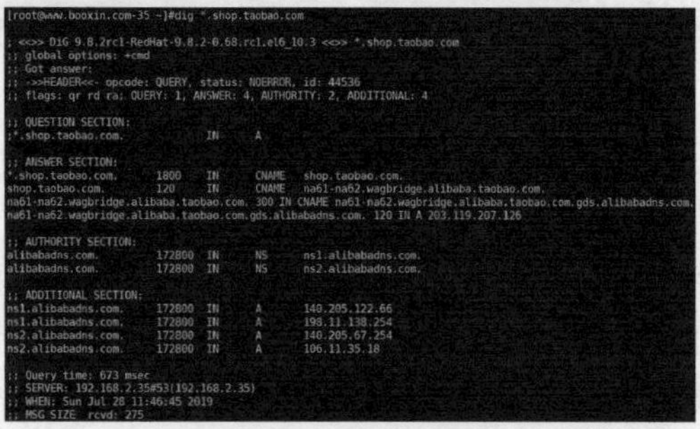

图 2-26 泛域名解析测试 1   图 2-27 泛域名解析测试 2

从图 2-27 中可以看到,不管采用什么样的主机名,只要以 ".shop.taobao.com" 结尾,地址都将解析为 shop.taobao.com。

3. 子域解析

子域是域名层次结构中的一个术语,是对某一个域进行细分时的下一级域。如 shop.taobao.com 是一个顶级域名,可以把 aaa.shop.taobao.com 配置为它的一个子域。

配置子域可以采用两种方式,一种是把子域配置放在另一台 DNS 服务器上,另一种是子域配置与父域配置放在一起,此时也称为虚拟子域,配置如下。

```
$ORIGIN shop.taobao.com.
*.shop.taobao.com.      1800    IN    CNAME    shop.taobao.com.
```

注意:子域一般适用于部门或子公司,方便管理。

2.5.2 直接域名解析实例

直接域名解析在企业环境中应用非常广泛,是架构师、Linux 高级工程师,以及 DevOps 和

2.5 用 BIND 实现域名解析

SRE 工程师必备技能之一,直接域名解析服务器环境信息如表 2-7 所示。

表 2-7 直接域名解析服务器环境信息

角色	操作系统	IP 地址	主机名和域名
DNS 服务器	CentOS 6.9 x86_64	192.168.2.33	DNS.m.booxin*.com
Web 服务器	CentOS 6.9 x86_64	192.168.2.37	bj-Web37
客户端	Windows 10 专业版	192.168.2.100	hanyw

1. DNS 服务器配置

192.168.2.33 作为 DNS 服务器,配置步骤如下。

(1) 配置域名解析文件,代码如下。

```
[root@www.DNS_master_33.com named]#cat booxin*.com.zone
$TTL 3H
$ORIGIN booxin*.com.
@       IN SOA  booxin*.com. 3128743705.qq.com. (
                                2       ; serial
                                1D      ; refresh
                                1H      ; retry
                                1W      ; expire
                                3H )    ; minimum
        IN      NS      ns1.booxin*.com.
        IN      NS      ns2.booxin*.com.
        IN      MX   5  mail.booxin*.com.

ns1     IN      A       192.168.2.33
ns2     IN      A       192.168.2.34
mail    IN      A       192.168.2.33
ftp             IN      A       192.168.2.33
ftp     IN      A       192.168.2.34
www     IN      A       192.168.2.37
;www    IN      A       192.168.2.38
ww      IN      CNAME   www.booxin*.com.
w       IN      CNAME   www.booxin*.com.
handuoduo       IN      CNAME       www.booxin*.com.
; 子域授权-韩艳威 2019/07/25
ops     IN      NS      ns.ops
ns.ops          IN      A       192.168.2.35

;ns.ops 有两个名称服务,再新增一条记录即可(主从配置)
;ops    IN      NS      ns2.ops
;ns2.ops IN     A       192.168.2.37
;blog           IN      NS      ns.blog
;ns.blog        IN      A       192.168.2.36
```

上述代码配置域名解析文件。

(2) 检测配置文件语法,代码如下。

```
[root@www.DNS_master_33.com named]#named-checkzone booxin*.com booxin*.com.zone
zone booxin*.com/IN: ops.booxin*.com/NS 'ns.ops.booxin*.com' (out of zone) has no
addresses records (A or AAAA)
zone booxin*.com/IN: loaded serial 2
OK
```

上述代码检测 booxin*.com 区域配置文件语法是否正确。

(3) 重启服务,代码如下。

```
[root@www.DNS_master_33.com ~]#/etc/init.d/named restart
停止 named:                                          [确定]
启动 named:                                          [确定]
```

上述代码重启 named 服务。

（4）查看 named 进程和端口是否存在，代码如下。

```
[root@www.DNS_master_33.com ~]#netstat -ntpl
Active Internet connections (only servers)
Proto Recv-Q Send-Q Local Address      Foreign Address     State    PID/Program name
tcp       0      0 0.0.0.0:42894 0.0.0.0:*      LISTEN    1283/rpc.statd
tcp       0      0 0.0.0.0:111   0.0.0.0:*      LISTEN    1261/rpcBIND
tcp       0      0 192.168.2.33:53 0.0.0.0:*    LISTEN    1620/named
tcp       0      0 127.0.0.1:53  0.0.0.0:*      LISTEN    1620/named
tcp       0      0 0.0.0.0:22    0.0.0.0:*      LISTEN    1355/sshd
tcp       0      0 127.0.0.1:953 0.0.0.0:*      LISTEN    1620/named
tcp       0      0 127.0.0.1:25  0.0.0.0:*      LISTEN    1434/master
tcp       0      0 :::46182      :::*           LISTEN    1283/rpc.statd
tcp       0      0 :::111        :::*           LISTEN    1261/rpcBIN
tcp       0      0 :::22         :::*           LISTEN    1355/sshd
tcp       0      0 ::1:953       :::*           LISTEN    1620/named
tcp       0      0 ::1:25        :::*           LISTEN    1434/master
[root@www.DNS_master_33.com ~]#ps -ef |grep named
named    1620    1  0 21:06 ?        00:00:00 /usr/sbin/named -u named -t /var/named/chroot
root     1631 1480  0 21:15 pts/0    00:00:00 grep named
```

上述代码中，named 进程和 53 端口已经存在。

（5）DNS 客户端设置，代码如下。

```
[root@www.DNS_master_33.com ~]#cat /etc/resolv.conf
nameserver 192.168.2.33
nameserver 8.8.8.8
```

（6）解析域名，代码如下。

```
[root@www.DNS_master_33.com ~]#nslookup  www.booxin*.com 192.168.2.33
Server:       192.168.2.33
Address:192.168.2.33#53

Name:    www.booxin*.com
Address: 192.168.2.37
```

上述代码使用 DNS 服务器（192.168.2.33）解析 www.booxin*.com 域名，返回该域名的 A 记录解析 IP 地址为 192.168.2.37。

2. 配置 Web 服务器

Web 服务器配置步骤如下。

（1）查看 IP 地址和主机名，代码如下。

```
[root@bj-Web-37 ~]#ip ro
192.168.2.0/24 dev eth0  proto kernel  scope link  src 192.168.2.37
169.254.0.0/16 dev eth0  scope link  metric 1002
default via 192.168.2.1 dev eth0
[root@bj-Web-37 ~]#uname -n
bj-Web-37
```

上述代码中，Web 服务器的 IP 地址为 192.168.2.37，主机名为 bj-Web-37。

（2）确认 SELinux 和防火墙状态，代码如下。

```
[root@bj-Web-37 ~]#sestatus
SELinux status:                 disabled
[root@bj-Web-37 ~]#/etc/init.d/iptables status
iptables: Firewall is not running.
```

上述代码表示 SELinux 和防火墙已经处于关闭状态。

（3）安装 epel 源，代码如下。

```
[root@bj-Web-37 ~]#yum install epel-release -y
Loaded plugins: fastestmirror
Setting up Install Process
Loading mirror speeds from cached hostfile
 * base: mirrors.neusoft.edu.cn
 * epel: mirrors.tuna.tsinghua.edu.cn
```

```
 * extras: mirrors.neusoft.edu.cn
 * updates: mirrors.tuna.tsinghua.edu.cn
Package epel-release-6-8.noarch already installed and latest version
Nothing to do
```

上述代码表示 epel 源已经安装成功。

（4）安装 Nginx，代码如下。

```
[root@bj-Web-37 ~]#yum install nginx -y
Loaded plugins: fastestmirror
Setting up Install Process
```

（中间代码略）

```
Dependency Updated:
  nginx-all-modules.noarch 0:1.10.3-1.el6              nginx-filesystem.noarch
0:1.10.3-1.el6            nginx-mod-http-geoip.x86_64 0:1.10.3-1.el6
  nginx-mod-http-image-filter.x86_64 0:1.10.3-1.el6
nginx-mod-http-perl.x86_64 0:1.10.3-1.el6       nginx-mod-http-xslt-filter.x86_64
0:1.10.3-1.el6
  nginx-mod-mail.x86_64 0:1.10.3-1.el6                 nginx-mod-stream.x86_64
0:1.10.3-1.el6

Complete!
```

（5）查看 Nginx 是否安装成功，代码如下。

```
[root@bj-Web-37 ~]#rpm -q nginx
nginx-1.10.3-1.el6.x86_64
```

上述代码表示 Nginx 已经安装成功。

（6）配置 Nginx，代码如下。

```
[root@bj-Web-37 ~]#cd /etc/nginx/conf.d/
[root@bj-Web-37 conf.d]#cat www.booxin*.com.conf
server {
    server_name  www.booxin*.com ww.booxin*.com w.booxin*.com;
    root         /usr/share/nginx/html;

    # Load configuration files for the default server block.
    include /etc/nginx/default.d/*.conf;

    location / {
    }

    error_page 404 /404.html;
        location = /40x.html {
    }

    error_page 500 502 503 504 /50x.html;
        location = /50x.html {
    }

}
```

上述代码设置 server_name 分别为 www.booxin*.com、ww.booxin*.com、w.booxin*.com。

（7）设置 Nginx 默认首页配置文件 index.html，代码如下。

```
[root@bj-Web-37 ~]#cat /usr/share/nginx/html/index.html
<!DOCTYPE html>
<html lang="en">
<!--
******************** 作者：韩艳威 ********************
******************** 邮箱：3128743705@qq.com ********************
******************** 说明：测试 DNS 解析 ********************
******************** 日期：2019/06/ ********************
******************** 版本：******************** 
******************** 补充：********************
-->
```

```html
<head>
    <meta charset="UTF-8">
    <title>韩艳威的 HTML 测试页</title>
    <meta charset="UTF-8">
    <title>显示系统时间</title>


    <style type="text/css">
        .time_sys {
            font-size: 60px;
            background-color: #ffd4dc;
        }
        #show_time {
            font-size: 36px;
            background-color: goldenrod;
        }
    </style>


    <script type="text/javascript">
        setInterval("fun(show_time)",1);
        function fun(timeID){
            var date = new Date();      //创建对象
            var y = date.getFullYear();         //获取年份
            var m =date.getMonth()+1;    //获取月份,返回0~11
            var d = date.getDate(); // 获取日
            var w = date.getDay();      //获取星期几,返回0~6     (0表示星期日)
            var ww = ' 星期'+'日一二三四五六'.charAt(new Date().getDay()) ;//星期几
            var h = date.getHours();    //时
            var minute = date.getMinutes();//分
            var s = date.getSeconds(); //秒
            var sss = date.getMilliseconds() ; //毫秒
            if(m<10){
                m = "0"+m;
            }
            if(d<10){
                d = "0"+d;
            }
            if(h<10){
                h = "0"+h;
            }
            if(minute<10){
                minute = "0"+minute;
            }
            if(s<10){
                s = "0"+s;
            }
            if(sss<10){
                sss = "00"+sss;
            }else if(sss<100){
                sss = "0"+sss;
            }
            document.getElementById(timeID.id).innerHTML
                = y+"-"+m+"-"+d+"    "+h+":"+minute+":"+s+"    "+" "+ww
        }
    </script>

</head>
<body>

<span class="time_sys">系统时间: </span>
<br >
<br />
```

```
<div ID="show_time"> </div>


</body>
</html>
```

上述代码在 Nginx 默认首页配置文件中使用浏览器访问时会动态显示当前操作系统时间。

3．Windows 10 客户端测试

（1）配置 DNS 服务器地址。

如图 2-28 所示，在"首选 DNS 服务器"文本框配置 DNS 服务器地址。

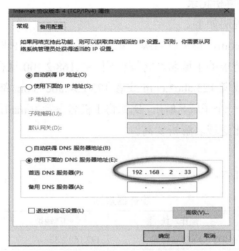

图 2-28　配置 DNS 服务器地址

（2）浏览器访问网站测试。

在浏览器中分别输入 www.booxin*.com、ww.booxin*.com、w.booxin*.com 域名进行访问测试，如图 2-29～图 2-31 所示。

图 2-29　www.booxin*.com 默认首页

图 2-30　ww.booxin*.com 默认首页

图 2-31　w.booxin*.com 默认首页

2.5.3 泛域名解析实例

1. 泛域名解析概念

泛域名：泛域名是指在一个域名下，以.hanyanwei*.com 的形式表示这个域名所有未建立的子域名。

泛域名解析是把.hanyanwei*.com 的 A 记录解析到某个 IP 地址上，我们通过访问任意的前缀.hanyanwei*.com 都能访问到解析的站点上。

2. 设置正向的泛域名解析记录

关于泛域名解析举例说明如下。

如用户的域名是 abc*.com，将子域名解析记录设置为 "*"，IP 地址解析到 192.168.2.100，"*" 是通配符，代表用户请求的子域名都将解析到 192.168.2.100 这台服务器。

如输入 bbs.abc*.com 或者 123.abc*.com 或者 123.234.abc*.com 都将解析为 192.168.2.100。

系统里面如果单独设置一个子域名解析，比如将主机名设置为 mail，单独解析到 192.168.2.118，那么该解析记录的优先级比泛域名解析要高。

3. 服务器规划

服务器规划如表 2-8 所示。

表 2-8　　　　　　　　　　　　服务器规划

角色	操作系统	IP 地址	主机名和域名
DNS 服务器	CentOS 6.9 x86_64	192.168.2.33	DNS.m.booxin*.com
Nginx 负载服务器	CentOS 6.9 x86_64	192.168.2.37	bj-Web37
客户端	Windows 10 专业版	192.168.2.100	hanyw

域名后缀也必须与本地域配置文件中定义的 "zone" 相同，代码如下。

```
*.d.booxin*.com.    IN A192.168.2.60
*.t.booxin*.com.    IN A192.168.2.61
*.q.booxin*.com.    IN A192.168.2.62
*.booxin*.com.      IN A192.168.2.63
```

- *.d.booxin*.com.表示匹配所有的以 booxin*.com.t 结尾的访问请求，这些请求都将转发到 192.168.2.60 这台服务器。
- *.t.booxin*.com.表示匹配所有的以 booxin*.com.d 结尾的访问请求，这些请求都将转发到 192.168.2.61 这台服务器。
- *.q.booxin*.com.表示匹配所有的以 booxin*.com.d 结尾的访问请求，这些请求都将转发到 192.168.2.62 这台服务器。
- *.booxin*.com.表示匹配所有的以 booxin*.com.d 结尾的访问请求，这些请求都将转发到 192.168.2.63 这台服务器。

4. 父域服务器设置

父域服务器设置如下。

```
[root@www.DNS_master_33.com named]#cat booxin*.com.zone
$TTL 3H
$ORIGIN booxin*.com.
@       IN SOA   booxin*.com. 3128743705.qq.com. (
                      2       ; serial
                      1D      ; refresh
                      1H      ; retry
```

```
                                1W       ; expire
                                3H )     ; minimum
                IN      NS      ns1.booxin*.com.
                IN      NS      ns2.booxin*.com.
                IN      MX  5   mail.booxin*.com.

ns1     IN      A       192.168.2.33
ns2     IN      A       192.168.2.34
mail    IN      A       192.168.2.33
ftp     IN      A       192.168.2.33
ftp     IN      A       192.168.2.34
www     IN      A       192.168.2.37
;www    IN      A       192.168.2.38
ww      IN      CNAME   www.booxin*.com.
w       IN      CNAME   www.booxin*.com.
handuoduo       IN      CNAME   www.booxin*.com.

; 子域授权-韩艳威 2019/07/25
;ops            IN      NS      ns.ops.
;ns.ops         IN      A       192.168.2.35

;ns.ops 有两个名称服务，再新增一条记录即可(主从配置)
;ops            IN      NS      ns2.ops
;ns2.opsIN      A       192.168.2.37
;blog           IN      NS      ns.blog
;ns.blog        IN      A       192.168.2.36

; 2019/07/31 泛域名解析
*.d.booxin*.com.   IN   A    192.168.2.60
;可以作为研发环境的 Nginx 代理服务器
*.t.booxin*.com.   IN   A    192.168.2.61
;可以作为测试环境的 Nginx 代理服务器
*.q.booxin*.com.   IN   A    192.168.2.62
;可以作为准生产环境的 Nginx 代理服务器
*.booxin*.com.     IN   A    192.168.2.63
;可以作为生产环境的 Nginx 代理服务器
```

上述代码定义 4 个环境的 Nginx 代理服务器 IP 地址，实际企业应用中一般会有 4 个环境的负载均衡器，分别为研发环境（开发人员开始调试代码的环境）、测试环境（测试人员测试代码的环境）、准生产环境（开发和测试联调后的稳定版程序运行环境）、生产环境（正式对外提供服务的最终版本服务运行环境）。

5. Nginx 负载均衡器设置

这里使用 192.168.2.37 模拟 Nginx 代理服务器，配置步骤如下。

（1）设置负载均衡器 IP 地址，代码如下。

```
[root@bj-Web-37 conf.d]#ifconfig
eth0    Link encap:Ethernet   HWaddr 00:0C:29:37:2B:C4
        inet addr:192.168.2.37  Bcast:192.168.2.255  Mask:255.255.255.0
        inet6 addr: fe80::20c:29ff:fe37:2bc4/64 Scope:Link
        UP BROADCAST RUNNING MULTICAST  MTU:1500  Metric:1
        RX packets:3795 errors:0 dropped:0 overruns:0 frame:0
        TX packets:1796 errors:0 dropped:0 overruns:0 carrier:0
        collisions:0 txqueuelen:1000
        RX bytes:309907 (302.6 KiB)  TX bytes:248035 (242.2 KiB)

lo      Link encap:Local Loopback
        inet addr:127.0.0.1  Mask:255.0.0.0
        inet6 addr: ::1/128 Scope:Host
        UP LOOPBACK RUNNING  MTU:65536  Metric:1
        RX packets:134 errors:0 dropped:0 overruns:0 frame:0
        TX packets:134 errors:0 dropped:0 overruns:0 carrier:0
        collisions:0 txqueuelen:0
        RX bytes:12358 (12.0 KiB)  TX bytes:12358 (12.0 KiB)
```

使用 ifconfig 指令，临时设置 eth0:0 网卡的 IP 地址为 192.168.2.60，代码如下。

```
[root@bj-Web-37 conf.d]#ifconfig eth0:0 192.168.2.60
[root@bj-Web-37 conf.d]#ifconfig
eth0      Link encap:Ethernet  HWaddr 00:0C:29:37:2B:C4
          inet addr:192.168.2.37  Bcast:192.168.2.255  Mask:255.255.255.0
          inet6 addr: fe80::20c:29ff:fe37:2bc4/64 Scope:Link
          UP BROADCAST RUNNING MULTICAST  MTU:1500  Metric:1
          RX packets:3869 errors:0 dropped:0 overruns:0 frame:0
          TX packets:1837 errors:0 dropped:0 overruns:0 carrier:0
          collisions:0 txqueuelen:1000
          RX bytes:316491 (309.0 KiB)  TX bytes:253591 (247.6 KiB)

eth0:0    Link encap:Ethernet  HWaddr 00:0C:29:37:2B:C4
          inet addr:192.168.2.60  Bcast:192.168.2.255  Mask:255.255.255.0
          UP BROADCAST RUNNING MULTICAST  MTU:1500  Metric:1

lo        Link encap:Local Loopback
          inet addr:127.0.0.1  Mask:255.0.0.0
          inet6 addr: ::1/128 Scope:Host
          UP LOOPBACK RUNNING  MTU:65536  Metric:1
          RX packets:134 errors:0 dropped:0 overruns:0 frame:0
          TX packets:134 errors:0 dropped:0 overruns:0 carrier:0
          collisions:0 txqueuelen:0
          RX bytes:12358 (12.0 KiB)  TX bytes:12358 (12.0 KiB)
```

使用 ping 指令测试 www.t.booxin*.com（测试环境）域名返回的 DNS 服务器 IP 地址（192.168.2.61），代码如下。

```
[root@bj-Web-37 conf.d]#ping www.t.booxin*.com
PING www.t.booxin*.com (192.168.2.61) 56(84) bytes of data.
^C
--- www.t.booxin*.com ping statistics ---
3 packets transmitted, 0 received, 100% packet loss, time 2703ms
```

使用 ping 指令测试 www.d.booxin*.com（研发环境）域名返回的 DNS 服务器 IP 地址（192.168.2.60），代码如下。

```
[root@bj-Web-37 conf.d]#ping www.d.booxin*.com
PING www.d.booxin*.com (192.168.2.60) 56(84) bytes of data.
64 bytes from 192.168.2.60: icmp_seq=1 ttl=64 time=0.010 ms
64 bytes from 192.168.2.60: icmp_seq=2 ttl=64 time=0.038 ms
64 bytes from 192.168.2.60: icmp_seq=3 ttl=64 time=0.023 ms
64 bytes from 192.168.2.60: icmp_seq=4 ttl=64 time=0.039 ms
64 bytes from 192.168.2.60: icmp_seq=5 ttl=64 time=0.039 ms
64 bytes from 192.168.2.60: icmp_seq=6 ttl=64 time=0.033 ms
64 bytes from 192.168.2.60: icmp_seq=7 ttl=64 time=0.074 ms
64 bytes from 192.168.2.60: icmp_seq=8 ttl=64 time=0.040 ms
64 bytes from 192.168.2.60: icmp_seq=9 ttl=64 time=0.134 ms
^C
--- www.d.booxin*.com ping statistics ---
9 packets transmitted, 9 received, 0% packet loss, time 8160ms
rtt min/avg/max/mdev = 0.010/0.047/0.134/0.035 ms
```

上述代码根据测试域名的不同，返回 DNS 服务器的 IP 地址也不相同。

（2）启动 Nginx 服务。

启动 Nginx 服务，并查看 DNS 服务器地址设置是否正确，代码如下。

```
[root@bj-Web-37 conf.d]#/etc/init.d/nginx start
Starting nginx:                                            [  OK  ]
[root@bj-Web-37 conf.d]#netstat -ntpl
Active Internet connections (only servers)
Proto Recv-Q Send-Q Local Address           Foreign Address         State       PID/Program name
tcp        0      0 0.0.0.0:80              0.0.0.0:*               LISTEN      1461/nginx
```

```
tcp        0      0 0.0.0.0:22              0.0.0.0:*               LISTEN      1210/sshd
tcp        0      0 127.0.0.1:25            0.0.0.0:*               LISTEN      1289/master
tcp        0      0 :::80                   :::*                    LISTEN      1461/nginx
tcp        0      0 :::22                   :::*                    LISTEN      1210/sshd
tcp        0      0 ::1:25                  :::*                    LISTEN      1289/master
[root@bj-Web-37 conf.d]#
[root@bj-Web-37 conf.d]#
[root@bj-Web-37 conf.d]#
[root@bj-Web-37 conf.d]#cat /etc/resolv.conf
; generated by /sbin/dhclient-script
nameserver 192.168.2.33
nameserver 202.106.195.68
```

上述代码表示 Nginx 服务已经启动，其中 80 端口已经被成功监听。

（3）ping 测试，代码如下。

```
[root@bj-Web-37 conf.d]#
[root@bj-Web-37 conf.d]#ping www.d.booxin*.com
PING www.d.booxin*.com (192.168.2.60) 56(84) bytes of data.
64 bytes from 192.168.2.60: icmp_seq=1 ttl=64 time=0.007 ms
64 bytes from 192.168.2.60: icmp_seq=2 ttl=64 time=0.040 ms
^C
--- www.d.booxin*.com ping statistics ---
2 packets transmitted, 2 received, 0% packet loss, time 1169ms
rtt min/avg/max/mdev = 0.007/0.023/0.040/0.017 ms
[root@bj-Web-37 conf.d]#
```

上述代码中测试 Nginx 负载均衡器返回 IP 地址正确。

（4）设置 DNS 服务器地址，代码如下。

```
[root@bj-Web-37 conf.d]#cat /etc/resolv.conf
; generated by /sbin/dhclient-script
nameserver 192.168.2.33
nameserver 202.106.195.68
```

上述代码设置 DNS 服务器地址。

6．客户端设置 DNS 服务器地址

Windows 10 客户端设置 DNS 服务器地址，如图 2-32 所示。

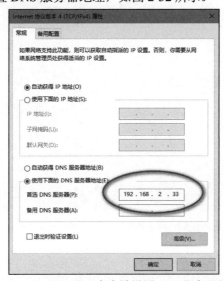

图 2-32　Windows 10 客户端设置 DNS 服务器地址

7. 浏览器访问测试

Windows 10 客户端使用浏览器访问 www.booxin*.com，如图 2-33 所示。

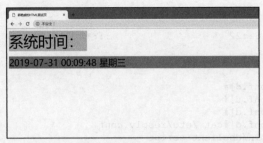

图 2-33　Windows 10 客户端使用浏览器访问 www.booxin*.com

第 3 章

DNS服务器核心应用与运维管理

3.1 构建企业级缓存 DNS 服务器

3.1.1 BIND 缓存基本实现

一个局域网中可能有很多的主机要访问互联网,这时就可以在局域网的出口处配置缓存服务器,加速网络访问。

也可以为缓存服务器配置一个上游 DNS 服务器地址,当缓存服务器无法完成解析或者想让局域网内的用户可以访问到除指定 DNS 配置之外的 DNS 服务器的情况下,会给客户端返回另一个更稳定或者信息更全的 DNS 服务器地址,客户端就可以通过这个 DNS 服务器继续完成查询,DNS 服务器配置代码如下。

```
[root@localhost etc]# vim /var/named/chroot/etc/named.conf
options
{
        directory "/var/named";
     forwarders {114.114.114.114; };

};
```

如果我们在 named.conf 里面指定 forward only,缓存服务器则不再进行任何的解析查询,而是直接把需求转发到上游 DNS 服务器,配置文件信息如下。

```
options
{
        directory "/var/named";
     forwarders {114.114.114.114; };
     forward only;
};
```

上述代码直接转发客户端所有 DNS 解析请求到 114.114.114.114 DNS 服务器。

3.1.2 DNS 转发器工作原理

1. DNS 转发器

DNS 服务器可以解析自己区域文件中的域名。那么,对本服务器上没有的域名的查询请求应如何处理呢?一种方法是直接转发查询请求到根域 DNS 服务器,进行迭代查询。另一种方法就是直接将请求转发给其他 DNS 服务器。

一个 DNS 服务器,能将本地 DNS 服务器无法解析的查询请求转发给网络上的其他 DNS 服务器,该 DNS 服务器即被指定为转发器。当 DNS 服务器将查询请求转发给转发器时,这种查询请求通常被称为递归查询。

对本地 DNS 服务器(在局域网中)无法解析的域名的查询,我们可以通过配置转发器来实现域名解析,DNS 转发器工作原理如图 3-1 所示。也可以在本地 DNS 服务器上设置转发器,指向互联网上的 DNS 服务器。

- 当本地 DNS 服务器(转发器)收到查询请求时,它会尝试使用本地数据解析该查询请求。
- 如果不能使用本地数据解析查询请求,此时本地 DNS 服务器作为客户端,会将查询请求转发给外网 DNS 服务器。
- 本地 DNS 服务器等待来自外网 DNS 服务器的应答。

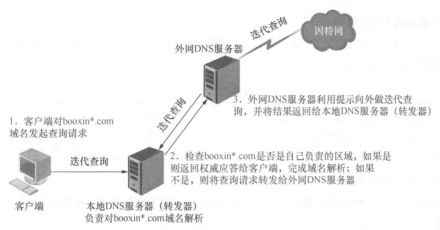

图 3-1　DNS 转发器工作原理

- 对于外网 DNS 服务器来说，它接收到的查询请求是迭代查询。此时，它自己需要向外做迭代查询找到"最终答案"并返回给转发器。
- 转发器将外网 DNS 服务器返回的查询结果送到客户端，完成解析过程。

2. DNS 转发类型

（1）全局转发。

全局转发将凡是本地没有通过"zone"定义的区域查询请求，全部转给某转发器。代码如下。

```
options {
   forwarders { ip; };
   forward only|first;
};
```

（2）局部转发。

局部转发仅转发对某特定区域的解析请求。代码如下。

```
options {
   forwarders { ip; };
   forward only|first;
forwarders { ip; };
};
```

参数说明如下。

- forwarders { ip; };：指明转发器。
- forward only|first;：only 表示仅转发；first 表示先进行转发。如果没查询到结果，那么它会根据提示向外做迭代查询。

3.1.3　使用 BIND 搭建缓存 DNS 服务器

1. 缓存 DNS 服务器应用场景

当我们在访问网页时，需要通过 DNS 服务器去解析网站地址。因为解析时间很短，所以当 DNS 访问量非常大的时候，会导致访问网页的响应时间变长，影响服务质量。

2. 缓存 DNS 服务器

当一台主机第一次访问网页时，该主机的 DNS 服务器会从其他 DNS 服务器获取数据（搭建的缓存 DNS 服务器中本来是不存在数据的），同时缓存到本地 DNS 服务器。当下次进行访问时，可以从该主机 DNS 服务器的缓存中直接获取数据，不必访问其他 DNS 服务器，大大减少访问网页的响应时间。

3. 配置缓存 DNS 服务器

我们使用一台虚拟机作为 DNS 服务器，在缓存配置好后分别使用另外两台服务器做测试，服务器规划如表 3-1 所示。

表 3-1　　　　　　　　　　　　　　　服务器规划

角色	操作系统	IP 地址	主机名和域名
DNS 服务器	CentOS 6.9 x86_64	192.168.2.38	DNS.m.booxin*.com
Nginx 负载服务器	CentOS 6.9 x86_64	192.168.2.39	bj-WEB39

4. 测试域名解析时间

在没有配置缓存 DNS 服务器时，我们分别在服务端和客户端使用 dig 指令观察访问网页的响应时间。

（1）服务端测试如下。

```
[root@DNS.m.booxin*.com ~]#dig www.jd*.com
-bash: dig: command not found
[root@DNS.m.booxin*.com ~]#
[root@DNS.m.booxin*.com ~]#yum install BIND-utils -y >/dev/null 2>&1
[root@DNS.m.booxin*.com ~]#dig www.jd*.com

; <<>> DiG 9.8.2rc1-RedHat-9.8.2-0.68.rc1.el6_10.3 <<>> www.jd*.com
;; global options: +cmd
;; Got answer:
;; ->>HEADER<<- opcode: QUERY, status: NOERROR, id: 52247
;; flags: qr rd ra; QUERY: 1, ANSWER: 3, AUTHORITY: 0, ADDITIONAL: 0

;; QUESTION SECTION:
;www.jd*.com.            IN    A

;; ANSWER SECTION:
www.jd*.com.         80  IN   CNAME    www.jd*.com.gslb.qianxun.com.
www.jd*.com.gslb.qianxun.com. 34  IN   CNAME    www.jdcdn*.com.
www.jdcdn*.com.      34  IN   A        111.206.231.1

;; Query time: 18 msec
;; SERVER: 114.114.114.114#53(114.114.114.114)
;; WHEN: Sat Aug  3 20:44:15 2019
;; MSG SIZE  rcvd: 106
```

上述代码表示服务端解析 www.jd*.com 的响应时间为 18ms。

（2）客户端测试如下。

使用 yum 指令安装 BIND-utils 软件，代码如下。

```
[root@nginx-39 ~]#yum install BIND-utils -y >/dev/null  2>&1
[root@nginx-39 ~]#dig www.jd*.com

; <<>> DiG 9.8.2rc1-RedHat-9.8.2-0.68.rc1.el6_10.3 <<>> www.jd*.com
;; global options: +cmd
;; Got answer:
;; ->>HEADER<<- opcode: QUERY, status: NOERROR, id: 63846
;; flags: qr rd ra; QUERY: 1, ANSWER: 3, AUTHORITY: 0, ADDITIONAL: 0

;; QUESTION SECTION:
;www.jd*.com.            IN    A

;; ANSWER SECTION:
www.jd*.com.         34  IN   CNAME    www.jd*.com.gslb.qianxun.com.
www.jd*.com.gslb.qianxun.com. 34  IN   CNAME    www.jdcdn*.com.
www.jdcdn*.com.      34  IN   A        111.206.231.1
```

```
;; Query time: 19 msec
;; SERVER: 114.114.114.114#53(114.114.114.114)
;; WHEN: Sat Aug  3 20:45:32 2019
;; MSG SIZE  rcvd: 106
```

上述代码表示客户端解析 www.jd*.com 的响应时间为 19ms。

虽然两台主机访问网页的响应时间都只有十几毫秒，但是如果 DNS 访问量非常大的话，那响应时间将变长。

下面我们来配置缓存 DNS 服务器。

5．配置缓存 DNS 服务器

（1）安装 BIND 软件。

安装 BIND 软件，代码如下。

```
[root@DNS.m.booxin*.com ~]#yum install BIND BIND-chroot -y >/dev/null  2>&1
```

建议读者直接使用 yum 指令安装 BIND 软件，方便快捷。

（2）查询 BIND 软件是否安装成功，代码如下。

```
[root@DNS.m.booxin*.com ~]#rpm -qa  |grep BIND
BIND-libs-9.8.2-0.68.rc1.el6_10.3.x86_64
BIND-9.8.2-0.68.rc1.el6_10.3.x86_64
BIND-chroot-9.8.2-0.68.rc1.el6_10.3.x86_64
```

上述代码表示 BIND 软件已经安装成功。

（3）设置网络可以访问 114.114.114.114，代码如下。

```
[root@DNS.m.booxin*.com ~]#cat /etc/resolv.conf
; generated by /sbin/dhclient-script
nameserver 192.168.2.38
nameserver 8.8.8.8
```

配置好 DNS 服务器 IP 地址并且能够正常上网，使服务端能够访问 114.114.114.114 DNS 服务器。

（4）测试 DNS 服务器，代码如下。

```
[root@DNS.m.booxin*.com ~]#ping 114.114.114.114 -c 3
PING 114.114.114.114 (114.114.114.114) 56(84) bytes of data.
64 bytes from 114.114.114.114: icmp_seq=1 ttl=67 time=17.2 ms
64 bytes from 114.114.114.114: icmp_seq=2 ttl=91 time=18.2 ms
64 bytes from 114.114.114.114: icmp_seq=3 ttl=88 time=18.2 ms

--- 114.114.114.114 ping statistics ---
3 packets transmitted, 3 received, 0% packet loss, time 2022ms
rtt min/avg/max/mdev = 17.248/17.913/18.264/0.495 ms
```

上述代码表示可以正常访问 114.114.114.114 DNS 服务器。

（5）全局转发配置。

配置 BIND 服务，代码如下。

```
[root@DNS.m.booxin*.com ~]#cat /etc/named.conf
options
{
   directory       "/var/named";        // "Working" directory
   dump-file       "data/cache_dump.db";
      statistics-file    "data/named_stats.txt";
      memstatistics-file  "data/named_mem_stats.txt";
   listen-on port 53   { any; };
   listen-on-v6 port 53   { any; };
   allow-query     { any; };
   recursion yes;
   #forward first;
   #本机不能解析的转发给 202.106.195.68 做解析
   #first:首先转发；转发器不响应时，自行迭代查询
   #only:只转发
```

```
            forwarders  {
              #202.106.195.68;
               114.114.114.114;
               8.8.8.8;
              };
        DNSsec-enable yes;
        DNSsec-validation no;
        DNSsec-lookaside auto;
        pid-file "/run/named/named.pid";
        Session-keyfile "/run/named/Session.key";
        managed-keys-directory "/var/named/dynamic";
};
logging
{
        channel default_debug {
              file "data/named.run";
              severity dynamic;
        };
     channel gsquery {
              file "data/query.log"    versions 3 size 20m;
              severity info;
              print-time yes;
              print-category yes;
              print-severity yes;
        };
              category queries { gsquery; };
};
```

上述代码配置即为全局转发配置。

（6）局部转发配置。

```
[root@DNS.m.booxin*.com ~]#tail -6  /etc/named.rfc1912.zones

zone "goo***.com.hk" IN {
   type forward;
   forward only;
   forwarders {8.8.8.8;};
   };
```

上述代码配置即为局部转发配置。

（7）检查语法并重启服务，代码如下。

```
[root@DNS.m.booxin*.com ~]#/etc/init.d/named configtest
[root@DNS.m.booxin*.com ~]#/etc/init.d/named restart
Stopping named:                                    [  OK  ]
Generating /etc/rndc.key:                          [  OK  ]
Starting named:                                    [  OK  ]
```

6. DNS 服务端测试

DNS 服务端测试代码如下。

```
[root@DNS.m.booxin*.com ~]#dig www.jd*.com

; <<>> DiG 9.8.2rc1-RedHat-9.8.2-0.68.rc1.el6_10.3 <<>> www.jd*.com
;; global options: +cmd
;; Got answer:
;; ->>HEADER<<- opcode: QUERY, status: NOERROR, id: 61001
;; flags: qr rd ra; QUERY: 1, ANSWER: 3, AUTHORITY: 13, ADDITIONAL: 0

;; QUESTION SECTION:
;www.jd*.com.              IN    A

;; ANSWER SECTION:
www.jd*.com.        68  IN  CNAME   www.jd*.com.gslb.qianxun.com.
www.jd*.com.gslb.qianxun.com. 34  IN  CNAME   www.jdcdn*.com.
www.jdcdn*.com.     34  IN  A       111.206.231.1
```

```
;; AUTHORITY SECTION:
.                       1582    IN      NS      c.root-servers.net.
.                       1582    IN      NS      e.root-servers.net.
.                       1582    IN      NS      j.root-servers.net.
.                       1582    IN      NS      l.root-servers.net.
.                       1582    IN      NS      k.root-servers.net.
.                       1582    IN      NS      d.root-servers.net.
.                       1582    IN      NS      b.root-servers.net.
.                       1582    IN      NS      g.root-servers.net.
.                       1582    IN      NS      i.root-servers.net.
.                       1582    IN      NS      h.root-servers.net.
.                       1582    IN      NS      a.root-servers.net.
.                       1582    IN      NS      m.root-servers.net.
.                       1582    IN      NS      f.root-servers.net.

;; Query time: 57 msec
;; SERVER: 192.168.2.38#53(192.168.2.38)
;; WHEN: Sat Aug  3 21:01:38 2019
;; MSG SIZE  rcvd: 317
```

第一次访问网页时的响应时间会比较长，因为本地 DNS 服务器要从其他 DNS 服务器去缓存数据。下面我们进行第二次访问，观察得出响应时间为 0ms，已经能够高速访问网页，代码如下。

```
[root@DNS.m.booxin*.com ~]#dig www.jd*.com

; <<>> DiG 9.8.2rc1-RedHat-9.8.2-0.68.rc1.el6_10.3 <<>> www.jd*.com
;; global options: +cmd
;; Got answer:
;; ->>HEADER<<- opcode: QUERY, status: NOERROR, id: 36691
;; flags: qr rd ra; QUERY: 1, ANSWER: 3, AUTHORITY: 13, ADDITIONAL: 0

;; QUESTION SECTION:
;www.jd*.com.                   IN      A

;; ANSWER SECTION:
www.jd*.com.            63      IN      CNAME   www.jd*.com.gslb.qianxun.com.
www.jd*.com.gslb.qianxun.com. 29  IN    CNAME   www.jdcdn*.com.
www.jdcdn*.com.         29      IN      A       111.206.231.1

;; AUTHORITY SECTION:
.                       1577 IN    NS      h.root-servers.net.
.                       1577 IN    NS      a.root-servers.net.
.                       1577 IN    NS      i.root-servers.net.
.                       1577 IN    NS      k.root-servers.net.
.                       1577 IN    NS      e.root-servers.net.
.                       1577 IN    NS      c.root-servers.net.
.                       1577 IN    NS      m.root-servers.net.
.                       1577 IN    NS      l.root-servers.net.
.                       1577 IN    NS      d.root-servers.net.
.                       1577 IN    NS      f.root-servers.net.
.                       1577 IN    NS      g.root-servers.net.
.                       1577 IN    NS      j.root-servers.net.
.                       1577 IN    NS      b.root-servers.net.

;; Query time: 0 msec
;; SERVER: 192.168.2.38#53(192.168.2.38)
;; WHEN: Sat Aug  3 21:01:43 2019
;; MSG SIZE  rcvd: 317
```

从上述代码可以看到，缓存的效果非常显著，没有延迟，立即响应并返回结果。

（1）测试端测试。

在配置文件 vim /etc/resolv.conf 中添加高速缓存服务器的地址，代码如下。

```
[root@nginx-39 ~]#cat /etc/resolv.conf
```

```
; generated by /sbin/dhclient-script
nameserver 192.168.2.38
nameserver 192.168.2.38
```

测试端依然可以实现高速访问网页，代码如下。

```
[root@nginx-39 ~]#dig www.jd*.com

; <<>> DiG 9.8.2rc1-RedHat-9.8.2-0.68.rc1.el6_10.3 <<>> www.jd*.com
;; global options: +cmd
;; Got answer:
;; ->>HEADER<<- opcode: QUERY, status: NOERROR, id: 5002
;; flags: qr rd ra; QUERY: 1, ANSWER: 3, AUTHORITY: 13, ADDITIONAL: 0

;; QUESTION SECTION:
;www.jd*.com.                   IN      A

;; ANSWER SECTION:
www.jd*.com.            35      IN      CNAME   www.jd*.com.gslb.qianxun.com.
www.jd*.com.gslb.qianxun.com. 46 IN     CNAME   www.jdcdn*.com.
www.jdcdn*.com.         34      IN      A       111.206.231.1

;; AUTHORITY SECTION:
.                       310     IN      NS      l.root-servers.net.
.                       310     IN      NS      k.root-servers.net.
.                       310     IN      NS      b.root-servers.net.
.                       310     IN      NS      a.root-servers.net.
.                       310     IN      NS      m.root-servers.net.
.                       310     IN      NS      c.root-servers.net.
.                       310     IN      NS      h.root-servers.net.
.                       310     IN      NS      j.root-servers.net.
.                       310     IN      NS      i.root-servers.net.
.                       310     IN      NS      d.root-servers.net.
.                       310     IN      NS      e.root-servers.net.
.                       310     IN      NS      f.root-servers.net.
.                       310     IN      NS      g.root-servers.net.

;; Query time: 36 msec
;; SERVER: 192.168.2.38#53(192.168.2.38)
;; WHEN: Sat Aug  3 21:17:03 2019
;; MSG SIZE  rcvd: 317
```

上述代码表示第一次访问会有 **36ms** 的响应时间。

```
[root@nginx-39 ~]#dig www.jd*.com

; <<>> DiG 9.8.2rc1-RedHat-9.8.2-0.68.rc1.el6_10.3 <<>> www.jd*.com
;; global options: +cmd
;; Got answer:
;; ->>HEADER<<- opcode: QUERY, status: NOERROR, id: 43213
;; flags: qr rd ra; QUERY: 1, ANSWER: 3, AUTHORITY: 13, ADDITIONAL: 0

;; QUESTION SECTION:
;www.jd*.com.                   IN      A

;; ANSWER SECTION:
www.jd*.com.            32      IN      CNAME   www.jd*.com.gslb.qianxun.com.
www.jd*.com.gslb.qianxun.com. 43 IN     CNAME   www.jdcdn*.com.
www.jdcdn*.com.         31      IN      A       111.206.231.1

;; AUTHORITY SECTION:
.                       307     IN      NS      k.root-servers.net.
.                       307     IN      NS      l.root-servers.net.
.                       307     IN      NS      m.root-servers.net.
.                       307     IN      NS      b.root-servers.net.
.                       307     IN      NS      d.root-servers.net.
.                       307     IN      NS      f.root-servers.net.
.                       307     IN      NS      a.root-servers.net.
```

```
.                   307  IN    NS    c.root-servers.net.
.                   307  IN    NS    g.root-servers.net.
.                   307  IN    NS    j.root-servers.net.
.                   307  IN    NS    i.root-servers.net.
.                   307  IN    NS    h.root-servers.net.
.                   307  IN    NS    e.root-servers.net.

;; Query time: 0 msec
;; SERVER: 192.168.2.38#53(192.168.2.38)
;; WHEN: Sat Aug  3 21:17:05 2019
;; MSG SIZE  rcvd: 317
```

上述代码中，响应时间为 0ms，表示缓存设置已经生效。

（2）服务端测试。

服务端测试代码每隔一段时间都需要重新获取最新的信息。尝试修改 DNS 服务器的 named.conf 主配置文件，代码如下。

```
[root@DNS.m.booxin*.com ~]#cat /etc/named.conf
options
{
    directory       "/var/named";        // "Working" directory
    dump-file       "data/cache_dump.db";
        statistics-file   "data/named_stats.txt";
        memstatistics-file "data/named_mem_stats.txt";
    listen-on port 53   { any; };
    listen-on-v6 port 53   { any; };
    allow-query     { any; };
    recursion yes;
    auth-nxdomain no;
    #forward first;
    #本机不能解析的转发给 202.106.195.68 做解析
    #first: 首先转发；转发器不响应时，自行迭代查询
    #only: 只转发

#    forwarders {
#        #202.106.195.68;
#        114.114.114.114;
#        8.8.8.8;
#     };
    DNSsec-enable yes;
    DNSsec-validation no;
    DNSsec-lookaside auto;
    pid-file "/run/named/named.pid";
    Session-keyfile "/run/named/Session.key";
    managed-keys-directory "/var/named/dynamic";
};
logging
{
        channel default_debug {
                file "data/named.run";
                severity dynamic;
        };
    channel gsquery {
                file "data/query.log"   versions 3 size 20m;
                severity info;
                print-time yes;
                print-category yes;
                print-severity yes;
        };
        category queries { gsquery; };
};
```

上述代码中注释掉了 forwarders   {};代码块。

（3）重新配置 DNS 服务端。

重启 named 服务，重新测试，代码如下。

```
[root@DNS.m.booxin*.com ~]#/etc/init.d/named restart
Stopping named:                                         [  OK  ]
Starting named:                                         [  OK  ]
```

上述代码表示 named 服务已经重启成功。

```
[root@DNS.m.booxin*.com ~]#dig www.jd*.com

; <<>> DiG 9.8.2rc1-RedHat-9.8.2-0.68.rc1.el6_10.3 <<>> www.jd*.com
;; global options: +cmd
;; Got answer:
;; ->>HEADER<<- opcode: QUERY, status: NOERROR, id: 28258
;; flags: qr rd ra; QUERY: 1, ANSWER: 3, AUTHORITY: 4, ADDITIONAL: 4

;; QUESTION SECTION:
;www.jd*.com.                   IN      A

;; ANSWER SECTION:
www.jd*.com.            120     IN      CNAME   www.jd*.com.gslb.qianxun.com.
www.jd*.com.gslb.qianxun.com. 60 IN     CNAME   www.jdcdn*.com.
www.jdcdn*.com.         60 IN   A       111.206.231.1

;; AUTHORITY SECTION:
jdcdn*.com.             720 IN  NS      ns3.jd*.com.
jdcdn*.com.             720 IN  NS      ns4.jd*.com.
jdcdn*.com.             720 IN  NS      ns1.jd*.com.
jdcdn*.com.             720 IN  NS      ns2.jd*.com.

;; ADDITIONAL SECTION:
ns3.jd*.com.            172800  IN      A       120.52.149.254
ns4.jd*.com.            172800  IN      A       106.39.177.32
ns2.jd*.com.            172800  IN      A       111.206.226.10
ns1.jd*.com.            172800  IN      A       111.13.28.10

;; Query time: 1400 msec
;; SERVER: 192.168.2.38#53(192.168.2.38)
;; WHEN: Sat Aug  3 21:40:08 2019
;; MSG SIZE  rcvd: 242
```

上述代码访问 www.jd*.com 耗时 1400ms。再次使用指令进行测试，测试结果如下。

```
[root@dns.m.booxin*.com ~]#dig www.jd*.com

; <<>> DiG 9.8.2rc1-RedHat-9.8.2-0.68.rc1.el6_10.3 <<>> www.jd*.com
;; global options: +cmd
;; Got answer:
;; ->>HEADER<<- opcode: QUERY, status: NOERROR, id: 33365
;; flags: qr rd ra; QUERY: 1, ANSWER: 3, AUTHORITY: 4, ADDITIONAL: 4

;; QUESTION SECTION:
;www.jd*.com.                   IN      A

;; ANSWER SECTION:
www.jd*.com.            112     IN      CNAME   www.jd*.com.gslb.qianxun.com.
www.jd*.com.gslb.qianxun.com. 52 IN     CNAME   www.jdcdn*.com.
www.jdcdn*.com.         52 IN   A       111.206.231.1

;; AUTHORITY SECTION:
jdcdn*.com.             712 IN  NS      ns4.jd.com.
jdcdn*.com.             712 IN  NS      ns3.jd.com.
jdcdn*.com.             712 IN  NS      ns1.jd.com.
jdcdn*.com.             712 IN  NS      ns2.jd.com.

;; ADDITIONAL SECTION:
ns3.jd*.com.            172792  IN      A       120.52.149.254
ns4.jd*.com.            172792  IN      A       106.39.177.32
ns2.jd*.com.            172792  IN      A       111.206.226.10
```

```
ns1.jd*.com.          172792  IN   A    111.13.28.10
;; Query time: 0 msec
;; SERVER: 192.168.2.38#53(192.168.2.38)
;; WHEN: Sat Aug  3 21:40:16 2019
;; MSG SIZE  rcvd: 242
```

上述代码使用 dig 指令解析 www.jd*.com 网站时，耗时 0ms。

3.2 BIND 实现智能 DNS 服务器

3.2.1 智能 DNS 服务器基础知识

1．智能 DNS 服务器的出现

普通的域名解析只为用户返回解析结果，不会考虑访问者 IP 地址的来源和类型。这样，所有的访问者都被解析到同样的 IP 地址上，容易出现由跨运营商或者跨地域访问引起的网络体验欠佳问题。

智能 DNS 服务器的智能线路解析功能支持按运营商、地域等维度区分访问者 IP 地址的来源和类型，对同一域名的访问请求做出不同的解析响应，指向不同服务器的 IP 地址。当联通用户访问时，智能 DNS 服务器返回联通服务器的 IP 地址；当电信用户访问时，返回电信服务器的 IP 地址，通过解决跨网访问慢的难题，实现高效解析。

除了按运营商线路及地域解析之外，智能 DNS 服务器还支持按 IP 网段划分访问者的自定义线路解析，可以更细粒度地设置解析线路，将访问者路由至不同的网站服务器。

对于部署了多台服务器的网站，如果想要实现解析的负载均衡，可以选择智能 DNS 服务器。

2．解决方案

智能 DNS 服务器策略解析很好地满足了上面的需求。智能 DNS 服务器策略解析最基本的功能是可以智能地判断访问网站的用户，然后根据不同的用户把域名分别解析成不同的 IP 地址。

如果访问者是联通用户，智能 DNS 服务器策略解析会把域名对应的联通 IP 地址解析给这个用户。如果访问者是电信用户，智能 DNS 服务器策略解析会把域名对应的电信 IP 地址解析给这个用户。智能 DNS 服务器策略解析还可以为多个主机实现负载均衡，这时来自各地的访问流量会比较平均地分布到每一台服务器。

3．基于 Linux 下的 BIND 9 来实现智能 DNS 服务器策略解析

（1）硬件需求。

- 网络环境：独立于联通节点和电信节点之外的线路，1000MB 独享接入。
- 服务器数目要求：3 台（DNS 服务器一主一从，另外一台备用机）。

（2）系统及软件需求。

- 系统：CentOS 6.9，64 位。
- 软件：BIND 9。

（3）其他需求。

一个可以注册为 NS 服务器的一级域名。

（4）实现双线自动选择。

根据 BIND 9 的设置和设定好的联通 IP 地址和电信 IP 地址的访问控制列表（Access Control

List, ACL), 按照用户来源进行解析, 处于电信 IP 地址范围内的用户自动解析为电信节点的服务器, 处于联通 IP 地址范围内的用户自动解析为联通节点的服务器, 不处于电信和联通 IP 地址范围内的, 进行轮询解析。

(5) 实现宕机检测。

DNS 服务器每隔 60s 对服务器进行一次宕机检测, 使用脚本实现。若发现某节点宕机, 则自动执行设定好的脚本, 修改 DNS 记录并刷新 DNS。

4. 构建智能 DNS 服务器核心知识

(1) 基本现状描述。

Linux 系统管理员把 www.booxin*.com 解析到 192.168.3.100, 那么世界上任何一台计算机在请求 www.booxin*.com 时, 解析到的都是这个 IP 地址。

注意: www.booxin*.com 域名仅供内网访问 (此 IP 地址只是基于本地局域网实验的, 而且也没有从域名供应商处注册并购买, 除了内网之外, 世界上任何一台计算机都访问不到)。

(2) 问题和解决方案。

假设有两个用户, 一个在电信, 另一个在联通, 他们都想访问同一个域名, 服务器要放在哪个运营商那里更好呢？现在提供两台服务器, 一台放在电信, 另一台放在联通, 然后让电信的用户查询 www.booxin*.com 时, 解析到电信的服务器上, 联通的用户查询 www.booxin*.com 时, 解析到联通的服务器上。这是 BIND 的高级功能: 视图查询。

5. 视图基础知识

BIND 视图: 在主配置文件中, 可以指定多个视图, 使用 BIND 提供的 view 指令可以实现根据不同的 IP 地址范围来对同一个域名进行解析 (启用了视图, 则所有的区域包括根区域都要定义在视图中, 视图是有先后次序的)。配置视图主要需要 3 个步骤。

(1) 收集 IP 地址的集合。

例如现在有 3 个 IP 地址的集合, 让每个集合能访问不同的地址。即某个集合中的一个 IP 地址在查询时, 提供一个解析结果, 另一个集合中的另一个 IP 地址查询时, 提供另一个结果。

(2) 给每个 IP 地址集合提供一个视图。

这个视图中要写清楚是哪个集合, 能请求到哪个域。

(3) 给每个视图的每个域提供一个解析文件。

3.2.2 构建智能 DNS 服务器基础环境

1. 构建智能 DNS 服务器基础环境信息

构建智能 DNS 服务器基础环境信息如表 3-2 所示。其中使用一台作为 DNS 服务器, 两台作为 Web 服务器, 两台作为电信和联通的测试客户端。

表 3-2 构建智能 DNS 服务器基础环境信息

角色	操作系统	IP 地址	主机名和域名
DNS 服务器	CentOS 6.9	192.168.2.38	dns.m.booxin*.com
联通机房 Web 服务器	CentOS 6.9	192.168.2.39	bj-web39
电信机房 Web 服务器	CentOS 6.9	192.168.2.40	bj-web40
电信客户端	Windows 10	192.168.2.100	client-1
联通客户端	Windows 10	192.168.3.100	client-2

2. 智能 DNS 服务器架构

当收到客户端的 DNS 查询请求时，DNS 分离配置能区分客户端的来源地址，为不同类别的客户端提供不同的解析结果（IP 地址）。不同的客户端解析同一个域名时，解析结果不同，可以为客户端提供最近的资源，智能 DNS 服务器架构如图 3-2 所示。

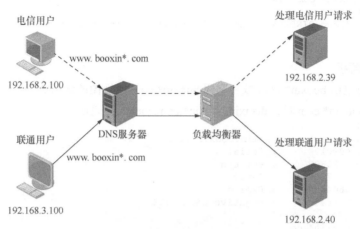

图 3-2　智能 DNS 服务器架构

在实现了 DNS 服务器主从同步、子域授权之后，还可以针对不同网络内的域名解析，请求 DNS 能够指向不同的主机地址，以实现分流。

假设图 3-2 中两台主机互为镜像，要实现来源不同的主机对 www.booxin*.com 域名的解析指向同网段内的镜像主机。例如 192.168.2.0/24 网段内主机对 www.booxin*.com 的解析指向 192.168.2.39 的服务器，192.168.3.0/24 网段内主机对 www.booxin*.com 的解析指向 192.168.2.40 的服务器，可以利用 BIND 的 ACL 规则和视图规则来实现。

3.2.3　智能 DNS 服务器实现核心步骤

下面，我们假设 192.168.2.0/24 这个网段是电信的，192.168.3.0/24 这个网段是联通的，对比配置视图。

1. IP 地址集合

（1）电信地址集合：dianxin.acl。代码如下。
```
acl "dianxin" {
    192.168.2.0/24;
};
```
（2）联通地址集合：liantong.acl。代码如下。
```
acl "liantong" {
    192.168.3.0/24;
};
```

2. 视图规则

（1）电信视图：dianxin.view。代码如下。
```
include "/etc/BIND/views/acls/dianxin.acl";
view "dianxin" {
    match-clients { "dianxin"; };
    zone "booxin*.com" IN {
        type master;
        file "/etc/BIND/views/zones/dianxin.booxin*.com.zone";
```

(2) 联通视图：liantong.view。代码如下。

```
include "/etc/BIND/views/acls/liantong.acl";
view "liantong" {
    match-clients { "liantong"; };
    zone "booxin*.com" IN {
        type master;
        file "/etc/BIND/views/zones/liantong.booxin*.com.zone";
    };
};
```

3. 配置域解析

最后配置对电信 booxin*.com 域和联通 booxin*.com 域视图的解析。

(1) 电信 booxin*.com 域：dianxin.booxin*.com.zone。代码如下。

```
$TTL 86400
@   IN   SOA booxin*.com. admin.booxin*.com. (
            2016090100    ; Serial
            28800         ; Refresh
            7200          ; Retry
            604800        ; Expire
            86400         ; Negative Cache TTL
)
@   IN   NS      booxin*.com.
@   IN   A       192.168.2.38
aaa IN   A       192.168.2.38
bbb IN   A       192.168.2.39
ccc IN   CNAME   bbb
```

(2) 联通 booxin*.com 域：liantong.booxin*.com.zone。代码如下。

```
$TTL 86400
@   IN   SOA booxin*.com. admin.booxin*.com. (
            2016090100    ; Serial
            28800         ; Refresh
            7200          ; Retry
            604800        ; Expire
            86400         ; Negative Cache TTL
)
@   IN   NS      booxin*.com.
@   IN   A       192.168.2.38
aaa IN   A       192.168.2.38
bbb IN   A       192.168.2.39
ccc IN   CNAME   bbb
```

上述代码中，配置了一套电信视图和一套联通视图。

3.2.4 智能 DNS 服务器核心构建步骤

1. yum 安装 BIND 软件

使用 yum 指令安装 BIND 软件，代码如下。

```
[root@DNS.m.booxin*.com ~]#yum -y install bind bind-utils >/dev/null 2>&1
[root@DNS.m.booxin*.com ~]#echo $?
0
```

上述代码安装 BIND 软件，返回值表示已经安装成功。

2. 定义 ACL 配置文件

定义 ACL 配置文件，步骤如下。

(1) 创建 ACL 控制目录，代码如下。

```
[root@DNS.m.booxin*.com ~]#mkdir -pv /etc/named/acl/{dianxin,liantong,other}
mkdir: created directory '/etc/named/acl'
mkdir: created directory '/etc/named/acl/dianxin'
```

```
mkdir: created directory '/etc/named/acl/liantong'
mkdir: created directory '/etc/named/acl/other'
```

上述代码创建 ACL 控制目录。

（2）将来源不同的两个网段定义到不同的 ACL 规则当中，代码如下。

```
[root@DNS.m.booxin*.com ~]#cat /etc/named/acl/dianxin/dianxin.conf
acl "dianxin" {
    192.168.2.7;
};
[root@DNS.m.booxin*.com ~]#cat /etc/named/acl/liantong/liantong.conf
acl "liantong" {
    192.168.2.6;
};

# 除上面两个网段之外的所有地址
[root@DNS.m.booxin*.com ~]#cat /etc/named/acl/other/other.conf
acl other {
! 192.168.2.7/24;
! 192.168.2.6/24;
any; };
```

上述代码定义 ACL 规则和对应的配置文件。

（3）修改权限。

修改 ACL 配置文件规则权限，代码如下。

```
[root@DNS.m.booxin*.com ~]#chown  -R named. /etc/named/acl/
[root@DNS.m.booxin*.com ~]#ll /etc/named/acl/
total 12
drwxr-xr-x 2 named named 4096 Aug   4 21:04 dianxin
drwxr-xr-x 2 named named 4096 Aug   4 21:05 liantong
drwxr-xr-x 2 named named 4096 Aug   4 21:07 other
```

3．配置视图

配置 BIND 视图，代码如下。

```
view "dianxin" {
   match-clients {         # 使用 match-clients 指令，指定匹配来自这些用户的 IP 地址
       dianxin;            # 写的是 ACL 配置文件定义的 aclname
   };
    zone   "booxin*.com"  {
     type master;
     file "booxin*.com.zone_dianxin";    #不同的匹配规则用不同的域名文件，方便管理
   };
};

view "liantong" {
   match-clients {
      liantong;
   };
    zone "booxin*.com"  {
     type master;
     file "booxin*.com.zone_liantong";
   };
};
```

上述代码表示电信和联通分别配置对应的规则。

4．引入 ACL 配置文件

引入 ACL 配置文件，代码如下。

```
[root@DNS.m.booxin*.com ~]#cat /etc/named.conf
options
{
   directory      "/var/named";
   dump-file      "data/cache_dump.db";
```

```
            statistics-file      "data/named_stats.txt";
            memstatistics-file "data/named_mem_stats.txt";
    listen-on port 53  { any; };
    listen-on-v6 port 53   { any; };
    allow-query     { any; };
    recursion yes;
    auth-nxdomain no;
    DNSsec-enable yes;
    DNSsec-validation no;
    DNSsec-lookaside auto;
    pid-file "/run/named/named.pid";
    Session-keyfile "/run/named/Session.key";
    managed-keys-directory "/var/named/dynamic";
};
logging
{
        channel default_debug {
                file "data/named.run";
                severity dynamic;
        };
    channel gsquery {
                file "data/query.log"   versions 3 size 20m;
                severity info;
                print-time yes;
                print-category yes;
                print-severity yes;
        };
        category queries { gsquery; };
};

include "/etc/named/acl/dianxin/dianxin.conf";
include "/etc/named/acl/liantong/liantong.conf";
include "/etc/named/acl/other/other.conf";
```

上述代码定义 ACL 规则。

5．创建区域数据文件

（1）创建电信区域数据文件。

在/var/named/目录下创建 booxin*.com.zone_dianxin 区域数据文件，代码如下。

```
[root@DNS.m.booxin*.com named]#cat booxin*.com.zone_dianxin
$TTL 600
@         IN    SOA    DNS.booxin*.com. admin.booxin*.com. (
                      2019030822
                      1H
                      10M
                      2D
                      1D)
@         IN    NS     DNS
@         IN    MX  10 mail
DNS       IN    A      192.168.2.38
www       IN    A      192.168.2.39
mailIN    A     192.168.2.88
```

上述代码创建电信区域数据文件。

（2）创建联通区域数据文件。

在/var/named/目录下创建 booxin*.com.zone_liantong 区域数据文件，代码如下。

```
[root@DNS.m.booxin*.com named]#cat  /var/named/booxin*.com.zone_liantong
$TTL 600
@         IN    SOA    DNS.booxin*.com. admin.booxin*.com. (
                      2019030822
                      1H
                      10M
```

```
                        2D
                        1D)
@       IN    NS        DNS
@       IN    MX   10   mail
DNS     IN    A         192.168.2.38
www     IN    A         192.168.2.40
mailIN  A     192.168.2.88
```

上述代码创建联通区域数据文件。

（3）修改区域数据文件的属主、属组及权限，代码如下。

```
[root@DNS.m.booxin*.com named]#chown -R root:named /var/named/
[root@DNS.m.booxin*.com named]#ll /var/named/
total 40
-rw-r--r-- 1 root named  371 Aug  4 21:26 booxin*.com.zone_dianxin
-rw-r--r-- 1 root named  368 Aug  4 21:27 booxin*.com.zone_liantong
drwxr-x--- 7 root named 4096 Aug  3 00:18 chroot
drwxrwx--- 2 root named 4096 Aug  4 18:17 data
drwxrwx--- 2 root named 4096 Aug  4 21:05 dynamic
-rw-r----- 1 root named 3289 Apr 11  2017 named.ca
-rw-r----- 1 root named  152 Dec 15  2009 named.empty
-rw-r----- 1 root named  152 Jun 21  2007 named.localhost
-rw-r----- 1 root named  168 Dec 15  2009 named.loopback
drwxrwx--- 2 root named 4096 Jun 19 00:19 slaves
```

上述代码修改区域数据文件的权限。

6. 查看完整配置文件

（1）查看 DNS 主配置文件，代码如下。

```
[root@DNS.m.booxin*.com ~]#cat /etc/named.conf
options
{
    directory       "/var/named";
    dump-file       "data/cache_dump.db";
        statistics-file     "data/named_stats.txt";
        memstatistics-file  "data/named_mem_stats.txt";
    listen-on port 53   { any; };
    listen-on-v6 port 53   { any; };
    allow-query     { any; };
    recursion yes;
    auth-nxdomain no;
    DNSsec-enable yes;
    DNSsec-validation no;
    DNSsec-lookaside auto;
    pid-file "/run/named/named.pid";
    Session-keyfile "/run/named/Session.key";
    managed-keys-directory "/var/named/dynamic";
};
logging
{
        channel default_debug {
                file "data/named.run";
                severity dynamic;
        };
    channel gsquery {
                file "data/query.log"   versions 3 size 20m;
                severity info;
                print-time yes;
                print-category yes;
                print-severity yes;
        };
        category queries { gsquery; };

};
```

```
view "dianxin" {
    match-clients   { dianxin; };
     zone  "booxin*.com"  {
        type master;
        file "booxin*.com.zone_dianxin";
    };

};

view "liantong" {
    match-clients   { liantong;  };
     zone "booxin*.com"  {
        type master;
        file "booxin*.com.zone_liantong";
    };
};


include "/etc/named/acl/dianxin/dianxin.conf";
include "/etc/named/acl/liantong/liantong.conf";
include "/etc/named/acl/other/other.conf";
```

上述代码为 DNS 主配置文件。

（2）查看 ACL 配置文件，代码如下。

```
[root@DNS.m.booxin*.com ~]#cat /etc/named/acl/dianxin/dianxin.conf
acl "dianxin" {
    192.168.2.7;
};
[root@DNS.m.booxin*.com ~]#cat /etc/named/acl/liantong/liantong.conf
acl "liantong" {
    192.168.2.6;
};
[root@DNS.m.booxin*.com ~]#
[root@DNS.m.booxin*.com ~]#cat /etc/named/acl/other/other.conf
acl other {
! 192.168.2.0/24;
! 192.168.3.0/24;
any; };
```

上述代码为 ACL 配置文件。

（3）查看区域配置文件，代码如下。

```
[root@DNS.m.booxin*.com ~]#cat /var/named/booxin*.com.zone_dianxin
$TTL 3H
$ORIGIN booxin*.com.
@       IN SOA   booxin*.com. 312874370*.qq.com. (
                                2       ; serial
                                1D      ; refresh
                                1H      ; retry
                                1W      ; expire
                                3H )    ; minimum
        IN      NS      ns1.booxin*.com.
        IN      NS      ns2.booxin*.com.
ns1     IN      A       192.168.2.38
ns2     IN      A       192.168.2.39
www     IN      A       192.168.2.39
www     IN      A       192.168.2.39
ww      IN      CNAME   www.booxin*.com.
w       IN      CNAME   www.booxin*.com.
handuoduo       IN      CNAME   www.booxin*.com.
[root@DNS.m.booxin*.com ~]#cat /var/named/booxin*.com.zone_liantong
$TTL 3H
$ORIGIN booxin*.com.
@       IN SOA   booxin*.com. 312874370*.qq.com. (
                                2       ; serial
```

```
                                    1D      ; refresh
                                    1H      ; retry
                                    1W      ; expire
                                    3H )    ; minimum
            IN      NS      ns1.booxin*.com.
            IN      NS      ns2.booxin*.com.

ns1         IN      A       192.168.2.38
ns2         IN      A       192.168.2.40
www         IN      A       192.168.2.40
www         IN      A       192.168.2.40
ww          IN      CNAME   www.booxin*.com.
w           IN      CNAME   www.booxin*.com.
handuoduo   IN      CNAME   www.booxin*.com.
```
上述代码为区域配置文件。

（4）检测语法并重启服务，代码如下。

```
[root@DNS.m.booxin*.com named]#/etc/init.d/named configtest
umount: /var/named/chroot/var/named: device is busy.
        (In some cases useful info about processes that use
        the device is found by lsof(8) or fuser(1))
[root@DNS.m.booxin*.com named]#
[root@DNS.m.booxin*.com named]#
[root@DNS.m.booxin*.com named]#/etc/init.d/named restart
Stopping named:                                     [  OK  ]
Starting named:                                     [  OK  ]
```

上述代码检测配置文件语法，并重启 BIND 服务。

（5）查看 DNS 服务端配置，代码如下。

```
[root@DNS.m.booxin*.com named]#cat /etc/resolv.conf
; generated by /sbin/dhclient-script
nameserver 192.168.2.38
nameserver 114.114.114.114
```

上述代码设置 DNS 服务端查询地址。

7．Nginx Web 服务器配置

（1）查看 Nginx Web 服务器（192.168.2.39）主配置文件，代码如下。

```
[root@nginx-39 ~]#cat /etc/nginx/conf.d/www.booxin*.conf
server {
    listen       80 ;
    server_name  www.booxin*.com;
    root         /usr/share/nginx/html;

    # Load configuration files for the default server block.
    include /etc/nginx/default.d/*.conf;

    location / {
    }

    error_page 404 /404.html;
        location = /40x.html {
    }

    error_page 500 502 503 504 /50x.html;
        location = /50x.html {
    }
}
```

上述代码为主配置文件内容。

默认首页配置文件代码如下。

```
[root@nginx-39 ~]#cat /usr/share/nginx/html/index.html
<h1> This is a 192.168.2.39 </h1>
```

查看 DNS 解析配置是否正确，代码如下。
```
[root@nginx-39 ~]#cat /etc/resolv.conf
; generated by /sbin/dhclient-script
nameserver 192.168.2.38
nameserver 8.8.8.8
```
查看 Nginx 配置文件语法是否正确，代码如下。
```
[root@nginx-39 ~]#/etc/init.d/nginx configtest
nginx: the configuration file /etc/nginx/nginx.conf syntax is ok
nginx: configuration file /etc/nginx/nginx.conf test is successful
```
重启 Nginx 服务，查看进程和端口是否存在，代码如下。
```
[root@nginx-39 ~]#/etc/init.d/nginx restart
Stopping nginx:                                            [  OK  ]
Starting nginx:                                            [  OK  ]
[root@nginx-39 ~]#netstat  -ntlp |grep 80
tcp        0      0 0.0.0.0:80              0.0.0.0:*               LISTEN      1603/nginx
tcp        0      0 :::80                   :::*                    LISTEN      1603/nginx
[root@nginx-39 ~]#ps -ef |grep nginx
root      1603     1  0 22:25 ?        00:00:00 nginx: master process /usr/sbin/nginx -c /etc/nginx/nginx.conf
nginx     1605  1603  0 22:25 ?        00:00:00 nginx: worker process
root      1609  1447  0 22:25 pts/3    00:00:00 grep nginx
```
上述代码中 Nginx 服务的运行端口为 80，Nginx 服务 master 进程的用户为 root，worker 进程的用户为 nginx。

（2）Nginx Web 服务器（192.168.2.40）配置。

查看主配置文件，代码如下。
```
[root@nginx-40 ~]#cat /etc/nginx/conf.d/www.booxin*.conf
#
# The default server
#

server {
    listen       80 ;
    server_name  www.booxin*.com;
    root         /usr/share/nginx/html;

    # Load configuration files for the default server block.
    include /etc/nginx/default.d/*.conf;

    location / {
    }

    error_page 404 /404.html;
        location = /40x.html {
    }

    error_page 500 502 503 504 /50x.html;
        location = /50x.html {
    }
}
```
查看默认首页配置文件，代码如下。
```
[root@nginx-40 ~]#cat /usr/share/nginx/html/index.html
<h1>This is 192.168.2.40 </h1>
```
检测配置文件语法，并重启 Nginx 服务，代码如下。
```
[root@nginx-40 ~]#/etc/init.d/nginx configtest
nginx: the configuration file /etc/nginx/nginx.conf syntax is ok
nginx: configuration file /etc/nginx/nginx.conf test is successful
```

```
[root@nginx-40 ~]#/etc/init.d/nginx restart
Stopping nginx:                                           [  OK  ]
Starting nginx:                                           [  OK  ]
```
查看进程和端口是否存在，代码如下。
```
[root@nginx-40 ~]#netstat  -tnpl |grep 80
tcp        0      0 0.0.0.0:80              0.0.0.0:*               LISTEN      1809/nginx
tcp        0      0 :::80                   :::*                    LISTEN      1809/nginx
[root@nginx-40 ~]#ps -ef |grep nginx
root      1809     1  0 22:27 ?        00:00:00 nginx: master process /usr/sbin/nginx -c /etc/nginx/nginx.conf
nginx     1811  1809  0 22:27 ?        00:00:00 nginx: worker process
root      1815  1678  0 22:27 pts/3    00:00:00 grep nginx
[root@nginx-40 ~]#
[root@nginx-40 ~]#cat /etc/resolv.conf
; generated by /sbin/dhclient-script
nameserver 192.168.2.38
nameserver 8.8.8.8
```
从上述代码中可以看到，Nginx 进程和端口都已经存在。

3.2.5　测试 BIND 视图

1. 电信客户测试

DNS 指向 192.168.2.38，电信客户访问结果如图 3-3 所示。

2. 联通客户测试

DNS 指向 192.168.2.38，联通客户访问结果如图 3-4 所示。

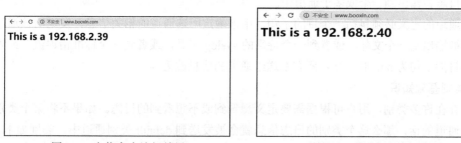

图 3-3　电信客户访问结果　　　　图 3-4　联通客户访问结果

3.3　BIND 日志配置

3.3.1　BIND 日志概念

BIND 日志中主要有两个概念：通道（Channel）和类别（Category）。通道指定应该向哪里发送日志数据，如发送给 syslog，还是写在一个文件里，是发送给 named 的标准错误输出，还是发送到位存储桶（Bit Bucket）。

通道允许根据不同的级别对消息进行过滤。

下面是不同级别的列表，按照严重性递减的顺序排列：critical、error、warning、notice、info、debug[level]、dynamic。最后两个级别是 BIND 8 和 BIND 9 特有的。debug 可以指定调试级别，默认是 1。如果指定 debug 3，就会看到第三级调试信息。如果规定 dynamic 这个级别，那么服务器就会记录所有与调试级别匹配的日志信息，默认的级别是 info。也就是说，除非你指定了级别，否则看不到任何调试信息。

3.3.2 logging 语句

logging 语句基本格式如下。

```
logging {
    [ channel channel_name {
      ( file path name
          [ versions ( number | unlimited ) ]
          [ size size spec ]
        | syslog syslog_facility
        | stderr
        | null );
      [ severity (critical | error | warning | notice |
                info | debug [ level ] | dynamic ); ]
      [ print-category yes or no; ]
      [ print-severity yes or no; ]
      [ print-time yes or no; ]
    }; ]
    [ category category_name {
      channel_name ; [ channel_name ; ... ]
    }; ]
    ...
};
```

上述代码规定 logging 语句的基础知识，下面进行分解。

1．通道基础知识

所有日志会输出到一个或多个通道中。

每个通道的定义必须包括一个目的子句，用来确定所选通道的相关信息，发送给该通道的信息将会被输出到一个文件，或者到一个特殊的 syslog 工具，或者到一个标准错误流，或者被忽略。当目的子句为 null 时，会使所有发送给通道的信息被丢弃。

2．类别基础知识

这里存在许多类别，用户可根据需要定义想看到或不想看到的日志。如果不将某个类别指定到某些通道的话，那么这个类别的日志信息就会被发送到 default 类别通道中。类别如下。

- default：默认类别，default 类别匹配所有未明确指定通道的类别。
- general：包括所有未明确类别的 BIND 消息，没有类别的日志都记录在此类别中。
- database：同 BIND 内部数据库相关的消息，用来存储区域数据和缓存记录。
- security：允许/拒绝的请求。
- config：配置文件分析和处理。
- resolver：DNS 解析，包括对来自解析器的递归查询的处理。
- xfer-in：接收区域传输。
- xfer-out：发送区域传输。
- notify：异步区变动通知。
- client：处理客户端请求。
- network：网络操作。
- update：动态更新事件。
- queries：客户端队列日志。

- DNSsec：DNSSEC 和 TSIG 协议处理。
- lame-servers：发现错误授权。

3.3.3 配置实例

1. 主配置文件配置日志格式

配置日志格式代码如下。

```
[root@www.DNS_master_33.com named]#cat /etc/named.conf
options {
    listen-on port 53 { 127.0.0.1; 192.168.2.33; };
    directory       "/var/named";
    dump-file       "/var/named/data/cache_dump.db";
        statistics-file "/var/named/data/named_stats.txt";
        memstatistics-file "/var/named/data/named_mem_stats.txt";
    allow-query     { 192.168.2.0/24; };
    recursion yes;
    DNSsec-enable no;
    DNSsec-validation no;

    BINDkeys-file "/etc/named.iscdlv.key";
    managed-keys-directory "/var/named/dynamic";
};



logging {
        channel default_debug {
                file "/var/named/log/named.run";
                severity dynamic;
        };
        channel query_info {
                file "/var/named/log/query.log" versions 1 size 100m;
                severity info;
                print-category yes;
                print-severity yes;
                print-time yes;
        };
        category queries {
                query_info;
                default_debug;
        };

        channel notify_info {
                file "/var/named/log/notify.log" versions 8 size 128m;
                severity info;
                print-category yes;
                print-severity yes;
                print-time yes;
        };

        category notify {
                notify_info;
                default_debug;
        };

        channel xfer_in_log {
                file "/var/named/log/xfer_in.log" versions 100 size 10m;
                severity info;
                print-category yes;
                print-severity yes;
                print-time yes;
        };
```

```
                channel xfer_out_log {
                        file "/var/named/log/xfer_out.log" versions 100 size 10m;
                        severity info;
                        print-category yes;
                        print-severity yes;
                        print-time yes;
                };

                category xfer-in { xfer_in_log; };
                category xfer-out { xfer_out_log; };
};
zone "." IN {
    type hint;
    file "named.ca";
};

zone "booxin*.com" IN {
    type master;
    file "booxin*.com.zone";
    also-notify { 192.168.2.34; };
        allow-transfer { 192.168.2.34; };
};

zone "2.168.192. in-addr.arpa" IN {
    type master;
    file "booxin*.com.rev";
    also-notify { 192.168.2.34; };
        allow-transfer { 192.168.2.34; };
};
include "/etc/named.rfc1912.zones";
include "/etc/named.root.key";
```

2. 日志格式说明

日志格式说明如下。

定义一个通道，名称叫作 default_debug，代码如下。
```
channel default_debug
```
通道的字句是 file，也就是将日志写到指定的文件中，代码如下。
```
file "/var/named/log/named.run"
```
日志级别设置成 dynamic，会观察所有的调试信息，代码如下。
```
severity dynamic;
```
定义一个 query_info 的通道。
```
channel query_info
file "/var/named/log/query.log" versions 1 size 100m;
```
上述代码中，日志记录位置是/var/named/log/query.log，versions 1 表示当此日志文件超过 100MB 时，会把当前的文件重命名，让后续的文件继续叫作 query.log 来接收日志，但是一旦又超过 100MB，就会将 query.log 覆盖掉。

定义日志级别为 info，默认也是这个级别。
```
severity info;
```
输出日志的类别，代码如下。
```
print-category yes;
```
输出日志级别，代码如下。
```
print-severity yes;
```
输出时间。
```
print-time yes;
```

3.3 BIND 日志配置

定义 queries 这个类别的日志分别写入 category queries/var/named/log/named.run 文件和 var/named/log/query.log 文件。

再定义一个名称为 notify_info 的通道，代码如下。

```
channel notify_info
```

写入/var/named/log/notify.log 文件，可以产生 8 份备份文件，日志大小为 128MB，不定义大小的话就是无限增长，代码如下。

```
file "/var/named/log/notify.log" versions 8 size 128m;
```

定义 notify 这个通知类别也会记录到两个通道里，代码如下。

```
category notify
```

定义名称为 xfer_in_log 的通道，代码如下。

```
channel xfer_in_log
```

定义名称为 xfer_out_log 的通道，代码如下。

```
channel xfer_out_log
```

3．创建日志目录并修改权限

创建日志目录并修改权限，代码如下。

```
[root@www.DNS_master_33.com named]#mkdir -pv /var/named/log/
mkdir: 已创建目录 "/var/named/log/"
[root@www.DNS_master_33.com named]#chown -R named. /var/named/log/
```

上述代码创建 BIND 日志目录，并修改权限。

4．重启 BIND 服务

重启 BIND 服务，代码如下。

```
[root@www.DNS_master_33.com named]#/etc/init.d/named restart
停止 named:                                                [确定]
启动 named:                                                [确定]
```

上述代码表示重启 BIND 服务成功。

5．DNS 客户端查询域名解析

DNS 客户端查询结果如下。

```
[root@www.DNS_slave_34.com ~]#dig www.booxin*.com

; <<>> DiG 9.8.2rc1-RedHat-9.8.2-0.68.rc1.el6_10.3 <<>> www.booxin*.com
;; global options: +cmd
;; Got answer:
;; ->>HEADER<<- opcode: QUERY, status: NOERROR, id: 24265
;; flags: qr aa rd ra; QUERY: 1, ANSWER: 2, AUTHORITY: 2, ADDITIONAL: 2

;; QUESTION SECTION:
;www.booxin*.com.               IN      A

;; ANSWER SECTION:
www.booxin*.com.        10800   IN      A       192.168.2.34
www.booxin*.com.        10800   IN      A       192.168.2.33

;; AUTHORITY SECTION:
booxin*.com.            10800   IN      NS      ns1.booxin*.com.
booxin*.com.            10800   IN      NS      ns2.booxin*.com.

;; ADDITIONAL SECTION:
ns1.booxin*.com.        10800   IN      A       192.168.2.33
ns2.booxin*.com.        10800   IN      A       192.168.2.34

;; Query time: 0 msec
;; SERVER: 192.168.2.33#53(192.168.2.33)
;; WHEN: Mon Aug  5 22:06:00 2019
;; MSG SIZE  rcvd: 132
```

```
[root@www.DNS_slave_34.com ~]#
[root@www.DNS_slave_34.com ~]#
[root@www.DNS_slave_34.com ~]#dig  +trace www.booxin*.com

; <<>> DiG 9.8.2rc1-RedHat-9.8.2-0.68.rc1.el6_10.3 <<>> +trace www.booxin*.com
;; global options: +cmd
.                       518374  IN      NS      c.root-servers.net.
.                       518374  IN      NS      m.root-servers.net.
.                       518374  IN      NS      i.root-servers.net.
.                       518374  IN      NS      b.root-servers.net.
.                       518374  IN      NS      j.root-servers.net.
.                       518374  IN      NS      d.root-servers.net.
.                       518374  IN      NS      h.root-servers.net.
.                       518374  IN      NS      f.root-servers.net.
.                       518374  IN      NS      k.root-servers.net.
.                       518374  IN      NS      e.root-servers.net.
.                       518374  IN      NS      g.root-servers.net.
.                       518374  IN      NS      a.root-servers.net.
.                       518374  IN      NS      l.root-servers.net.
;; Received 508 bytes from 192.168.2.33#53(192.168.2.33) in 3239 ms

com.                    172800  IN      NS      a.gtld-servers.net.
com.                    172800  IN      NS      b.gtld-servers.net.
com.                    172800  IN      NS      c.gtld-servers.net.
com.                    172800  IN      NS      d.gtld-servers.net.
com.                    172800  IN      NS      e.gtld-servers.net.
com.                    172800  IN      NS      f.gtld-servers.net.
com.                    172800  IN      NS      g.gtld-servers.net.
com.                    172800  IN      NS      h.gtld-servers.net.
com.                    172800  IN      NS      i.gtld-servers.net.
com.                    172800  IN      NS      j.gtld-servers.net.
com.                    172800  IN      NS      k.gtld-servers.net.
com.                    172800  IN      NS      l.gtld-servers.net.
com.                    172800  IN      NS      m.gtld-servers.net.
;; Received 492 bytes from 199.7.91.13#53(199.7.91.13) in 3701 ms

booxin*.com.        172800    IN    NS    DNS13.hichina.com.
booxin*.com.        172800    IN    NS    DNS14.hichina.com.
;; Received 392 bytes from 192.12.94.30#53(192.12.94.30) in 343 ms

www.booxin*.com.        600   IN   A    115.29.28.114
;; Received 48 bytes from 140.205.41.26#53(140.205.41.26) in 29 ms
```

上述代码首先从当前 DNS 服务器发起查询请求，然后从根域 DNS 服务器发起查询请求。

6. 查看 BIND 日志记录

查看 BIND 日志记录，代码如下。

```
[root@www.DNS_master_33.com named]#tailf /var/named/log/query.log
05-Aug-2019 22:05:49.047 queries: info: client 192.168.2.34#34148: query:
www.booxin*.com IN A + (192.168.2.33)
05-Aug-2019 22:06:01.639 queries: info: client 192.168.2.34#51517: query:
. IN NS - (192.168.2.33)
05-Aug-2019 22:06:01.640 queries: info: client 192.168.2.34#44036: query:
c.root-servers.net IN A + (192.168.2.33)
05-Aug-2019 22:06:01.640 queries: info: client 192.168.2.34#44036: query:
c.root-servers.net IN AAAA + (192.168.2.33)
05-Aug-2019 22:06:02.722 queries: info: client 192.168.2.34#34958: query:
m.root-servers.net IN A + (192.168.2.33)
05-Aug-2019 22:06:02.723 queries: info: client 192.168.2.34#34958: query:
m.root-servers.net IN AAAA + (192.168.2.33)
05-Aug-2019 22:06:02.733 queries: info: client 192.168.2.34#39168: query:
i.root-servers.net IN A + (192.168.2.33)
05-Aug-2019 22:06:02.733 queries: info: client 192.168.2.34#39168: query:
i.root-servers.net IN AAAA + (192.168.2.33)
```

```
05-Aug-2019 22:06:03.100 queries: info: client 192.168.2.34#45510: query:
b.root-servers.net IN A + (192.168.2.33)
05-Aug-2019 22:06:03.100 queries: info: client 192.168.2.34#45510: query:
b.root-servers.net IN AAAA + (192.168.2.33)
05-Aug-2019 22:06:04.007 queries: info: client 192.168.2.34#47680: query:
j.root-servers.net IN A + (192.168.2.33)
05-Aug-2019 22:06:04.008 queries: info: client 192.168.2.34#47680: query:
j.root-servers.net IN AAAA + (192.168.2.33)
05-Aug-2019 22:06:07.955 queries: info: client 192.168.2.34#59783: query:
j.gtld-servers.net IN A + (192.168.2.33)
05-Aug-2019 22:06:07.955 queries: info: client 192.168.2.34#59783: query:
j.gtld-servers.net IN AAAA + (192.168.2.33)
05-Aug-2019 22:06:08.079 queries: info: client 192.168.2.34#34165: query:
k.gtld-servers.net IN A + (192.168.2.33)
05-Aug-2019 22:06:08.079 queries: info: client 192.168.2.34#34165: query:
k.gtld-servers.net IN AAAA + (192.168.2.33)
05-Aug-2019 22:06:08.322 queries: info: client 192.168.2.34#34895: query:
l.gtld-servers.net IN A + (192.168.2.33)
05-Aug-2019 22:06:08.322 queries: info: client 192.168.2.34#34895: query:
l.gtld-servers.net IN AAAA + (192.168.2.33)
05-Aug-2019 22:06:08.493 queries: info: client 192.168.2.34#41292: query:
m.gtld-servers.net IN A + (192.168.2.33)
05-Aug-2019 22:06:08.493 queries: info: client 192.168.2.34#41292: query:
m.gtld-servers.net IN AAAA + (192.168.2.33)
05-Aug-2019 22:06:12.751 queries: info: client 192.168.2.34#38553: query:
DNS13.hichina.com IN A + (192.168.2.33)
05-Aug-2019 22:06:12.751 queries: info: client 192.168.2.34#38553: query:
DNS13.hichina.com IN AAAA + (192.168.2.33)
05-Aug-2019 22:06:12.936 queries: info: client 192.168.2.34#40090: query:
DNS14.hichina.com IN A + (192.168.2.33)
05-Aug-2019 22:06:12.936 queries: info: client 192.168.2.34#40090: query:
DNS14.hichina.com IN AAAA + (192.168.2.33)
```

上述代码查看客户端访问日志。为了丢弃一个类别中的所有信息，可以设定 null 通道，代码如下。

```
category "xfer-out" { "null"; };
category "notify" { "null"; };
```

3.4 DNS 与 CDN 企业级缓存架构

3.4.1 DNS 安全问题

1. DNS 放大攻击

放大攻击又叫杠杆攻击，攻击者向大量开放的 DNS 服务器发送大范围域名查询的 DNS 请求，并将该 DNS 请求的原 IP 地址伪造成想要攻击的目标 IP 地址。由于请求数据比响应数据小得多，攻击者可以利用该技术"放大"掌握的带宽资源和攻击流量。

2. DDoS 攻击

DDoS 大流量分布式攻击会对 DNS 服务器造成致命的打击，直接导致 DNS 服务器拒绝提供服务，目前没有特别有效的解决方案应对 DDoS 攻击。

3. DNS 劫持

DNS 劫持是通过篡改 DNS 服务器上的数据并返回给用户一个错误的查询结果来实现的。DNS 劫持是在劫持的网络范围内先拦截域名解析的请求，随后分析请求的域名，最后返回假的 IP 地址或者使请求失去响应。

4. DNS 缓存污染

DNS 缓存污染（DNS Cache Pollution），又称 DNS 缓存投毒（DNS Cache Poisoning），是指一些用户刻意或无意中制造出来的域名服务器数据包，把域名指向不正确的 IP 地址。一般来说，在互联网上都有可信赖的 DNS 服务器，但为降低网络上的流量压力，一般的域名服务器都会把从上游的域名服务器获得的解析记录暂存起来，待下次有其他机器要求解析域名时，可以立即提供服务。一旦有关 DNS 的局域域名服务器的缓存受到污染，就会把 DNS 内的计算机引向错误的服务器或服务器的网址。

DNS 污染与 DNS 劫持的区别在于，DNS 劫持修改了服务器 DNS 的解析结果，而 DNS 污染是不经过 DNS 服务器，直接返回错误信息。

5. DNS 信息被攻击者修改

攻击者攻击根域名或者顶级域名服务器中的某几台服务器，篡改解析信息，导致解析到恶意网站，给个人或企业带来巨大的损失。

3.4.2 CDN 基础知识

1. CDN 是什么

CDN 能够根据网络流量和各节点的连接、负载状况以及到用户的距离和响应时间等综合信息实时地将用户的请求重新导向到离用户最近的服务节点上。其目的是使用户可就近取得所需内容，解决网络拥挤的问题，提高用户访问网站的响应速度。典型的 CDN 系统由下面 3 个部分组成。

（1）分发服务系统。

分发服务系统最基本的工作单元就是缓存设备，缓存（边缘缓存）设备负责直接响应最终用户的访问请求，把缓存在本地的内容快速地提供给用户。同时缓存设备还负责与源站点进行内容同步，把更新的内容以及本地没有的内容从源站点获取并保存在本地。缓存设备的数量、规模、总服务能力是衡量一个 CDN 系统服务能力的最基本的指标之一。

（2）负载均衡系统。

负载均衡系统的主要功能是负责对所有发起服务请求的用户进行访问调度，确定提供给用户的最终实际访问地址。两级调度体系分为全局负载均衡（Global Server Load Balance，GSLB）和本地负载均衡（Server Load Balance，SLB）。GSLB 主要根据用户就近原则，通过对每个服务节点进行"最优"判断，确定向用户提供服务的缓存设备的物理位置。SLB 主要负责节点内部的设备负载均衡。

（3）运营管理系统。

运营管理系统又分为运营管理子系统和网络管理子系统，负责处理业务层面的与外界系统交互所必需的收集、整理、交付工作，包含客户管理、产品管理、计费管理、统计分析等功能。

2. 配置 DNS 解析指向 CDN

使用 CDN 的方法很简单，只需要修改自己的 DNS 解析，设置一个 CNAME 记录指向 CDN 服务器即可。

3. CDN 缓存的基本原理

- 用户单击网站页面上的 URL，URL 经过本地 DNS 解析，DNS 最终会将域名的解析权交给 CNAME 记录指向的 CDN 专用 DNS 服务器。
- CDN 的 DNS 服务器将 CDN 的全局负载均衡设备 IP 地址返回给用户。
- 用户向 CDN 的全局负载均衡设备发起内容 URL 访问请求。
- CDN 全局负载均衡设备根据用户 IP 地址和用户请求的内容 URL，选择一台用户所属区域的区域负载均衡设备，并告诉用户向这台设备发起请求。
- 区域负载均衡设备会为用户选择一台合适的缓存服务器提供服务，选择的思路为：根据用户 IP 地址，判断哪一台服务器距离用户最近；根据用户所请求的 URL 中携带的内容名称，判断哪一台服务器上有用户所需内容；查询各台服务器当前的负载情况，判断哪一台服务器尚有服务能力。基于以上这些条件的综合分析，区域负载均衡设备会向全局负载均衡设备返回一台缓存服务器的 IP 地址。
- 全局负载均衡设备把服务器的 IP 地址返回给用户。
- 用户向缓存服务器发起请求，缓存服务器响应用户请求，将用户所需内容传送到用户终端。如果这台缓存服务器上并没有用户想要的内容，而区域负载均衡设备依然将它分配给了用户，那么这台服务器就要向它的上一级缓存服务器请求内容，直至追溯到网站的源服务器，将内容返回到本地。

4．CDN 的优势

使用 CDN 的优势如下。
- 本地缓存加速，加快访问速度。
- 镜像服务，消除运营商之间互联的瓶颈影响，保证不同网络的用户都能得到良好的访问质量。
- 远程加速，自动选择缓存服务器。
- 带宽优化，分担网络流量，减轻压力。
- 集群抗攻击。
- 节约成本。

3.5 DNS 服务运维技巧

3.5.1 CNAME 记录和 A 记录

1．什么是 CNAME 记录

CNAME 记录被称为别名记录，这种记录允许用户将多个域名映射到同一台计算机，通常用于同时提供 WWW 和 MAIL 服务的计算机。例如，有一台计算机名为 laohan（A 记录），它同时提供 WWW 和 MAIL 服务，为了方便用户访问服务，可以为该计算机设置两个别名（CNAME 记录）：WWW 和 MAIL。

2．CNAME 记录和 A 记录的区别

A 记录解析域名到 IP 地址，CNAME 记录解析域名到另外一个域名。

A 记录与 CNAME 记录功能差不多，CNAME 记录将几个主机名指向一个别名，其实跟指向 IP 地址是一样的，因为这个别名也要做一个 A 记录。但是使用 CNAME 记录可以很方便地变更 IP 地址。如果一台服务器有 100 个网站，他们都设置了别名，该服务器变更 IP 地址时，只需要变更别名的 A 记录就可以了。

一台服务器可以同时运行多个网站，也可以有多个域名，每个网站对应一个单独的域名。解析格式如下所示。

- 域名 A→A 记录→1.1.1.1（IP 地址）。
- 域名 B（也叫 A 别名）。
- 域名 C（也叫 A 别名）。

如果 B、C 此时也是通过 A 记录的形式解析到真正的服务器，那么哪一天真正的服务器修改了，A、B、C 都得修改；而如果把 B、C 都解析到 A，那么如果修改 A 服务器的 IP 地址，只需要修改 A 地址的解析记录即可，如何实现这一功能呢？只需要把 B、C 通过 CNAME 记录解析到 A 记录即可。

3．使用 CNAME 记录的优势

使用 CNAME 记录的优势如下。

（1）降低多域名、多服务器、多业务的运维成本。

（2）多个域名、多种业务解析到同一台主机，同时不影响搜索引擎收录。

（3）解决多线和 CDN 分发加速问题。

（4）解决高并发下的负载问题。

3.5.2　CNAME 解析运维技巧

需要在公司的服务器 A 上配置 30 个网站，每个网站的域名都不一样，部分网站采用 A 记录绑定了很多域名，加上二级域名共有 80 条解析记录。

有一天由于电信线路故障导致服务器 A 无法对外提供服务，此时系统管理员就需要决定是否要把域名解析到服务器 B（服务器 B 是用来做备份的，每周备份一次）上。此时需要管理员手动更改 80 个域名的解析，等管理员更改完，服务器 A 已经恢复正常了，此种方式会浪费非常多的时间，且业务在短期内无法恢复正常。

基本信息描述。服务器 A：88.88.88.88。服务器 B：99.99.99.99。域名若干。

解决方案。将同一个服务器的域名全部配置 CNAME 记录到某个域名上，如 a.com，再为 a.com 配置 A 记录到 88.88.88.88，需要切换到服务器 B 的时候，只需要更改 a.com 的 A 记录到 99.99.99.99 即可，是不是方便了很多？遇上这种突发情况不用苦等故障恢复，且业务基本不受影响，可以快速恢复 DNS 解析。

3.6　DNS 管理工具之 rndc

3.6.1　rndc 基本环境描述

1．rndc 基本说明

rndc 管理工具基本说明如表 3-3 所示。

3.6 DNS 管理工具之 rndc

表 3-3　　　　　　　　　　　　　rndc 管理工具基本说明

协议	端口	说明
TCP/UDP	53 端口	用于 DNS 服务
TCP	953 端口	用于 rndc 管理 DNS 服务

主要功能：检查缓存状态、清空缓存、查询 DNS 服务运行状态详情、重启服务，只支持 HMAC-MD5 认证算法，在通信两端使用共享密钥。rndc 在连接通道中发送命令时，必须使用经过服务器认可的密钥加密。

2. 安装说明

安装 BIND 软件，rndc 工具包含在 BIND 软件包中，代码如下。

```
[root@www.DNS_master_33.com ~]yum install bind
```

上述代码中 BIND 软件包包含 rndc 工具。

```
[root@www.DNS_master_33.com ~]#rpm -ql BIND |grep rndc
/etc/rndc.conf
/etc/rndc.key
/usr/sbin/rndc
/usr/sbin/rndc-confgen
/usr/share/doc/BIND-9.8.2/arm/man.rndc-confgen.html
/usr/share/doc/BIND-9.8.2/arm/man.rndc.conf.html
/usr/share/doc/BIND-9.8.2/arm/man.rndc.html
/usr/share/man/man5/rndc.conf.5.gz
/usr/share/man/man8/rndc-confgen.8.gz
/usr/share/man/man8/rndc.8.gz
```

上述代码查看 BIND 软件包是否有 rndc 工具。

3. 配置文件说明

rndc 工具主要配置文件如下。

- /etc/rndc.conf：客户端密钥文件。
- /etc/named.conf：服务器端 DNS 配置文件，也是 rndc 被管理端配置文件。

3.6.2 配置 rndc

配置 rndc 工具步骤如下。

1. 生成密钥配置文件

生成密钥配置文件，代码如下。

```
[root@www.DNS_master_33.com ~]#rndc-confgen -r /dev/urandom
# Start of rndc.conf
key "rndc-key" {
    algorithm hmac-md5;
    secret "TYuBgIlCKSHVsLQ/BrB+SQ==";
};

options {
    default-key "rndc-key";
    default-server 127.0.0.1;
    default-port 953;
};
# End of rndc.conf

# Use with the following in named.conf, adjusting the allow list as needed:
# key "rndc-key" {
#     algorithm hmac-md5;
#     secret "TYuBgIlCKSHVsLQ/BrB+SQ==";
# };
```

```
#
# controls {
#    inet 127.0.0.1 port 953
#        allow { 127.0.0.1; } keys { "rndc-key"; };
# };
# End of named.conf
```

2. 配置文件说明

下面代码表示客户端密钥文件。

```
secret "TYuBgIlCKSHVsLQ/BrB+SQ==";
```

下面代码表示密钥名称。

```
default-key "rndc-key";
```

下面代码表示被管理服务器的目标 IP 地址。

```
default-server 127.0.0.1;
```

下面代码表示连接 DNS 服务器的目标端口。

```
default-port 953;
```

下面代码表示服务器密钥。

```
secret "TYuBgIlCKSHVsLQ/BrB+SQ==";
```

下面代码表示服务器监听的 IP 地址和端口。

```
inet 127.0.0.1 port 953
```

下面代码表示允许 IP 地址为 127.0.0.1 的主机连接管理本服务器。

```
allow { 127.0.0.1; } keys { "rndc-key"; };
```

3.6.3 配置 rndc 本地管理

1. 配置本地/etc/rndc.conf

系统中不存在/etc/rndc.conf 文件就自行创建，代码如下。

```
[root@www.DNS_master_33.com ~]#rndc-confgen -r /dev/urandom >/etc/rndc.conf
[root@www.DNS_master_33.com ~]#cat /etc/rndc.conf
# Start of rndc.conf
key "rndc-key" {
    algorithm hmac-md5;
    secret "jo+HvubCMTEyJ6GX5TwzaA==";
};

options {
    default-key "rndc-key";
    default-server 127.0.0.1;
    default-port 953;
};
# End of rndc.conf
```

2. 配置本地 rndc

过滤 rndc 配置信息，代码如下。

```
[root@www.DNS_master_33.com ~]#grep -v "#" /etc/rndc.conf
key "rndc-key" {
    algorithm hmac-md5;
    secret "jo+HvubCMTEyJ6GX5TwzaA==";
};

options {
    default-key "rndc-key";
    default-server 127.0.0.1;
    default-port 953;
};
```

配置本机服务端的/etc/named.conf，添加代码如下。

```
key "rndc-key" {
    algorithm hmac-md5;
```

```
    secret "jo+HvubCMTEyJ6GX5TwzaA==";
};

controls {
inet 127.0.0.1 port 953
allow { 127.0.0.1; } keys { "rndc-key"; };
};
```

/etc/named.conf 中核心配置如图 3-5 所示。

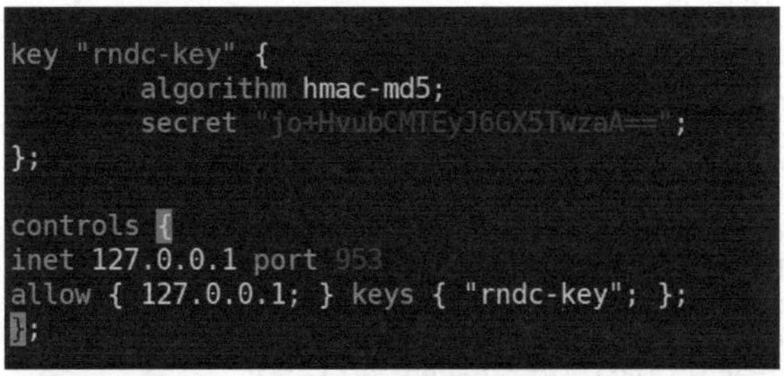

```
key "rndc-key" {
        algorithm hmac-md5;
        secret "jo+HvubCMTEyJ6GX5TwzaA==";
};

controls {
inet 127.0.0.1 port 953
allow { 127.0.0.1; } keys { "rndc-key"; };
};
```

图 3-5 核心配置

3. 重启服务

检测 DNS 配置语法，代码如下。

```
[root@www.DNS_master_33.com ~]#/etc/init.d/named configtest
zone duoduo*.com/IN: loaded serial 222
zone mingze*.com/IN: loaded serial 222
zone booxin*.com/IN: loaded serial 21
zone 2.168.192.\032in-addr.arpa/IN: loaded serial 2
zone localhost.localdomain/IN: loaded serial 0
zone localhost/IN: loaded serial 0
zone 1.0.0.0.0.0.0.0.0.0.0.0.0.0.0.0.0.0.0.0.0.0.0.0.0.0.0.0.0.0.0.0.ip6.arpa
/IN: loaded serial 0
zone 1.0.0.127.in-addr.arpa/IN: loaded serial 0
zone 0.in-addr.arpa/IN: loaded serial 0
```

检测语法通过后，可以通过如下指令重启 BIND 服务，代码如下。

```
 [root@www.DNS_master_33.com ~]#/etc/init.d/named restart
停止 named: .umount: /var/named/chroot/var/named: device is busy.
        (In some cases useful info about processes that use
         the device is found by lsof(8) or fuser(1))
                                                           [确定]
启动 named:                                                [确定]
[root@www.DNS_master_33.com ~]#ss -ntplu |grep 953
tcp    LISTEN     0      128     127.0.0.1:953      *:*    users:(("named",2250,22))
```

上述代码表示 BIND 服务重启成功，953 端口已经被监听，可以提供服务。

3.6.4 配置 rndc 远程管理

1. 基本环境信息描述

- DNS 服务器 IP 地址：192.168.2.33。
- 客户端 IP 地址：192.168.2.34。

2. 服务端配置

服务端配置/etc/named.conf，添加代码如下。

```
#服务端密钥
      key "rndc-key" {
algorithm hmac-md5;
secret "TYuBgIlCKSHVsLQ/BrB+SQ==";
 };
```

```
        controls {
inet 192.168.2.33 port 953      #服务器监听的 IP 地址和端口
allow { 192.168.2.33; 192.168.2.34; } keys { "rndc-key"; };   #允许本地和远程客户端
192.168.2.34 远程管理
};
```

3. 检测配置文件

检测配置文件，代码如下。

```
[root@www.DNS_master_33.com ~]#/etc/init.d/named configtest
zone duoduo*.com/IN: loaded serial 222
zone mingze*.com/IN: loaded serial 222
zone booxin*.com/IN: loaded serial 21
zone 2.168.192.\032in-addr.arpa/IN: loaded serial 2
zone localhost.localdomain/IN: loaded serial 0
zone localhost/IN: loaded serial 0
zone 1.0.0.0.0.0.0.0.0.0.0.0.0.0.0.0.0.0.0.0.0.0.0.0.0.0.0.0.0.0.0.0.ip6.arpa
/IN: loaded serial 0
zone 1.0.0.127.in-addr.arpa/IN: loaded serial 0
zone 0.in-addr.arpa/IN: loaded serial 0
```

4. 本地客户端配置/etc/rndc.conf

本地客户端配置/etc/rndc.conf，代码如下。

```
[root@www.DNS_master_33.com ~]#cat /etc/rndc.conf
key "rndc-key" {
    algorithm hmac-md5;
    secret "jo+HvubCMTEyJ6GX5TwzaA==";
};

options {
    default-key "rndc-key";
    default-server 127.0.0.1,192.168.2.34;
    default-port 953;
};
```

5. 远程客户端配置/etc/rndc.conf

远程客户端配置/etc/rndc.conf，代码如下。

```
[root@www.DNS_slave_34.com ~]#cat /etc/rndc.conf
 key "rndc-key" {
algorithm hmac-md5;
secret "jo+HvubCMTEyJ6GX5TwzaA==";   #客户端密钥
 };

options {
default-key "rndc-key";       #密钥名
default-server 192.168.2.33;   #被管理服务器的目标 IP 地址
default-port 953;              #连接 DNS 服务器的目标端口
};
```

6. 重启 DNS 服务

重启 DNS 服务，代码如下。

```
[root@www.DNS_master_33.com ~]#/etc/init.d/named configtest
zone duoduo*.com/IN: loaded serial 222
zone mingze*.com/IN: loaded serial 222
zone booxin*.com/IN: loaded serial 21
zone 2.168.192.\032in-addr.arpa/IN: loaded serial 2
zone localhost.localdomain/IN: loaded serial 0
zone localhost/IN: loaded serial 0
zone 1.0.0.0.0.0.0.0.0.0.0.0.0.0.0.0.0.0.0.0.0.0.0.0.0.0.0.0.0.0.0.0.ip6.arpa
/IN: loaded serial 0
zone 1.0.0.127.in-addr.arpa/IN: loaded serial 0
zone 0.in-addr.arpa/IN: loaded serial 0
[root@www.DNS_master_33.com ~]#/etc/init.d/named restart
```

```
停止 named: umount: /var/named/chroot/var/named: device is busy.
       (In some cases useful info about processes that use
        the device is found by lsof(8) or fuser(1))
                                                            [确定]
启动 named:                                                  [确定]
[root@www.DNS_master_33.com ~]#ss -ntplu |grep 953
tcp    LISTEN    0     128    192.168.2.33:953    *:*    users:(("named",2429,22))
```

上述代码先检测 DNS 配置文件语法，接着重启 BIND 服务。

查看端口和服务是否存在，代码如下。

```
[root@www.DNS_master_33.com ~]#netstat  -ntpl
Active Internet connections (only servers)
Proto Recv-Q Send-Q Local Address        Foreign Address      State      PID/Program name
tcp       0      0 0.0.0.0:111          0.0.0.0:*            LISTEN     1233/rpcBIND
tcp       0      0 192.168.2.33:5       0.0.0.0:*            LISTEN     2891/named
tcp       0      0 127.0.0.1:53         0.0.0.0:*            LISTEN     2891/named
tcp       0      0 0.0.0.0:22           0.0.0.0:*            LISTEN     1326/sshd
tcp       0      0 192.168.2.33:953     0.0.0.0:*            LISTEN     2891/named
tcp       0      0 127.0.0.1:25         0.0.0.0:*            LISTEN     1405/master
tcp       0      0 0.0.0.0:56795        0.0.0.0:*            LISTEN     1255/rpc.statd
tcp       0      0 :::42028             :::*                 LISTEN     1255/rpc.statd
tcp       0      0 :::111               :::*                 LISTEN     1233/rpcBIND
tcp       0      0 :::22                :::*                 LISTEN     1326/sshd
tcp       0      0 ::1:25               :::*                 LISTEN     1405/master
```

7. 基本管理指令

DNS 服务器执行指令，代码输出如下。

```
[root@www.DNS_master_33.com ~]#rndc status
WARNING: key file (/etc/rndc.key) exists, but using default configuration file (/etc/rndc.conf)
version: 9.8.2rc1-RedHat-9.8.2-0.68.rc1.el6_10.3
CPUs found: 1
worker threads: 1
number of zones: 23
debug level: 0
xfers running: 0
xfers deferred: 0
soa queries in progress: 0
query logging is ON
recursive clients: 0/0/1000
tcp clients: 2/100
server is up and running
```

客户端执行指令，代码输出如下。

```
[root@www.DNS_slave_34.com ~]# rndc  -s 192.168.2.33 status
version: 9.8.2rc1-RedHat-9.8.2-0.68.rc1.el6_10.3
CPUs found: 1
worker threads: 1
number of zones: 23
debug level: 0
xfers running: 0
xfers deferred: 0
soa queries in progress: 0
query logging is ON
recursive clients: 0/0/1000
tcp clients: 2/100
server is up and running
```

客户端可以使用下面这条指令启动 rndc，效果和 rndc –s 相同。

```
[root@www.DNS_master_33.com ~]# rndc -c /etc/rndc.conf status
version: 9.8.2rc1-RedHat-9.8.2-0.68.rc1.el6_10.3
CPUs found: 1
worker threads: 1
number of zones: 23
debug level: 0
```

```
xfers running: 0
xfers deferred: 0
soa queries in progress: 0
query logging is ON
recursive clients: 0/0/1000
tcp clients: 2/100
server is up and running
```

3.6.5 rndc 管理工具常用选项和指令

rndc 管理工具常用选项如表 3-4 所示。

表 3-4　　　　　　　　　　　　rndc 管理工具常用选项

选项	说明
-b	绑定 rndc 客户端使用的源地址，一个网卡可有多个地址
-c	指定连接时使用的配置文件，而不是默认的/etc/rndc.conf
-s	指定要连接的服务器的 IP 地址
-p	指定要连接的服务器的端口
-k	指定连接时使用的密钥文件，而不是默认的/etc/rndc.key
-y	指定要使用的密钥标识，必须与服务器一致

rndc 管理工具常用指令如表 3-5 所示。

表 3-5　　　　　　　　　　　　rndc 管理工具常用指令

指令	说明
status	显示 BIND 服务器的工作状态
reload	重新加载配置文件和区域文件
reload zone_name	重新加载指定区域
reconfig	重读配置文件并加载新增的区域
querylog	关闭或开启查询日志
dumpdb	将高速缓存转储到转储文件（named_dump.db）
freeze	暂停更新所有动态区域
freeze zone [class [view]]	暂停更新一个动态区域
flush [view]	刷新服务器的所有高速缓存
flushname name	为某一视图刷新服务器的高速缓存
stats	将服务器统计信息写入统计文件中
stop	停止服务器运行
trace	打开 debug，每执行一次提升一次级别
trace LEVEL	指定 debug 的级别，trace 0 表示关闭 debug
notrace	将调试级别设置为 0
restart	重启服务器（尚未实现）
tsig-list	查询当前有效的 TSIG 列表

3.6.6 管理 DNS 注意事项

在使用 rndc 连接 BIND 前一定要确保两台 Linux 服务器的时间同步，否则会报错，可以手动修改系统时间，也可以与 NTP 服务器同步。若实验用的服务器无法连接外网，又不想手动配置时间，则可以将其中一台服务器配置为 NTP，供另一台服务器同步。

同步时间指令如下。

```
[root@www.DNS_master_33.com ~]#ntpdate ntp1.aliyun.com
14 Sep 10:16:07 ntpdate[2939]: step time server 120.25.115.20 offset 43046.342760 sec
[root@www.DNS_master_33.com ~]#ntpdate ntp1.aliyun.com >/dev/null 2>&1
```

将同步时间指令加入 Linux 操作系统定时任务计划中，代码输出如下。

```
[root@www.DNS_master_33.com ~]#crontab -e
crontab: installing new crontab
[root@www.DNS_master_33.com ~]#
[root@www.DNS_master_33.com ~]#crontab -l
# ntpdate time by hanyw 20190914
*/30 * * * * ntpdate ntp1.aliyun.com >/dev/null 2>&1
```

上述代码表示，每隔 30min 执行一次 ntpdate ntp1.aliyun.com >/dev/null 2>&1 指令。

3.7 TTL 值配置

3.7.1 TTL 值基础知识

1．什么是 TTL 值？

DNS 服务器不可能永久保存缓存数据。若永久保存缓存数据，当 DNS 服务器列表数据发生变更的时候，记录无法被及时更新和传达。因此，需要通过定义 TTL，来定义数据在缓存中的存放时间。TTL 一到期，DNS 服务器就丢弃原有的缓存数据并从其他服务器获取新的数据。

2．常见 TTL 值设置

TTL 通常以秒为单位，一些常见的 TTL 值设置如下。

```
300 seconds = 5 minutes = "Very Short"
3600 seconds = 1 hour = "Short"
86400 seconds = 24 hours = "Long"
604800 seconds = 7 days = "Insanity"
```

3．TTL 值生效时间解析

修改完域名的 TTL 值后已经过去很久还没有生效，可能存在以下几个原因。

- 浏览器缓存。浏览器缓存将文件保存在客户端，在同一个会话过程中会检查缓存的副本是否足够新，在后退网页时，访问过的资源可以从浏览器缓存中取出使用。网络管理员通过减少服务器处理请求的数量，用户将获得更快的体验。该缓存并**不遵循 DNS 服务器的 TTL 值**，在此不过多介绍。
- 具有更多 DNS 服务器的复杂内部网络产生比预期更长的 DNS 更改时间。

这是大多数服务声明的原因。注意：修改 DNS 服务器需要 0～72h 的全球生效时间，如果发现某些地方记录没有生效，并且修改时间还不到 72h，请耐心等待。

3.7.2 TTL 值最佳配置实战

1．短时间设置 TTL 值使用场景

（1）域名会频繁更改记录。

域名频繁更新的 TTL 值设置，建议将 TTL 值设置得相对小，避免 TTL 值影响 DNS 服务器的性能。

（2）一些重要的域名，一旦发生记录不可达则损失很大，这时候建议将 TTL 值设置得小

一些，可以及时完成变更。（一些本地 DNS 会对 TTL 值进行默认设置，所以在灾难恢复的时候时间不可控。）

（3）如果对 DNS 记录进行增加或者修改时，碰巧输错了记录，这时候最好的操作方法是，增加或修改记录时，先修改为一个小的 TTL 值，然后对记录进行修改。

2．长时间设置 TTL 值使用场景

（1）考虑到一定成本的时候，例如，DNSpod 免费托管域名的最小 TTL 值是 600。而在 DNSpod，TTL 值越小意味着套餐价格越高，也就是客户承担更高的成本。

（2）一个大的 TTL 值可以缩短查询时间。

（3）MX、DKIM、SPF、TXT 记录，这几个记录可以设置更大的 TTL 值。

（4）当根域或者 ISP 级别的 DNS 服务器发生 DDoS 时，如果攻击事件在 TTL 值内，部分用户会无感知。

3．SOA TTL 使用场景

在每个 DNS 服务器区域的顶部，有几个 TTL 值在 DNS 服务器中发挥重要作用。

（1）refresh TTL。

从 DNS 服务器向主 DNS 服务器刷新的时间。notify 参数可以设置主 DNS 服务器发生改变时主动向从 DNS 服务器更新，关闭 notify 时会采用这个 refresh TTL。

（2）retry TTL。

refresh TTL 失败后的重试时间。

（3）expiry TTL。

当 refresh TTL 和 retry TTL 都失败时，从 DNS 服务器无法和主 DNS 服务器建立连接，过了 expiry TTL 后将无法提供权威解析，从而会删除自己的副本。一般这几个 TTL 值不建议自行修改，除非明确知道自己在做什么。

4．TTL 值设置建议

TTL 值过大，修改 DNS 解析后等待生效的时间就会过长；TTL 值过小，域名解析的稳定性和解析速度就会受到影响。那么 TTL 值多大合适，就要根据具体的网站来决定，没有统一的标准答案。

下面给出一些常见网站类型的 TTL 推荐值，可以按照以下建议进行 TTL 值设置，如表 3-6 所示。

表 3-6　　　　　　　　　　DNS 解析多场景 TTL 推荐值参照

IP 地址频繁更新	动态 IP 地址	宕机检测	服务器架构	建议 TTL 值
否	否	不使用	单服务器	600
否	否	使用	多服务器	180
是	否	不使用	单服务器	300
是	是	不限	不限	120
否	否	是	大型商业网站	60
否	否	否	热备、容灾、IP 固定	3600

第 4 章

DHCP 服务器运维实战

4.1 DHCP 服务器详解

4.1.1 DHCP 服务器基础

1. DHCP 简介

DHCP 的英文全称为 Dynamic Host Configuration Protocol，即动态主机配置协议。它是由因特网工程任务组（Internet Engineering Task Force，IETF）设计开发的，专门用于为 TCP/IP 网络中的计算机自动分配 TCP/IP 参数。

DHCP 通常被应用在大型的局域网环境中，主要作用是集中管理 IP 地址，使网络环境中的主机动态获得 IP 地址、网关地址、DNS 服务器地址等信息，并能够提升 IP 地址的使用效率。

DHCP 工作在开放系统互连（Open System Interconnection，OSI）的**应用层**，可以帮助计算机从指定的 DHCP 服务器获取配置信息的协议，主要包括 IP 地址、子网掩码、网关和 DNS 等。

DHCP 采用 C/S 模型，主机地址的动态分配任务由网络主机驱动。当 DHCP 服务器接收到来自网络主机申请地址的信息时，才会向网络主机发送相关的地址配置等信息，以实现网络主机地址信息的动态配置。DHCP 具有以下功能。

- 保证任何 IP 地址在同一时刻只能由一个 DHCP 客户端使用。
- 给用户分配永久固定的 IP 地址。
- 可以同用其他方法获得 IP 地址的主机共存（如手动配置 IP 地址的主机）。
- 向现有的引导协议（Boot Strap Protocol，BOOTP）客户端提供服务。

互联网是目前世界上用户最多的服务之一，有几十亿人在使用，为每位用户分配一个固定的 IP 地址的做法不仅造成了 IP 地址的浪费，还会给 ISP 服务商带来高额的维护成本。

2. 使用 DHCP 服务器的优点

使用 DHCP 服务器的优点如下。

- 减少管理员的工作量。
- 避免输入错误的可能。
- 避免 IP 地址冲突。
- 提高 IP 地址的利用率。
- 方便客户端的配置。

3. DHCP 服务器主机配置的特点

DHCP 服务器主机配置方式最重要的特征之一是整个配置过程自动实现，而且所有配置信息在一个地方集中控制，这就是 DHCP 服务器的作用。

最初的 BOOTP/DHCP 是在同一个子网中使用广播方式实现的，无法穿越路由器扩展到不同的子网中，也就是要使用 DHCP 的每一个网络（广播域）中必须配置一台 DHCP 服务器。为了弥补这一缺陷，网络管理员采用了 DHCP 中继服务器的方式，使 BOOTP/DHCP 能够穿透路由器，实现跨了网传输。

DHCP 一般采用终端的硬件地址（如果是以太网，则为 MAC 地址）来作为一个终端设备的唯一标识。

通过设置 IP 地址使用租期，可以达到 IP 地址的时分复用效果，解决 IP 地址资源短缺的问题。

DHCP 基本上是一个单向驱动协议，服务器完全是被动的，其动作、行为基本由客户端的请求行为激发，即服务器无法主动控制客户端。因此其交互性和安全性就没有点对点协议（Point-to-Point Protocol, PPP）那么完善，这是 DHCP 的一个风险点。除了分配 IP 地址，DHCP 还可以帮助客户端指定如下网络环境信息。

- 默认网关（Router）。
- 默认子网掩码（Netmask）。
- DNS 服务器（DNS Server）。

4. DHCP 分配 IP 地址方式

DHCP 有如下 3 种方式分配 IP 地址。

（1）自动分配（Automatic Allocation）方式。

DHCP 服务器为主机指定一个永久性的 IP 地址，一旦 DHCP 客户端第一次成功从 DHCP 服务器租用到 IP 地址后，就可以永久地使用该地址。

（2）动态分配（Dynamic Allocation）方式。

DHCP 服务器给主机指定一个具有时间限制的 IP 地址，当时间到期或主机明确表示放弃该地址时，该地址可以被其他主机使用。

（3）手动分配（Manual Allocation）方式。

客户端的 IP 地址是由网络管理员指定的，DHCP 服务器只是将指定的 IP 地址告诉客户端主机。

上述 3 种 IP 地址分配方式中，只有动态分配方式可以重复使用客户端不再需要的地址。

5. DHCP 服务器常用端口

DHCP 服务器有如下 3 个常用端口。

（1）UDP67：DHCP 服务器服务端口。

（2）UDP68：DHCP 客户端服务端口。

（3）UDP546：用于 DHCP v6，而不用于 DHCP v4。

4.1.2 DHCP 运行机制

1. DHCP 运行方式

客户端发送广播包给整个网段内的所有主机或服务器，局域网内有 DHCP 服务器时，才会响应 DHCP 客户端的 IP 寻址请求，因此 DHCP 服务器与 DHCP 客户端应该在同一个网段内。DHCP 客户端与 DHCP 服务器连接过程如图 4-1 所示。

DHCP 客户端与 DHCP 服务器数据交互步骤如下。

（1）DHCP 客户端利用广播包发送搜索 DHCP 服务器的包。（DHCP Discover。）

（2）DHCP 服务器提供 DHCP 客户端

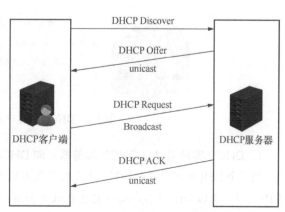

图 4-1 DHCP 客户端与 DHCP 服务器连接过程

网络相关的租约选择。（DHCP Offer，unicast。）

（3）DHCP 客户端选择 DHCP 服务器提供的网络参数租约并汇报给服务器。（DHCP Request，Broadcast。）

（4）DHCP 服务器记录这次租约并回报给 DHCP 客户端相关的封包信息。（DHCP ACK，unicast。）

2. DHCP 服务器基础概念

DHCP 服务器基础概念解释如下。

（1）DHCP 客户端：通过 DHCP 请求 IP 地址的客户端。DHCP 客户端是接口级的概念，如果一台主机有多个以太网接口，则该主机上的每个接口可以配置成一个 DHCP 客户端。交换机上的每个 VLAN 接口也可以配置成一个 DHCP 客户端。

（2）DHCP 服务器：负责为 DHCP 客户端提供 IP 地址，并且负责管理分配的 IP 地址。

（3）DHCP 中继器：DHCP 客户端跨网段申请 IP 地址的时候，实现 DHCP 报文的转发功能。

（4）DHCP 安全特性：实现合法用户 IP 地址表的管理功能。

（5）DHCP 监听：记录通过二层设备申请到 IP 地址的用户信息。

4.1.3 DHCP 服务器工作原理

DHCP 的实现过程涉及 DHCP 服务器和 DHCP 客户端之间的多种报文的交互，可分为客户端首次向服务器申请地址、客户端再次接入网络申请地址和续约等。

DHCP 客户端在启动时，会搜寻网络中是否存在 DHCP 服务器。如果存在，则给 DHCP 服务器发送一个请求。DHCP 服务器接到 DHCP 客户端请求后，为 DHCP 客户端选择 TCP/IP 配置的参数，并把这些参数发送给 DHCP 客户端。如果 DHCP 服务器已开启冲突检测功能，则 DHCP 服务器在将租约中的地址提供给 DHCP 客户端之前，先使用 ping 指令测试作用域中每个可用地址的连通性，以确保提供给 DHCP 客户端的每个 IP 地址都没有被手动配置给 TCP/IP 的另一台非 DHCP 客户端使用。

根据 DHCP 客户端是否第一次登录网络，DHCP 服务器的工作形式会有所不同。DHCP 客户端从 DHCP 服务器上获取 IP 地址的所有过程可以分为以下几个步骤，DHCP 新客户端的租约过程如图 4-2 所示。

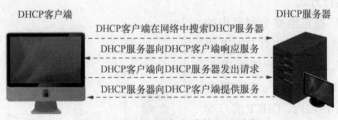

图 4-2　DHCP 新客户端的租约过程

1. DHCP 客户端寻找 DHCP 服务器，向 DHCP 服务器请求 IP 地址

当一个 DHCP 客户端启动时，计算机发现本机上没有任何 IP 地址设定，DHCP 客户端还没有 IP 地址，所以 DHCP 客户端要通过 DHCP 获取一个合法的 IP 地址，此时 DHCP 客户端以广播方式（因为 DHCP 服务器的 IP 地址对 DHCP 客户端来说是未知的）发送 DHCP Discover 请

求来寻找 DHCP 服务器。即向保留地址 255.255.255.255（网络设备）发送特定的广播请求，广播请求中包含 DHCP 客户端的 MAC 地址和主机名，以便 DHCP 服务器确定是哪个客户端发送的请求。网络上每一台安装了 TCP/IP 的主机都会接收到该广播信息，但只有 DHCP 服务器才会做出响应，寻找 DHCP 服务器如图 4-3 所示。

图 4-3　寻找 DHCP 服务器

2. DHCP 服务器响应，分配 IP 地址

当 DHCP 服务器接收到来自 DHCP 客户端请求 IP 地址的信息时，它在自己的 IP 地址池中查找是否有合法的 IP 地址可以提供给 DHCP 客户端。如果有，DHCP 服务器就将此 IP 地址添加标记，并加入 DHCP Offer 的消息，然后 DHCP 服务器就会广播一则 DHCP Offer 消息，其响应和分配过程如图 4-4 所示。

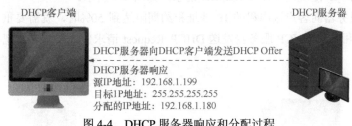

图 4-4　DHCP 服务器响应和分配过程

3. DHCP 客户端选择并接收 IP 地址

当 DHCP 客户端从第一个 DHCP 服务器接收 DHCP Offer 消息并提取了 IP 地址后，DHCP 客户端将 DHCP Request 消息广播到所有的 DHCP 服务器，表明它接受提供的内容。DHCP Request 消息包括为 DHCP 客户端提供 IP 地址配置的服务器的服务标识符（服务器 IP 地址），DHCP 服务器查看服务器标识符字段，以确定提供的 IP 地址是否被接受。如果 DHCP Offer 被接受，发出 IP 地址的 DHCP 服务器将该地址保留，这样该地址就不能提供给另一个 DHCP 客户端。如果 DHCP Offer 被拒绝，则 DHCP 服务器将会取消并保留其 IP 地址以提供给下一个 IP 地址租约的请求，如图 4-5 所示。

图 4-5　DHCP 客户端选择并接受 IP 地址

4. DHCP 服务器确认租约

DHCP 服务器接收到 DHCP Request 消息后,以 DHCP ACK 消息的形式向客户端广播租约成功的确认,该消息包含 IP 地址的有效租约和其他可配置的信息。此时,虽然服务器确认了客户端的租约请求,但是客户端还没有接收到服务器的 DHCP ACK 消息。当客户端收到 DHCP ACK 消息时,DHCP 客户端便将其 TCP/IP 与网卡绑定,此时客户端就配置了 IP 地址,完成 TCP/IP 寻址的初始化动作。另外,除了 DHCP 客户端选中的 DHCP 服务器外,其他的 DHCP 服务器也将收回曾经提供的 IP 地址,如图 4-6 所示。

图 4-6　DHCP 服务器确认租约

5. DHCP 客户端重新登录

DHCP 客户端每次重新登录网络时,不需要再发送 DHCP Discover 请求,而是直接发送包含前一次所分配的 IP 地址的 DHCP Request 请求,如图 4-7 所示。

当 DHCP 服务器向客户端出租的 IP 地址租约期限达到 50%时,就需要重新更新租约,客户端直接向提供租约的 DHCP 服务器发送 DHCP Request 请求包,要求更新现有的地址租约,如图 4-7 所示。

图 4-7　向 DHCP 服务器发送 DHCP Request 请求包

如果客户端 DHCP Request 内的 IP 地址在服务器端没有被使用,DHCP 服务器回复 DHCP ACK 消息继续使用 IP 地址,交互过程如图 4-8 所示。

图 4-8　DHCP ACK 消息

如果客户端 DHCP Request 内的 IP 地址在服务器已被使用,DHCP 服务器回复 DHCP NACK 消息告诉客户端 IP 地址已被使用,如图 4-9 所示。

图 4-9　DHCP NACK 消息

客户端重新开始向 DHCP 服务器发送 DHCP Discover 请求，交互过程如图 4-10 所示。

图 4-10　重新发送 DHCP Discover 请求

6. 更新租约

DHCP 服务器向 DHCP 客户端出租的 IP 地址一般都有一个租约期限，期满后 DHCP 服务器便会收回出租的 IP 地址。如果 DHCP 客户端要延长其 IP 租约，则必须更新其 IP 租约信息。DHCP 客户端启动和 IP 租约期限达到租约的 50%时，DHCP 客户端会自动向 DHCP 服务器发送更新其 IP 租约的信息。

DHCP 客户端和服务器租约交互过程大致可以分为以下几个阶段，如图 4-11 所示。

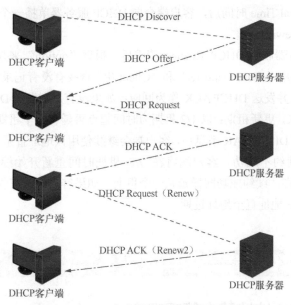

图 4-11　DHCP 客户端与服务器租约交互过程

7. IP 地址分配优先级

本地局域网的网络环境中，IP 地址分配的优先级如下。

- DHCP 服务器的数据库与客户端 MAC 地址静态绑定的固定 IP 地址。
- 客户端曾经使用过的 IP 地址，在客户端再次申请 IP 地址时，会在发送的 DHCP Request

请求中的"option"中加入"option 50: Requested IP Address=192.168.1.168"（192.168.1.168 为上次使用的 IP 地址）。
- 在 DHCP 地址池中按顺序查找可供分配的 IP 地址，首先查到的将被分配。

若上述几步都未找到可用的 IP 地址，DHCP 服务器会依次**查询租约过期**、**发生冲突**的 IP 地址，找到后则立即分配给 DHCP 客户端。

8．DHCP 租期和续约机制

（1）租约期限（简称"租期"）的选择。

DHCP 动态分配 IP 地址需要一个租期时间，以提高动态 IP 地址的重复利用率，DHCP 服务端与客户端租期机制如下。

- 如果客户端既没有指定请求租期，也没有指定 IP 地址，那么服务器将把租期指定为本机设置的默认租期。
- 如果客户端没有指定请求租期，但指定了 IP 地址，那么服务器将查询以前的租约信息，将以前该 IP 地址的租期再次指定给该 IP 地址。
- 如果客户端在 DHCP Discover 请求中指定了租期，则无论是否指定了 IP 地址，服务器都将把租期指定为请求租期。

（2）续约机制。

DHCP 服务器与客户端续约机制如下。

- 当 DHCP 客户端在收到 DHCP ACK 消息后，会根据租约设置两个续约的时间点：Renewal Time 1/2 租期和 Rebinding Time 7/8 租期。
- 当到达 Renewal Time 时间后，客户端会向 DHCP 服务器单播一个 DHCP Request 请求，同时进入 Renewing 状态。
- 当 DHCP 服务器收到 DHCP Request 请求后，根据"option"来判断是否为续约请求。如果是，则利用"client identifier"和"Client IP"查找有没有记录相应的租约。如果有，则更新租约，并发送 DHCP ACK 作为回应；如果没有，则发送 DHCP NAK。客户端收到 DHCP ACK，更新租期；以 1/2 租期为时间起点再延长一个租期时间，继续使用该 IP 地址。而收到 DHCP NAK 消息后，客户端会继续使用该地址直至 Rebinding Time 时间。
- 如果第一次续约不成功，客户端将在 7/8 租期时间重新开始续约。同时客户端进入 Rebinding 状态，续约成功同样延长一个租期。如果客户端一直收不到任何响应，它将继续使用该 IP 地址直至地址过期。

4.2 DHCP 服务应用场景

4.2.1 网络与 IP 地址基本管理理念

顺畅的网络依赖于良好的 IP 地址管理。随着网络和 IT 技术（例如 VoIP、云计算、服务器虚拟化、桌面虚拟化、IPv6 和服务自动化等）复杂度的提高，网络团队需要选择使用自动化 IP 地址管理（IP Address Management，IPAM）工具，自动化 IP 地址管理工具可以让管理员管理域名、分配子网，分配、追踪、回收、审计 IP 地址，解析域名，以及提供对网络的可视性。

IP 地址管理的基本理念：不同级别的用户是否有权限接入网络，使用哪一台终端，使用哪一个 IP 地址接入网络，是否存在地址欺骗和冲突的现象，对上述地址分配进行记录、审计和定位。因此其涉及实名制分配、终端 MAC 地址、IP 地址资源管理、域名管理、IP 地址接入授权和地址审计等方面。

4.2.2 DHCP 应用场景解析

1．DHCP 在办公网的应用

DHCP 自动获取 IP 地址在企业级办公网中的应用场景如下。

（1）企业员工办公。

在企业员工办公过程中，需要对网络流量进行限制，尤其是下载业务，会严重占用企业出口带宽，影响企业员工正常办公，因此需要针对员工的 IP 地址和 MAC 地址进行绑定，限制其网络下载速度。

（2）访客网络权限。

企业一般会有访客前来洽谈业务等，因此，需要对访客设置临时权限，即需要网络管理人员设置对应的 IP 地址分发权限，比如是否可以访问企业内部的其他服务或系统。提前规划好访客权限，避免访客计算机感染病毒导致企业内网信息泄露等安全问题。

2．DHCP 在家庭环境中的使用

家用网络一般会使用路由器或"光纤猫"加路由器的方式连接网络，家庭中也有许多设备（如智能电视、智能灯具、智能监控等）需要连接网络，因此需要在路由器上开启 DHCP 自动获取 IP 地址的功能，如图 4-12 和图 4-13 所示。

图 4-12　家用路由器 DHCP 自动获取 IP 地址

另外，现在家用高端路由器都自带防火墙，可根据实际需求选购。

3．DHCP 在 IDC 机房中的应用

DHCP 在 IDC 机房中的应用场景如下。

DHCP 一般在网络中的 3 层设备核心交换机进行设置。尤其是新建 IDC 时，比如要批量安装 100 多台服务器时，需要使用自动化部署操作系统的服务，如使用 Kickstart+PXE 自动化批量部署操作系统的场景。

图 4-13　DHCP 服务器信息

另外需要注意，在安装完操作系统后关闭 DHCP 功能，以免对其他服务器重启操作系统后自动获取 IP 地址造成影响，因为 IDC 机房内的服务器都是设置的静态 IP 地址，不需要设置动态 IP 地址。

4.3　DHCP 数据包格式

4.3.1　DHCP 的封装

1. DHCP 数据包基本组成

学习 DHCP 不能单纯学习协议本身，还必须了解它的数据包结构。因为 DHCP 是特殊的"服务发现"类型的协议，它用于客户端获取某种资源，所以它的数据包结构具有一定的特殊性。DHCP 数据包结构如图 4-14 所示。

图 4-14　DHCP 数据包结构

DHCP 数据包结构的说明如下。

（1）链路层头：承载报文的链路层消息头，常见的有 Ethernet_II 格式、802.1Q 格式、IEEE 802.3 格式、令牌环链路层头格式等。

（2）IP 头：包括 SrcIP、DstIP 等信息。

（3）UDP 头：包括 SrcPort、DstPort、报文长度及 UDP 校验和等信息。

（4）DHCP 报文：具体的 DHCP 报文内容。

2. DHCP 报文封装规则

DHCP 是初始化协议，即让终端获取 IP 地址的协议。如果终端连 IP 地址都没有，怎么发出 IP 报文呢？服务器给客户端回送的报文该怎么封装呢？为了解决这些问题，DHCP 报文的封装采取了如下规则。

（1）首先链路层的封装必须是广播形式，即让在同一子网中的所有主机都能够收到这个报文。

（2）由于终端目前没有 IP 地址，IP 头中的 SrcIP 规定全填为 0。

（3）当终端发出 DHCP 请求报文时，它并不知道 DHCP 服务器的 IP 地址，因此 IP 头中的 DstIP 填为有限的子网广播 IP——全 1（广播），以保证 DHCP 服务器的 IP 栈不丢弃这个报文。

（4）上面的措施保证了 DHCP 服务器能够收到终端的请求报文，但仅凭链路层和 IP 层信息，DHCP 服务器无法区分出 DHCP 报文。因此终端发出的 DHCP 请求报文的 UDP 头中的 SrcPort 为 68，DstPort 为 67，即 DHCP 服务器通过端口 67 来判断一个报文是否是 DHCP 报文。

（5）DHCP 服务器给终端的响应报文将会根据 DHCP 报文中的内容决定是广播还是单播。一般都是广播形式，广播封装时规则如下。

- 链路层的封装必须是广播形式。
- IP 头中的 DstIP 填为有限的子网广播 IP——全 1（广播）。

单播封装时规则如下。

- 链路层的封装是单播形式（因为客户端在发送 DHCP 报文的时候在链路层帧头填写了 sourceMAC）。
- IP 头中的 DstIP 填为有限的子网广播 IP（全 1），或者是即将分配给用户的 IP 地址（当终端能够接收这样的 IP 报文时）。

两种封装方式中的 UDP 头都是相同的，SrcPort 为 67，DstPort 为 68。终端通过端口 68 来判断一个报文是否是 DHCP 服务器的响应报文。

4.3.2　DHCP 数据包本身的报文格式

在 DHCP 交互过程中，DHCP 数据包本身的报文格式，如图 4-15 和图 4-16 所示。

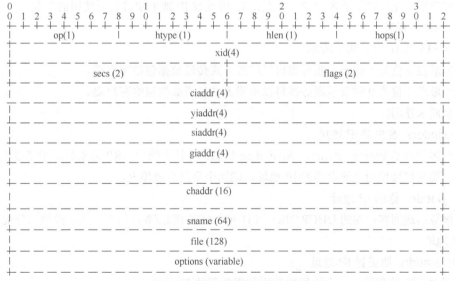

图 4-15　DHCP 数据包报文格式

1．op：封包

（1）客户端发送给服务器的数据包，设为 1。

（2）服务器发送给客户端的数据包，设为 2。

2．htype：硬件地址

表示 10MB/s 的以太网的硬件地址。

```
 0        8        16        24        31
┌─────────┬─────────┬─────────┬─────────┐
│ 封包(op)│硬件类别 │硬件地址 │ 跳数    │
│         │(htype)  │长度(hlen)│(hops)  │
├─────────┴─────────┼─────────┴─────────┤
│          事务ID（xid）                │
├───────────────────┬───────────────────┤
│  请求耗时（secs） │   标志（flags）   │
├───────────────────┴───────────────────┤
│         客户端IP地址（ciaddr）        │
├───────────────────────────────────────┤
│          你的IP地址（yiaddr）         │
├───────────────────────────────────────┤
│         服务器IP地址（siaddr）        │
├───────────────────────────────────────┤
│       中继代理IP地址（giaddr）        │
├───────────────────────────────────────┤
│   客户端硬件地址（chaddr）16字节      │
├───────────────────────────────────────┤
│    服务器的主机名（sname）64字节      │
├───────────────────────────────────────┤
│       启动文件名（file）128字节       │
├───────────────────────────────────────┤
│      允许厂商选项（options）不定长    │
└───────────────────────────────────────┘
```

图 4-16　DHCP 数据包报文格式封装说明

3．hlen：硬件地址长度

以太网为 6 字节。

4．hops：跳数

若封包需经过网关传送，则每"站"跳数加 1；若在同一网内，则跳数为 0（客户端的初始设置为 0）。

5．xid：事务 ID

xid 由客户端设置并由服务器返回，为 32 位整数，以作发送 DHCP Reply 时的依据。

6．secs：请求耗时

由客户端填充，表示从客户端开始获得 IP 地址或 IP 地址续借后所使用的秒数。

7．flags：标志

从 0～15 共 16 位，如下所示。

（1）最左 1 位为 1 时表示服务器将以广播方式传送数据包给客户端。

（2）最左 1 位为 0 时表示服务器将以单播方式传送数据包给客户端。

其余尚未使用。

8．ciaddr：客户端 IP 地址

只有客户端是 Bound、Renew、Rebinding 状态，并且能响应 ARP 请求时，才能被填充。即如果客户端想继续使用之前取得的 IP 地址，则这个字段会被填充。

9．yiaddr：你的 IP 地址

从服务器送回客户端的 DHCP Offer 与 DHCP ACK 消息/数据包中，此字段填写分配给客户端的 IP 地址。

10．siaddr：服务器 IP 地址

表明 DHCP 流程的下一个阶段要使用的服务器的 IP 地址。

11．giaddr：中继代理 IP 地址

若需跨网域进行 DHCP 获取，此栏为中继代理的地址，否则为 0。

12．chaddr：客户端硬件地址

客户端必须设置它的"chaddr"字段。UDP 数据包中的以太网帧首部也有该字段，但通常通过查看 UDP 数据包来确定以太网帧首部中的该字段和获取该值比较困难或者说不可能，而在

4.3 DHCP 数据包格式

UDP 承载的 DHCP 报文中设置该字段，用户进程就可以很容易地获取该值。

13．sname：服务器的主机名

由 DHCP 服务器填写，以 0x00 结尾。

14．file：启动文件名

DHCP Offer 请求中提供有效的目录路径全名。

15．options：允许厂商自定义选项

允许厂商自定义选项以提供更多的设定信息。

格式为 CODE（占 1 字节）+LEN（占 1 字节）+VALUE（长度由 LEN 决定）。

从数据包报文格式中可以看到，DHCP 服务器除了进行 IP 地址分配之外，还会负责给客户端发送 DNS 服务器 IP 地址、默认网关 IP 地址、默认子网掩码之类的信息。

> **注意：** 上述代码中的 n 代表 Option：选项，不定长度，是 DHCP 报文中比较重要的字段。

DHCP 从 BOOTP 拓展而来，DHCP 报文也是由 BOOTP 报文发展而来的。但是 DHCP 在 BOOTP 之上添加了许多功能，其报文也需要有一定的拓展。如果 BOOTP 报文不能满足的内容，就以 Option 的形式存在于 DHCP 报文中。

DHCP 协议其实就是携带许多 Option 的 BOOTP。

DHCP 有许多类型的 Option，长度不一（但都是整数字节）。Option 遵循以下格式。

- 如果 Option 没有值，只有标志位之类的内容，则以一个字节表示。
- 如果 Option 有值，则 Opiton 需要以多个字节表示。

DHCP 支持大量的 Option（BOOTP 也支持其中的一部分）。下面列举一些常用的 DHCP Option（见表 4-1）。

表 4-1　　　　　　　　　　　　　DHCP Option

Option	名称	描述
0	Pad	填充位
1	Subnet Mask	子网掩码
3	Router Address	路由器地址
6	DNS	DNS 服务器
15	DN	域名
50	Requested IP Address	请求的 IP 地址
51	Address Lease Time	地址租约时间
53	DHCP Message Type	DHCP 消息类型，如 Discover、Request、Offer、ACK
54	Server Identifier	服务器标识
55	Parameter Request List	参数请求列表
56	DHCP Error Message	DHCP 错误消息
58	Lease Renewal Time	租约续期时间
59	Lease Rebinding Time	租约重新设定的时间
61	Client Identifier	客户标识
119	Domain Search List	域名查找列表
255	End	结束

在 DHCP Option 中，我们着重看一下 DHCP Message Type。DHCP Message Type 标识 DHCP 报文类型，主要有以下类型，如表 4-2 所示。

表 4-2　DHCP 报文类型

类型	对应的 Option 值
DHCP Discover	1
DHCP Offer	2
DHCP Request	3
DHCP Decline	4
DHCP ACK	5
DHCP NAK	6
DHCP Release	7
DHCP Inform	8

4.3.3　DHCP 报文类型简析

1. DHCP Discover

DHCP 客户端请求地址时，并不知道 DHCP 服务器的位置，因此 DHCP 客户端会在本地网络内以广播方式发送请求报文，这个报文称为 DHCP Discover 报文，目的是发现网络中的 DHCP 服务器。所有收到 DHCP Discover 报文的 DHCP 服务器都会发送 DHCP Offer 报文，DHCP 客户端据此可以知道网络中存在的 DHCP 服务器的位置。

2. DHCP Offer

DHCP 服务器收到 DHCP Discover 报文后，就会在所配置的地址池中查找一个合适的 IP 地址，加上相应的租约期限和其他配置信息（如网关、DNS 服务器等），构造一个 DHCP Offer 报文，发送给 DHCP 客户端（既可以广播，也可以单播），告知 DHCP 客户端本服务器可以为其提供 IP 地址。（注意，只是告诉 DHCP 客户端可以提供 IP 地址，是预分配，还需要 DHCP 客户端通过 ARP 检测该 IP 地址是否重复。）

3. DHCP Request

客户端会在两种情况下发送 DHCP Request 报文。

（1）DHCP 客户端可能会收到来自 DHCP 服务器的很多 DHCP Offer 报文，所以必须在这些 DHCP Offer 报文中选择一个。DHCP 客户端通常选择第一个回应 DHCP Offer 报文的 DHCP 服务器作为自己的目标服务器，并回应一个广播 DHCP Request 报文，通告选择的服务器。注意，"DHCP 客户端通常选择第一个回应 DHCP Offer 报文的 DHCP 服务器作为自己的目标服务器"，这里存在一个安全问题，如果伪 DHCP 服务器能比原始 DHCP 服务器先发送 DHCP Offer 报文，就能达到欺骗的目的，从而劫持目标用户的流量。

（2）DHCP 客户端成功获取 IP 地址后，在地址使用租期过去 1/2 时，会向 DHCP 服务器发送单播 DHCP Request 报文续延租期。如果没有收到 DHCP ACK 报文，在租期过去 3/4 时，发送广播 DHCP Request 报文续延租期。

4. DHCP ACK

DHCP 服务器收到 DHCP Request 报文后，根据 DHCP Request 报文中携带的 MAC 地址来查找有没有相应的租约记录（即之前的预分配过程中登记的那个 MAC 地址），如果有则发送 DHCP ACK 报文作为回应，通知 DHCP 客户端可以使用分配的 IP 地址。

5. DHCP NAK

如果 DHCP 服务器收到 DHCP Request 报文后，没有发现相应的租约记录或者由于某些原因无法正常分配 IP 地址，则发送 DHCP NAK 报文作为回应，通知 DHCP 客户端无法分配合适的 IP 地址。

6. DHCP Release

当 DHCP 客户端不再需要分配 IP 地址时，就会"主动"向 DHCP 服务器发送 DHCP Release 报文，告知 DHCP 服务器 DHCP 客户端不再需要分配 IP 地址。DHCP 服务器会释放被绑定的租约（在数据库中清除某个 MAC 地址对某个 IP 地址的租约记录，这样，这个 IP 地址就可以分配给下一个请求租约的 MAC 地址）。

7. DHCP Decline

DHCP 客户端收到 DHCP 服务器回应的 DHCP ACK 报文后，通过地址冲突检测发现 DHCP 服务器分配的地址冲突或者由于其他原因不能使用，则发送 DHCP Decline 报文，通知 DHCP 服务器所分配的 IP 地址不可用。在手动设置静态 IP 地址或者 DHCP 分配中有时会遇到"检测到 IP 地址冲突"的提示，就是因为客户端利用 ARP 机制在当前内网中确认当前指定的 IP 地址是否已经被占用。

8. DHCP Inform

DHCP 客户端如果需要从 DHCP 服务器获取更为详细的配置信息，则发送 DHCP Inform 报文向服务器进行请求，服务器收到该报文后，将根据租约进行查找，找到相应的配置信息后，发送 DHCP ACK 报文回应 DHCP 客户端。

4.4 DHCP 服务器部署规划

4.4.1 准备 DHCP 服务器基础环境

采用 3 台虚拟机模拟企业实战环境。第 1 台作为 DHCP 服务器，第 2 台和第 3 台作为 DHCP 客户端，其中，第 3 台采用 fixed 的方式固定设置 IP 地址。Linux 操作系统搭建 DHCP 服务器基础环境信息如表 4-3 所示。

表 4-3　　　　　　　　　　DHCP 服务器基础环境信息

角色	操作系统	IP（MAC）地址	主机名
DHCP 服务器	CentOS 6.9	192.168.1.199	dhcp-server
DHCP 客户端	CentOS 6.9	暂无（等待开机后 DHCP 自动分配）	bj-web00
DHCP 客户端	CentOS 6.9	192.168.2.99（00:0C:29:3A:22:B6）	bj-web01

操作系统和内核版本信息如下。

```
[root@dhcp-server ~]# cat /etc/redhat-release
CentOS release 6.9 (Final)
[root@dhcp-server ~]# uname -r
2.6.32-696.el6.x86_64
[root@dhcp-server ~]# getconf LONG_BIT
64
```

实验环境 DHCP 服务器地址池相关信息如图 4-17 所示。

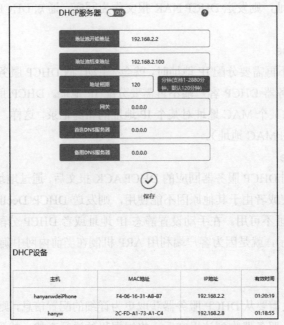

图 4-17 DHCP 服务器地址池相关信息

配置 2 台或 3 台 Linux 虚拟机，第 1 台命名为 www.dhcp-server.com 作为 DHCP 服务器，第 2 台和第 3 台分别命名为 www.bj-web00.com 和 www.bj-web01.com 作为 DHCP 客户端。DHCP 服务器实验架构如图 4-18 所示。

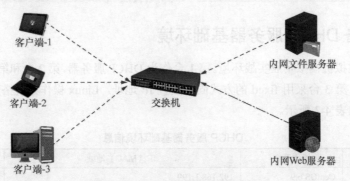

图 4-18 DHCP 服务器实验架构

4.4.2 配置网络环境与防火墙

1. 设置 DHCP 服务器主机名

主机名设置信息如下。

```
tail -1 /etc/hosts
```

查看主机名和 IP 地址最后一行设置信息，代码如下。

```
tail -1 /etc/sysconfig/network
```

查看主机名文件最后一行的设置，代码如下。

```
[root@dhcp-server ~]# tail -1 /etc/hosts
192.168.1.105 www.dhcp-server.com dhcp-server
[root@dhcp-server ~]# tail -1 /etc/sysconfig/network
HOSTNAME=www.dhcp-server.com
```

```
[root@dhcp-server ~]# hostname
www.dhcp-server.com
```

从上述代码执行返回结果中可以看到,当前主机名已经被成功设置为 www.dhcp-server.com。

2. 设置 DHCP 服务器网卡为静态 IP 地址

(1) 确认当前服务器网卡 MAC 地址,代码如下。

```
[root@dhcp-server ~]#ifconfig  |grep HWaddr| awk '{print $NF}'
00:0c:29:cb:ab:82
```

从上述代码可以看到,当前服务器的网卡 MAC 地址为 00:0C:29:CB:AB:82。

(2) 根据网卡配置文件信息确认 MAC 地址,代码如下。

```
[root@dhcp-server ~]#grep 'address' -color
/etc/udev/rules.d/70-persistent-net.rules  |cut -d',' -f4|cut -d'"' -f2
00:0c:29:cb:ab:82
```

上述代码根据网卡的配置文件信息确定当前服务器的网卡 MAC 地址。

```
[root@dhcp-server ~]#awk -F',' '{print $NF,$4}'
/etc/udev/rules.d/70-persistent-net.rules|tail -1|tr '=' ' '|awk -F'"' '{print
"网卡的名称是:================>"$2 "\n" "网卡的MAC地址是:"$4}'
网卡的名称是:================>eth0
网卡的MAC地址是:00:0c:29:cb:ab:82
```

上述代码确定当前操作系统上的网卡名称和 MAC 地址。

(3) 建立网卡的配置文件。

DHCP 服务器需要向客户端分配 IP 地址,服务器自身需要使用一个静态的 IP 地址,代码如下。

```
[root@dhcp-server ~]#cat /etc/sysconfig/network-scripts/ifcfg-eth0
DEVICE=eth0
TYPE=Ethernet
UUID=69d6ec39-9137-4deb-9fcb-f4d7a1945981
ONBOOT=yes
NM_CONTROLLED=yes
BOOTPROTO=static
IPADDR=192.168.1.199
NETMASK=255.255.255.0
GATEWAY=192.168.1.1
DNS1=114.114.114.114
DNS2=8.8.8.8
HWADDR=00:0C:29:CB:AB:82
IPV4_FAILURE_FATAL=yes
NAME="System eth0"
```

eth0 网卡配置文件如上述代码所示。查看并确认 eth0 网卡信息,代码如下。

```
[root@dhcp-server ~]#ifconfig
eth0      Link encap:Ethernet  HWaddr 00:0C:29:CB:AB:82
          inet addr:192.168.1.199  Bcast:192.168.1.255  Mask:255.255.255.0
          inet6 addr: fe80::20c:29ff:fecb:ab82/64 Scope:Link
          UP BROADCAST RUNNING MULTICAST  MTU:1500  Metric:1
          RX packets:547 errors:0 dropped:0 overruns:0 frame:0
          TX packets:307 errors:0 dropped:0 overruns:0 carrier:0
          collisions:0 txqueuelen:1000
          RX bytes:44839 (43.7 KiB)  TX bytes:31445 (30.7 KiB)

lo        Link encap:Local Loopback
          inet addr:127.0.0.1  Mask:255.0.0.0
          inet6 addr: ::1/128 Scope:Host
          UP LOOPBACK RUNNING  MTU:65536  Metric:1
          RX packets:0 errors:0 dropped:0 overruns:0 frame:0
          TX packets:0 errors:0 dropped:0 overruns:0 carrier:0
          collisions:0 txqueuelen:0
          RX bytes:0 (0.0 b)  TX bytes:0 (0.0 b)
```

上述代码中,192.168.1.199 即为 DHCP 服务器的静态 IP 地址。

(4) 重启网络服务。

重启网络服务，使网络配置信息生效，代码如下。

```
[root@dhcp-server ~]#/etc/init.d/network restart
Shutting down interface eth0:                              [  OK  ]
Shutting down loopback interface:                          [  OK  ]
Bringing up loopback interface:                            [  OK  ]
Bringing up interface eth0:  Determining if ip address 192.168.1.199 is already
in use for device eth0...
                                                           [  OK  ]
```

从上述代码可以看到，当前网络配置信息已经生效。

(5) 修改网卡 DNS 服务器地址，代码如下。

```
[root@dhcp-server ~]# cat /etc/resolv.conf
# Generated by NetworkManager
search 8.8.8.8
nameserver 114.114.114.114
```

上述代码已经将 DNS 服务器地址设置为 114.114.114.114。

3. 测试 DHCP 服务器网络连通性

测试 DHCP 服务器网络连通性代码如下。

```
ping 192.168.1.199 -c10
```

使用 ping 指令测试 DHCP 服务器网络的连通性，连续发送 10 次通信探测包，-c10 表示使用 ping 指令 ping10 次。

```
[root@dhcp-server ~]#ping 192.168.1.199 -c10
PING 192.168.1.199 (192.168.1.199) 56(84) bytes of data.
64 bytes from 192.168.1.199: icmp_seq=1 ttl=64 time=0.009 ms
64 bytes from 192.168.1.199: icmp_seq=2 ttl=64 time=0.039 ms
64 bytes from 192.168.1.199: icmp_seq=3 ttl=64 time=0.046 ms
64 bytes from 192.168.1.199: icmp_seq=4 ttl=64 time=0.229 ms
64 bytes from 192.168.1.199: icmp_seq=5 ttl=64 time=0.032 ms
64 bytes from 192.168.1.199: icmp_seq=6 ttl=64 time=0.046 ms
64 bytes from 192.168.1.199: icmp_seq=7 ttl=64 time=0.028 ms
64 bytes from 192.168.1.199: icmp_seq=8 ttl=64 time=0.035 ms
64 bytes from 192.168.1.199: icmp_seq=9 ttl=64 time=0.032 ms
64 bytes from 192.168.1.199: icmp_seq=10 ttl=64 time=0.039 ms

--- 192.168.1.199 ping statistics ---
10 packets transmitted, 10 received, 0% packet loss, time 8999ms
rtt min/avg/max/mdev = 0.009/0.053/0.229/0.059 ms
```

上述代码使用 ping 指令测试 192.168.1.199 的连通性。

4. 测试 DNS 是否正常工作

使用 nslookup 指令测试 DHCP 服务器 DNS 解析是否正常，代码如下。

```
[root@dhcp-server ~]#nslookup www.booxin*.vip
Server:         114.114.114.114
Address:114.114.114.114#53

Non-authoritative answer:
Name:   www.booxin*.vip
Address: 122.114.79.8
[root@dhcp-server ~]#nslookup   www.ziroom*.com
Server:         114.114.114.114
Address:114.114.114.114#53

Non-authoritative answer:
www.ziroom*.com   canonical name = csdc.ziroom*.com.
csdc.ziroom*.com canonical name = gtm-cn-v0h0r3hu405.gtm-a1b2.com.
Name:   gtm-cn-v0h0r3hu405.gtm-a1b2.com
Address: 119.254.76.126
Name:   gtm-cn-v0h0r3hu405.gtm-a1b2.com
```

```
       Address: 124.251.119.64
       Name:    gtm-cn-v0h0r3hu405.gtm-a1b2.com
       Address: 119.254.83.228
       Name:    gtm-cn-v0h0r3hu405.gtm-a1b2.com
       Address: 124.251.119.63
       Name:    gtm-cn-v0h0r3hu405.gtm-a1b2.com
       Address: 119.254.89.46
       Name:    gtm-cn-v0h0r3hu405.gtm-a1b2.com
       Address: 119.254.89.45
       Name:    gtm-cn-v0h0r3hu405.gtm-a1b2.com
       Address: 124.251.119.65
       Name:    gtm-cn-v0h0r3hu405.gtm-a1b2.com
       Address: 124.251.119.62
       Name:    gtm-cn-v0h0r3hu405.gtm-a1b2.com
       Address: 119.254.76.105
```

若系统提示无 nslookup 指令，代码如下。

```
[root@dhcp-server ~]#nslookup www.booxin*.vip
-bash: nslookup: command not found
```

则可以使用如下指令进行相关软件包的安装，然后再使用 nslookup 指令即可。

```
[root@dhcp-server ~]#yum -y install bind-utils
Loaded plugins: fastestmirror
Setting up Install Process
Determining fastest mirrors
epel/metalink
| 9.4 kB     00:00
 * base: mirrors.aliyun.com
 * epel: mirrors.yun-idc.com
 * extras: mirrors.tuna.tsinghua.edu.cn
 * updates: mirrors.aliyun.com
base
//中间代码略

Installed:
  bind-utils.x86_64 32:9.8.2-0.68.rc1.el6_10.1

Dependency Installed:
  bind-libs.x86_64 32:9.8.2-0.68.rc1.el6_10.1

Complete!
```

确定 nslookup 指令所在位置，并确定其属于哪个软件包，代码如下。

```
[root@dhcp-server ~]#which nslookup
/usr/bin/nslookup
[root@dhcp-server ~]#rpm -qf /usr/bin/nslookup
bind-utils-9.8.2-0.68.rc1.el6_10.1.x86_64
```

查看 bind-utils 软件包安装了哪些实用工具，代码如下。

```
[root@dhcp-server ~]#rpm -ql bind-utils|head
/usr/bin/dig
/usr/bin/host
/usr/bin/nslookup
/usr/bin/nsupdate
/usr/share/man/man1/dig.1.gz
/usr/share/man/man1/host.1.gz
/usr/share/man/man1/nslookup.1.gz
/usr/share/man/man1/nsupdate.1.gz
```

5．设置 DHCP 防火墙

（1）关闭 SELinux，代码如下。

```
[root@dhcp-server ~]# grep --color '^\<SELINUX\>' /etc/sysconfig/selinux
SELINUX=disabled
[root@dhcp-server ~]# sestatus
SELINUX status:              disabled
```

上述代码关闭 SELinux 并确认 SELinux 状态信息。

(2) 关闭防火墙，代码如下。
```
[root@dhcp-server ~]#  /etc/init.d/iptables stop
[root@dhcp-server ~]# chkconfig --level 2345 iptables off
[root@dhcp-server ~]# chkconfig --list |grep iptables
iptables        0:off   1:off   2:off   3:off   4:off   5:off   6:off
```
上述代码确认防火墙是否关闭。

6. DHCP 服务器规划总结

在安装 DHCP 服务器之前，需要规划以下信息。
- 确定 DHCP 服务器应分发给客户端的 IP 地址范围。
- 为客户端确定正确的子网掩码。
- 确定 DHCP 服务器不应向客户端分发的所有 IP 地址。（保留一些固定 IP 地址提供给服务器等使用。）
- 确定 IP 地址的租用期限，默认值为 8 天。

通常，租用期限应等于该客户端的平均活动时间。例如，如果客户端是很少关闭的桌面计算机，理想的期限可以比 8 天长；如果客户端是经常离开网络或在子网之间移动的移动设备，该期限可以少于 8 天。

4.4.3 配置 DHCP 客户端环境信息

根据规划提前配置好 DHCP 客户端的环境信息，如下所示。

1. 客户端 1 基础环境信息

客户端 1 基础环境信息如下。
```
[root@www.bj-web01.com ~]#ifconfig  |grep HW
eth0      Link encap:Ethernet   HWaddr 00:0C:29:24:ED:44
[root@www.bj-web01.com ~]#hostname
www.bj-web01.com
[root@www.bj-web01.com ~]#sestatus
SELinux status:                 disabled
[root@www.bj-web01.com ~]#/etc/init.d/iptables status
iptables: Firewall is not running.
[root@www.bj-web01.com ~]#
[root@www.bj-web01.com ~]#cat /etc/redhat-release
CentOS release 6.9 (Final)
[root@www.bj-web01.com ~]#uname -rnm
www.bj-web01.com 2.6.32-696.el6.x86_64 x86_64
```
上述代码设置客户端 1 的主机名和防火墙等基本信息。

2. 客户端 2 基础环境信息

客户端 2 基础环境信息如下。
```
[root@www.bj-web00.com ~]#ifconfig  |grep HW
eth0      Link encap:Ethernet   HWaddr 00:0C:29:01:68:35
[root@www.bj-web00.com ~]#hostname
www.bj-web00.com
[root@www.bj-web00.com ~]#sestatus
SELinux status:                 disabled
[root@www.bj-web00.com ~]#/etc/init.d/iptables status
iptables: Firewall is not running.
[root@www.bj-web00.com ~]#cat /etc/redhat-release
CentOS release 6.9 (Final)
[root@www.bj-web00.com ~]#uname -rnm
www.bj-web00.com 2.6.32-696.el6.x86_64 x86_64
```
上述代码设置客户端 2 的主机名和防火墙等基本信息。

3. 配置客户端网络信息

DHCP 客户端分别配置 eth0 网卡，采用 dhcp 自动获取 IP 地址。

```
[root@www.bj-web01.com ~]#grep dhcp /etc/sysconfig/network-scripts/ifcfg-eth0
BOOTPROTO=dhcp
[root@www.bj-web00.com ~]#grep dhcp /etc/sysconfig/network-scripts/ifcfg-eth0
BOOTPROTO=dhcp
```

从上述代码可以看到，两个客户端都已经开通了 DHCP，网卡将动态获取 IP 地址。

4.5 CentOS 搭建 DHCP 服务器实战

4.5.1 DHCP 服务器基本配置

1. DHCP 服务器基本环境信息

搭建 DHCP 服务器基本环境信息如下。

- 作用：自动为客户分配 TCP/IP 参数。
- 确认已经安装了 epel 第三方 YUM 源。
- 软件：dhcp.x86_64、dhcp-devel.x86_64。
- 端口：UDP 67（DHCP 服务器）、68（DHCP 客户端）。
- 配置文件：/etc/dhcp/dhcpd.conf。
- 服务名称：dhcpd。

2. 安装 DHCP 软件

（1）检查 DHCP 软件包是否安装。

安装之前首先使用 rpm –qa | grep dhcp 查看系统中是否已安装了 DHCP 软件包，代码如下。

```
[root@dhcp-server ~]# rpm -qa | grep dhcp
dhcp-common-4.1.1-53.P1.el6.centos.x86_64
```

首先需要安装 DHCP 的软件包。用 yum 指令进行安装，即通过 YUM 服务器安装 DHCP 服务器软件，代码如下。

```
yum -y install dhcp
```

上述代码表示采用自动应答方式安装 DHCP 软件包。

```
[root@dhcp-server ~]# yum -y install dhcp
Loaded plugins: fastestmirror
Setting up Install Process
Loading mirror speeds from cached hostfile
epel/metalink                                              | 6.7 kB     00:00
 * base: mirror.bit.edu.cn
 * epel: mirrors.tuna.tsinghua.edu.cn
 * extras: mirror.bit.edu.cn
 * updates: mirror.bit.edu.cn
base                                                       | 3.7 kB     00:00
extras                                                     | 3.4 kB     00:00
updates                                                    | 3.4 kB     00:00
//中间代码略

Installed:
  dhcp.x86_64 12:4.1.1-53.P1.el6.centos.1

Dependency Updated:
  dhclient.x86_64 12:4.1.1-53.P1.el6.centos.1      dhcp-common.x86_64
```

```
12:4.1.1-53.P1.el6.centos.1

Complete!
```

(2) 确认 DHCP 软件包是否安装，代码如下。

```
[root@dhcp-server ~]# rpm -qa | grep 'dhcp*'
dhcp-4.1.1-53.P1.el6.centos.1.x86_64
dhcp-common-4.1.1-53.P1.el6.centos.1.x86_64
dhclient-4.1.1-53.P1.el6.centos.1.x86_64
```

从上述代码可以看出，DHCP 软件包已经被成功安装。

3. 查看 DHCP 软件包组件信息

查看 DHCP 软件包安装了哪些组件，代码如下。

```
[root@dhcp-server ~]# rpm -ql dhcp|nl
     1  /etc/dhcp
     2  /etc/dhcp/dhcpd.conf
     3  /etc/dhcp/dhcpd6.conf
     4  /etc/openldap/schema/dhcp.schema
     5  /etc/portreserve/dhcpd
     6  /etc/rc.d/init.d/dhcpd
     7  /etc/rc.d/init.d/dhcpd6
     8  /etc/rc.d/init.d/dhcrelay
     9  /etc/rc.d/init.d/dhcrelay6
    10  /etc/sysconfig/dhcpd
    11  /etc/sysconfig/dhcpd6
    12  /etc/sysconfig/dhcrelay
    13  /etc/sysconfig/dhcrelay6
    14  /usr/bin/omshell
    15  /usr/sbin/dhcpd
    16  /usr/sbin/dhcrelay
    17  /usr/share/doc/dhcp-4.1.1
    18  /usr/share/doc/dhcp-4.1.1/3.0b1-lease-convert
    19  /usr/share/doc/dhcp-4.1.1/IANA-arp-parameters
    20  /usr/share/doc/dhcp-4.1.1/README.ldap
    21  /usr/share/doc/dhcp-4.1.1/api+protocol
    22  /usr/share/doc/dhcp-4.1.1/dhclient-tz-exithook.sh
    23  /usr/share/doc/dhcp-4.1.1/dhcpd-conf-to-ldap
    24  /usr/share/doc/dhcp-4.1.1/dhcpd.conf.sample
    25  /usr/share/doc/dhcp-4.1.1/dhcpd6.conf.sample
    26  /usr/share/doc/dhcp-4.1.1/draft-ietf-dhc-ldap-schema-01.txt
    27  /usr/share/doc/dhcp-4.1.1/ms2isc
    28  /usr/share/doc/dhcp-4.1.1/ms2isc/Registry.perlmodule
    29  /usr/share/doc/dhcp-4.1.1/ms2isc/ms2isc.pl
    30  /usr/share/doc/dhcp-4.1.1/ms2isc/readme.txt
    31  /usr/share/doc/dhcp-4.1.1/sethostname.sh
    32  /usr/share/doc/dhcp-4.1.1/solaris.init
    33  /usr/share/man/man1/omshell.1.gz
    34  /usr/share/man/man5/dhcpd.conf.5.gz
    35  /usr/share/man/man5/dhcpd.leases.5.gz
    36  /usr/share/man/man8/dhcpd.8.gz
    37  /usr/share/man/man8/dhcrelay.8.gz
    38  /var/lib/dhcpd
    39  /var/lib/dhcpd/dhcpd.leases
    40  /var/lib/dhcpd/dhcpd6.leases
```

从上述代码第 1~34 行可以看到，安装 DHCP 软件包后，组件众多，下面将重点介绍和 DHCP 服务器有关的核心组件。

4. DHCP 的相关配置文件说明

DHCP 的相关配置文件如下。

（1）DHCP 服务器端口号信息。

- IPv4：UDP 67（源端口：接收客户端请求）、UDP 68（目标端口：向客户端发送请求成功或失败的回应)。
- IPv6：UDP 546、UDP 547。

(2) DHCP 服务器名称。

DHCP 服务器名称分别为 dhcpd 和 dhcrelay。

(3) DHCP 主配置文件。

DHCP 默认主配置文件为/etc/dhcp/dhcpd.conf。

(4) DHCP 模板配置文件。

DHCP 模板配置文件为/usr/share/doc/dhcp-4.1.1/dhcpd.conf.sample。

(5) DHCP 中继配置文件。

DHCP 中继配置文件为/etc/sysconfig/dhcrelay。

(6) DHCP 执行程序。

DHCP 执行程序为/usr/sbin/dhcpd 与/usr/sbin/dhcrelay。

(7) DHCP 服务器脚本。

DHCP 服务器脚本为/etc/rc.d/init.d/dhcpd 与/etc/rc.d/init.d/dhcrelay。

(8) DHCP 执行参数配置文件。

DHCP 执行参数配置文件为/etc/sysconfig/dhcpd。

(9) DHCP 服务器租约文件。

DHCP 服务器租约文件为/var/lib/dhcpd/dhcpd.leases。

(10) DHCP 服务器日志记录文件。

DHCP 服务器默认日志记录文件为/var/log/messages。

5. 配置 DHCP 服务器

(1) 查看 DHCP 配置文件，代码如下。

```
[root@dhcp-server ~]# cat /etc/dhcp/dhcpd.conf
#
# DHCP Server Configuration file.
#   see /usr/share/doc/dhcp*/dhcpd.conf.sample
#   see 'man 5 dhcpd.conf'
#
```

DHCP 服务器主要配置文件是 dhcpd.conf，默认情况下该文件不存在，用户安装完 DHCP 服务器后会自动生成该文件。

(2) 查看 DHCP 服务器模板配置文件。

查看 DHCP 服务器中所有包含 sample 的文件和详细路径，代码如下。

```
[root@dhcp-server ~]# rpm -ql dhcp|grep sample
/usr/share/doc/dhcp-4.1.1/dhcpd.conf.sample
/usr/share/doc/dhcp-4.1.1/dhcpd6.conf.sample
```

可以看到有 3 个样例文件，后面将采用样例文件作为模板和 DHCP 服务器的配置文件，并根据企业实际业务需求进行调整和修改。

(3) 复制 DHCP 模板文件作为 DHCP 配置文件。

复制 DHCP 模板文件来配置 DHCP，代码如下。

```
/bin/cp -av /usr/share/doc/dhcp-4.1.1/dhcpd.conf.sample  /etc/dhcp/dhcpd.conf
```

将/usr/share/doc/dhcp-4.1.1/dhcpd.conf.sample 配置文件去掉注释和空行并重定向到/etc/dhcp/dhcpd.conf 文件中，具体代码如下。

```
[root@dhcp-server ~]# cp -av /usr/share/doc/dhcp-4.1.1/dhcpd.conf.sample
/usr/share/doc/dhcp-4.1.1/dhcpd.conf.sample_bak
'/usr/share/doc/dhcp-4.1.1/dhcpd.conf.sample' -> '/usr/share/doc/dhcp-4.1.1
/dhcpd.conf.sample_bak
[root@localhost ~]# egrep -v "#|^$" /usr/share/doc/dhcp-4.1.1/dhcpd.conf.
sample > /etc/dhcp/dhcpd.conf
```

使用 cp 指令成功将样例文件复制为 DHCP 服务器主配置文件。

6. 编辑 DHCP 配置文件

(1) 简化/etc/dhcp/dhcpd.conf 配置文件，代码如下。

```
ddns-update-style interim;
subnet 192.168.1.0 netmask 255.255.255.0{
range 192.168.1.120 192.168.1.198;
default-lease-time  600;
max-lease-time 7200;
option routers 192.168.1.1;
option broadcast-address 192.168.1.255;
option domain-name-servers 8.8.8.8;
option domain-name "www.booxin*.vip";
#next-server 192.168.1.1;
#filename "pxelinux.0";
host www.bj-web01.com
   {
        hardware   ethernet   00:0C:29:24:ED:44;
        fixed-address 192.168.1.198;
   }
}
```

上述代码为 DHCP 服务器的简化配置文件。

(2) 查看 dhcpd.conf 配置文件，代码如下。

```
[root@dhcp-server ~]#cat /etc/dhcp/dhcpd.conf
ddns-update-style interim;
subnet  192.168.2.0  netmask 255.255.255.0{
range 192.168.2.3  192.168.2.99;
default-lease-time   600;
max-lease-time  7200;
option routers 192.168.2.1;
option broadcast-address 192.168.2.255;
#option domain-name-servers 8.8.8.8;
#option domain-name "www.booxin*.vip";
#next-server 192.168.1.1;
#filename "pxelinux.0";
host www.bj-web00.com
   {
        hardware   ethernet   00:0C:29:3A:22:B6;
        fixed-address 192.168.2.99;
   }
}
```

上述代码为 DHCP 服务器最终的配置文件。

上述代码提供的 IP 地址段为 192.168.2.3~192.168.2.100，其中 192.168.2.100 为 DHCP 服务器地址，192.168.2.99 为 DHCP 服务器分配给 DHCP 客户端（MAC 地址为 00:0C:29:3A: 22:B6）的静态 IP 地址。

(3) 检测 dhcpd.conf 配置文件语法，代码如下。

```
[root@dhcp-server ~]# /etc/init.d/dhcpd configtest
Syntax: OK
```

从上述代码可以看到，DHCP 服务主配置文件语法准确无误。

7. DHCP 服务配置参数

DHCP 服务常用配置参数如表 4-4 所示。

表 4-4　DHCP 服务常用配置参数

参数	说明
ddns-update-style	定义所支持的 DNS 动态更新类型
range	指定可分配的 IP 地址池
option domain-name-servers	设置指定域名服务器
option routers	设置网关地址
option broadcast-address	设置广播地址
option subnet-mask	设置客户端的子网掩码
host	保留主机，作用于单个主机
default-lease-time	设置默认的租约
max-lease-time	最大的租约时间
hardware ethernet	指定对应主机的 MAC 地址
fixed-address	指定为该主机保留的 IP 地址

说明：如果想为 DHCP 服务器配置不同网段的 IP 地址，前提是必须有一个和本地 IP 地址同网段的声明。

4.5.2　DHCP 服务器常用操作

1. 启动和关闭 DHCP 服务器

（1）设定系统重启后自动启动 DHCP 服务器。

在当前 Runlevel 运行级别上打开 DHCP 服务开关，代码如下。

```
chkconfig dhcpd on
chkconfig  dhcpd on/off
```

开机自动启动或关闭自动启动 DHCP 服务，验证设置结果是否正确。

```
[root@dhcp-server ~]# chkconfig dhcpd on
[root@dhcp-server ~]# chkconfig --list |grep dhcp
dhcpd           0:off   1:off   2:on    3:on    4:on    5:on    6:off
dhcpd6          0:off   1:off   2:off   3:off   4:off   5:off   6:off
```

从上述代码可以看到，DHCP 服务已经成功加入开机自动启动服务中，会随着服务器电源的打开而自动启动 DHCP 服务。

（2）启动 DHCP 服务器，代码如下。

```
[root@dhcp-server ~]# /etc/init.d/dhcpd start
Starting dhcpd:                                            [  OK  ]
```

从上述代码可以看到，已经成功启动 DHCP 服务器。

（3）关闭 DHCP 服务器，代码如下。

```
[root@dhcp-server ~]# service dhcpd stop
Shutting down dhcpd:                                       [  OK  ]
```

从上述代码可以看到，已经成功关闭 DHCP 服务器。

（4）重启 DHCP 服务器，代码如下。

```
[root@dhcp-server ~]# service dhcpd restart
Starting dhcpd:                                            [  OK  ]
```

从上述代码可以看到，已经成功重启 DHCP 服务器。

2. 查看 DHCP 服务器监听端口

查看 DHCP 服务器监听端口，代码如下。

```
[root@dhcp-server ~]# netstat -aunp| grep 67
udp        0      0 0.0.0.0:67              0.0.0.0:*                           1575/dhcpd
```

从上述代码可以看到，DHCP 服务器已经成功监听 UDP 服务的 67 端口。

3. DHCP 服务器启动日志

查看 DHCP 服务器启动日志信息，代码如下。

```
[root@dhcp-server ~]#/etc/init.d/dhcpd start;tailf /var/log/messages
Starting dhcpd:                                              [  OK  ]
[root@dhcp-server ~]#tailf /var/log/messages
Sep 15 20:21:20 dhcp-server dhcpd: All rights reserved.
Sep 15 20:21:20 dhcp-server dhcpd: For info, please visit
Sep 15 20:21:20 dhcp-server dhcpd: WARNING: Host declarations are global.
They are not limited to the scope you declared them in.
Sep 15 20:21:20 dhcp-server dhcpd: Not searching LDAP since ldap-server,
ldap-port and ldap-base-dn were not specified in the config file
Sep 15 20:21:20 dhcp-server dhcpd: Wrote 0 deleted host decls to leases file.
Sep 15 20:21:20 dhcp-server dhcpd: Wrote 0 new dynamic host decls to leases file.
Sep 15 20:21:20 dhcp-server dhcpd: Wrote 2 leases to leases file.
Sep 15 20:21:20 dhcp-server dhcpd: Listening on
LPF/eth0/00:0c:29:49:00:5c/192.168.2.0/24
Sep 15 20:21:20 dhcp-server dhcpd: Sending on
LPF/eth0/00:0c:29:49:00:5c/192.168.2.0/24
Sep 15 20:21:20 dhcp-server dhcpd: Sending on
Socket/fallback/fallback-net
```

从上述代码可以看到，DHCP 服务器已经成功启动，并等待客户端进一步的响应。

4. DHCP 服务器日志

查看 DHCP 服务器日志，内容如下。

```
[root@dhcp-server ~]#tailf /var/log/messages
Sep 15 20:21:20 dhcp-server dhcpd: All rights reserved.
Sep 15 20:21:20 dhcp-server dhcpd: For info, please visit
Sep 15 20:21:20 dhcp-server dhcpd: WARNING: Host declarations are global.
They are not limited to the scope you declared them in.
Sep 15 20:21:20 dhcp-server dhcpd: Not searching LDAP since ldap-server,
ldap-port and ldap-base-dn were not specified in the config file
Sep 15 20:21:20 dhcp-server dhcpd: Wrote 0 deleted host decls to leases file.
Sep 15 20:21:20 dhcp-server dhcpd: Wrote 0 new dynamic host decls to leases file.
Sep 15 20:21:20 dhcp-server dhcpd: Wrote 2 leases to leases file.
Sep 15 20:21:20 dhcp-server dhcpd: Listening on LPF/eth0/00:0c:29:49:00:5c
/192.168.2.0/24
Sep 15 20:21:20 dhcp-server dhcpd: Sending on   LPF/eth0/00:0c:29:49:00:5c
/192.168.2.0/24
Sep 15 20:21:20 dhcp-server dhcpd: Sending on   Socket/fallback/fallback-net

Sep 15 20:21:35 dhcp-server dhcpd: DHCPREQUEST for 192.168.2.4 from 00:0c:
29:3a:22:b6 via eth0: lease 192.168.2.4 unavailable.
Sep 15 20:21:35 dhcp-server dhcpd: DHCPNAK on 192.168.2.4 to 00:0c:29:3a:22:
b6 via eth0
Sep 15 20:21:35 dhcp-server dhcpd: DHCPDISCOVER from 00:0c:29:3a:22:b6 via eth0
Sep 15 20:21:35 dhcp-server dhcpd: DHCPoffer on 192.168.2.99 to 00:0c:29:3a:22:
b6 via eth0
Sep 15 20:21:35 dhcp-server dhcpd: Dynamic and static leases present for
192.168.2.99.
Sep 15 20:21:35 dhcp-server dhcpd: Remove host declaration www.bj-web00.com
or remove 192.168.2.99
Sep 15 20:21:35 dhcp-server dhcpd: from the dynamic address pool for 192.168.
2.0/24
Sep 15 20:21:35 dhcp-server dhcpd: DHCPREQUEST for 192.168.2.99 (192.168.2.100)
from 00:0c:29:3a:22:b6 via eth0
Sep 15 20:21:35 dhcp-server dhcpd: DHCPACK on 192.168.2.99 to 00:0c:29:3a:22:
b6 via eth0
Sep 15 20:24:18 dhcp-server dhcpd: DHCPREQUEST for 192.168.2.3 from 00:0c:29:
2b:2a:62 via eth0
Sep 15 20:24:18 dhcp-server dhcpd: DHCPACK on 192.168.2.3 to 00:0c:29:2b:2a:
62 (www) via eth0
Sep 15 20:26:19 dhcp-server dhcpd: Dynamic and static leases present for
```

```
                                    192.168.2.99.
    Sep 15 20:26:19 dhcp-server dhcpd: Remove host declaration www.bj-web00.com
    or remove 192.168.2.99
    Sep 15 20:26:19 dhcp-server dhcpd: from the dynamic address pool for
    192.168.2.0/24
    Sep 15 20:26:19 dhcp-server dhcpd: DHCPREQUEST for 192.168.2.99 from 00:0c:29:
    3a:22:b6 via eth0
    Sep 15 20:26:19 dhcp-server dhcpd: DHCPACK on 192.168.2.99 to 00:0c:29:3a:22:
    b6 via eth0
    Sep 15 20:30:08 dhcp-server dhcpd: Dynamic and static leases present for
    192.168.2.99.
    Sep 15 20:30:08 dhcp-server dhcpd: Remove host declaration www.bj-web00.com
    or remove 192.168.2.99
    Sep 15 20:30:08 dhcp-server dhcpd: from the dynamic address pool for 192.168.
    2.0/24
    Sep 15 20:30:08 dhcp-server dhcpd: DHCPREQUEST for 192.168.2.99 from 00:0c:29:
    3a:22:b6 via eth0
    Sep 15 20:30:08 dhcp-server dhcpd: DHCPACK on 192.168.2.99 to 00:0c:29:3a:22:
    b6 via eth0
```

上述代码记录 DHCP 服务器与客户端交互的详细信息，查看 DHCP 服务器日志，有助于理解 DHCP 服务器与 DHCP 客户端的交互。

4.6 DHCP 客户端测试

4.6.1 DHCP 客户端测试注意事项

1. 确保 DHCP 已经开启

客户端确保 DHCP 已经开启，并核对 MAC 地址与服务器配置是否一致、是否准确无误。

2. 观察 DHCP 服务器日志

实时观察 DHCP 服务器日志，注意服务器与客户端的交互过程。

```
1   [root@dhcp-server ~]#tailf /var/log/messages
2   Sep 15 20:21:20 dhcp-server dhcpd: All rights reserved.
3   Sep 15 20:21:20 dhcp-server dhcpd: For info, please visit
4   Sep 15 20:21:20 dhcp-server dhcpd: WARNING: Host declarations are global.
    They are not limited to the scope you declared them in.
5   Sep 15 20:21:20 dhcp-server dhcpd: Not searching LDAP since ldap-server,
    ldap-port and ldap-base-dn were not specified in the config file
6   Sep 15 20:21:20 dhcp-server dhcpd: Wrote 0 deleted host decls to leases file.
7   Sep 15 20:21:20 dhcp-server dhcpd: Wrote 0 new dynamic host decls to leases file.
8   Sep 15 20:21:20 dhcp-server dhcpd: Wrote 2 leases to leases file.
9   Sep 15 20:21:20 dhcp-server dhcpd: Listening on LPF/eth0/00:0c:29:49:00:5c/
    192.168.2.0/24
10  Sep 15 20:21:20 dhcp-server dhcpd: Sending on   LPF/eth0/00:0c:29:49:00:5c/
    192.168.2.0/24
11  Sep 15 20:21:20 dhcp-server dhcpd: Sending on   Socket/fallback/fallback-net
12
13  Sep 15 20:21:35 dhcp-server dhcpd: DHCPREQUEST for 192.168.2.4 from 00:0c:
    29:3a:22:b6 via eth0: lease 192.168.2.4 unavailable.
14  Sep 15 20:21:35 dhcp-server dhcpd: DHCPNAK on 192.168.2.4 to 00:0c:29:3a:
    22:b6 via eth0
15  Sep 15 20:21:35 dhcp-server dhcpd: DHCPDISCOVER from 00:0c:29:3a:22:b6 via eth0
16  Sep 15 20:21:35 dhcp-server dhcpd: DHCPoffer on 192.168.2.99 to 00:0c:29:
    3a:22:b6 via eth0
17  Sep 15 20:21:35 dhcp-server dhcpd: Dynamic and static leases present for
    192.168.2.99.
18  Sep 15 20:21:35 dhcp-server dhcpd: Remove host declaration
    www.bj-web00.com or remove 192.168.2.99
```

```
19  Sep 15 20:21:35 dhcp-server dhcpd: from the dynamic address pool for
    192.168.2.0/24
20  Sep 15 20:21:35 dhcp-server dhcpd: DHCPREQUEST for 192.168.2.99
    (192.168.2.100) from 00:0c:29:3a:22:b6 via eth0
21  Sep 15 20:21:35 dhcp-server dhcpd: DHCPACK on 192.168.2.99 to 00:0c:29:
    3a:22:b6 via eth0
```

上述代码中第 6~21 行为客户端与服务器交互信息。

3. 观察 DHCP 客户端日志

实时观察 DHCP 客户端日志，注意与服务器的交互过程，根据日志信息判断是否交互成功。

4.6.2 DHCP 客户端测试步骤

1. 修改网卡配置文件

修改网卡配置文件为自动获取，代码如下：

```
1   [root@www.bj-web01.com ~]#cat /etc/sysconfig/network-scripts/ifcfg-eth0
2   DEVICE=eth0
3   TYPE=Ethernet
4   UUID=69d6ec39-9137-4deb-9fcb-f4d7a1945981
5   ONBOOT=yes
6   BOOTPROTO=dhcp
7   NETMASK=255.255.255.0
8   GATEWAY=192.168.2.1
9   DNS1=114.114.114.114
10  DNS2=8.8.8.8
11  IPV4_FAILURE_FATAL=yes
12  NAME="System eth0"
```

上述代码的第 5 行设置网卡开机启动，第 6 行配置自动获取 IP 地址。

2. 重启网络服务

重启网络服务，代码如下：

```
[root@www.bj-web01.com ~]#/etc/init.d/network restart
Shutting down interface eth0:                              [  OK  ]
Shutting down loopback interface:                          [  OK  ]
Bringing up loopback interface:                            [  OK  ]
Bringing up interface eth0:
Determining IP information for eth0... done.
                                                           [  OK  ]
```

上述代码重启网络服务，以便向 DHCP 服务器请求动态 IP 地址。

3. 验证

DHCP 客户端自动获取 IP 地址验证如图 4-19 所示。

DHCP 客户端获取 DNS 服务器信息如图 4-20 所示。

图 4-19　DHCP 客户端自动获取 IP 地址验证 　　图 4-20　DHCP 客户端获取 DNS 服务器信息

4.6.3 DHCP 运维常用文件/程序/脚本

1. DHCP 软件包的主要文件

在主配置文件 dhcpd.conf 中查看详细信息，代码如下。

```
[root@linuxidc ~]# ls -ltr /etc/dhcp/dhcpd.conf
-rw-r--r--. 1 root root 3260 Apr 12 15:10 /etc/dhcp/dhcpd.conf
```

2. DHCP 执行程序

DHCP 执行程序/usr/sbin/dhcpd、/usr/sbin/dhcrelay，代码如下。

```
[root@dhcp-server ~]# ls -lhrt /usr/sbin/{dhcpd,dhcrelay}
-rwxr-xr-x 1 root root 459K Oct  9  2018 /usr/sbin/dhcrelay
-rwxr-xr-x 1 root root 811K Oct  9  2018 /usr/sbin/dhcpd
```

3. DHCP 服务脚本

DHCP 服务脚本/etc/init.d/dhcpd、/etc/init.d/dhcrelay，代码如下。

```
[root@dhcp-server ~]# ls -lhrt /etc/init.d/{dhcpd,dhcrelay}
-rwxr-xr-x 1 root root 2.5K Sep 12  2018 /etc/init.d/dhcrelay
-rwxr-xr-x 1 root root 3.1K Sep 12  2018 /etc/init.d/dhcpd
```

上述代码查看 DHCP 服务脚本。

第 5 章

vsftpd 服务

5.1　FTP 基础知识

文件传送协议（File Transfer Protocol，FTP）是一个非常经典并且应用十分广泛的文件传输协议。

FTP 底层通过 TCP 来作为传输协议，所以 FTP 是一种可靠的文件传输方式。FTP 服务提供了两个端口号：20 和 21。20 端口是数据端口，提供数据之间的传输；21 端口是命令端口，提供命令之间的传输与交互。

FTP 服务器与客户端连接一般有两种模式：主动模式（Active Mode）和被动模式（Passive Mode）。

5.1.1　FTP 服务主动模式

FTP 服务主动模式工作流程如图 5-1 所示。

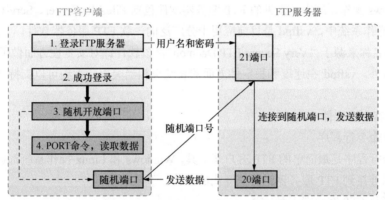

图 5-1　FTP 服务主动模式工作流程

主动模式下，FTP 客户端首先会向服务器的 21 端口发出一个连接命令，请求与服务器建立连接。此时服务器响应客户端，并要求客户端发送一个用于传送数据的端口，该端口号要大于 1023，此时服务端的 20 端口会与该数据端口主动建立连接，进行客户端与服务器的数据传送。

5.1.2　FTP 服务被动模式

FTP 服务被动模式工作流程如图 5-2 所示。

与主动模式不同的是，在被动模式下，客户端首先与服务器的 21 端口建立连接，服务器开启一个大于 1023 的数据传送端口，并返回给客户端。此时客户端也会开启一个大于 1023 的端口，然后客户端会主动与服务器的数据传输端口建立连接，两者之间进行数据的传送。

主动模式与被动模式的区别就在于究竟是服务器的 20 端口主动与客户端建立连接，还是服务器开放一个随机端口，等待客户端与其主动建立连接。

实际生产环境中，通常使用的是被动模式。因为服务器都有防火墙，而防火墙对于内网连接外网的端口一般是放行的，而对于外网连接内网的端口则一般是有限制的，所以如果使用主动模式连接，端口可能被防火墙拦截，从而不能为我们提供 FTP 服务。

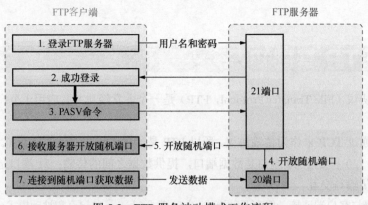

图 5-2　FTP 服务被动模式工作流程

5.1.3　FTP 软件种类

1．FTP 服务器软件

在 Windows 操作系统中，常见的 FTP 服务器软件包括 FileZilla Server、Serv-U、IIS 等。在 Linux/UNIX 操作系统中，vsftpd 是目前应用十分广泛的一款 FTP 服务器软件。

vsftpd 的名称来源于"Very Secure FTP Daemon"，该软件针对安全性方面做了大量的设计。除了安全性以外，vsftpd 在速度和稳定性方面的表现也相当突出，大约可以支持 15000 个用户并发连接。

2．FTP 客户端软件

（1）FTP 命令行程序。

FTP 命令行程序是最简单的 FTP 客户端工具，Windows 和 Linux 操作系统都拥有 FTP 命令行程序，可以连接到 FTP 服务器进行交互式的上传、下载通信。

（2）图形化 FTP 客户端工具。

Windows 操作系统中较常用的图形化工具包括 CuteFTP、FlashFXP、LeapFTP、FileZilla 等。在图形化的客户端程序中，用户通过鼠标和菜单即可访问、管理 FTP 资源，而不需要掌握 FTP 交互命令，更易于使用。

（3）FTP 下载工具。

FTP 下载工具包含 FlashGet、Wget 等，包括大多数网页浏览器程序，都支持通过 FTP 下载文件，但因为不具备 FTP 上传管理功能，通常不被称为 FTP 客户端工具。

5.1.4　FTP 服务器与客户端选型

1．FTP 服务器选型

（1）Serv-U。

Serv-U 是一种被广泛运用的 FTP 服务器软件，它具有非常完备的安全特性，支持 SSL FTP 传输，支持在多个 Serv-U 和 FTP 客户端中通过 SSL 加密连接保护数据安全等。

（2）FileZilla。

FileZilla 是一个免费开源的 FTP 软件，分为客户端版本和服务器版本，具备所有的 FTP 软件功能。可控、有条理的界面和管理多站点的简化方式使得 FileZilla 客户端成为一个方便、高效的工具，而 FileZilla 服务器则是一个小巧并且可靠的支持 FTP 和 SFTP 的 FTP 服务器软件。

如图 5-3 所示，FileZilla 可以支持多个平台，Windows、Linux、macOS 等。

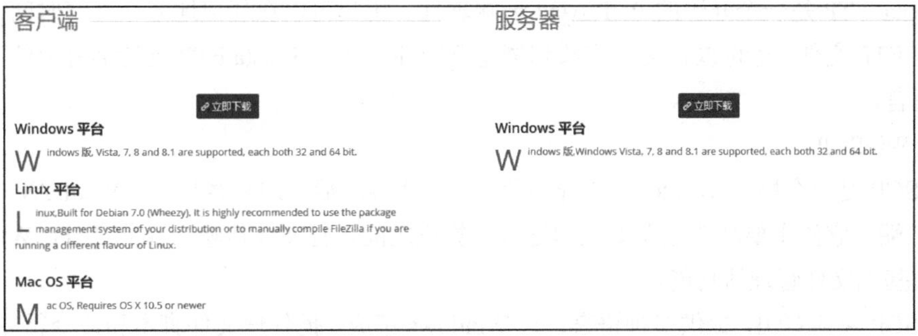

图 5-3　FileZilla 官网支持版本

FileZilla 客户端是一个快速、可信赖的、跨平台的 FTP、FTPS 和 SFTP 客户端。具有图形用户界面（Graphical User Interface，GUI）和很多有用的特性。相比其他 FTP 客户端，FileZilla 包含如下特性。

- 易于使用。
- 支持 FTP，FTP 支持 SSL/TLS（FTPS）协议，支持 SSH 文件传输协议（SFTP）。
- 跨平台。支持在 Windows、Linux、*BSD、macOS 和其他平台下运行。
- 支持 IPv6 协议。
- 多种可用的语言（包含中文）。
- 断点续传且支持容量大于 4GB 的文件。
- 多标签用户界面。
- 功能强大的站点管理器（Site Manager）和传输队列管理。
- 书签功能。
- 拖曳功能支持。
- 支持传输限速功能。
- 文件名过滤器。
- 文件夹比较功能。
- 网络设置向导。
- 远程文件编辑功能。
- 保持链接功能。
- 支持 HTTP/1.1、SOCKS5 和 FTP 代理（FTP-Proxy）。
- 登录到文件功能。
- 同步文件夹浏览。
- 远程查找文件。

2. FTP 客户端选型

（1）FileZilla。

FileZilla 是大多数用户的第一选择。

（2）FireFTP。

FireFTP 是一个用于 Mozilla Firefox 的免费、安全的跨平台 FTP/SFTP 客户端，可以轻松、直观地访问 FTP/SFTP 服务器。

(3) Monsta FTP。

Monsta FTP 是一个开源的 PHP/Ajax 云端软件。用户可以在任何地方、任何时间，在浏览器中管理 FTP 文件，也可以将文件拖放到浏览器中并上传。Monsta FTP 支持屏幕上的文件编辑和多种语言。

(4) WinSCP。

WinSCP 是一个用于 Windows 的开源免费 SFTP 客户端、FTP 客户端、WebDAV 客户端和 SCP 客户端。它的主要功能是在本地和远程计算机之间进行文件传输。除此之外，WinSCP 提供脚本和基本文件管理器功能。

WinSCP 包括 GUI，提供多种语言，与 Windows 集成，批处理文件脚本和命令行界面集成，以及提供各种其他有用的功能。

(5) OneButton FTP。

OneButton FTP 是用于 macOS 的图形 FTP 客户端，强调简单性，支持拖放和文件排队。它允许用户轻松地从远程服务器传输文件，只需将文件拖放到计算机中即可。

OneButton FTP 无须任何费用，它是完全免费的。它包含英语、法语、德语、意大利语、日语、西班牙语和瑞典语的本地化，支持未加密的 FTP 和 FTP over SSL 传输。

(6) gFTP。

gFTP 是一个基于 Nix 的免费多线程文件传输客户端。它支持 FTP、FTPS（仅控制连接）、HTTP、HTTPS、SSH 和 FSP。其上传和编辑文件的方法类似于 FileZilla。

FTP 客户端建议使用 FileZilla。

5.2 搭建 vsftpd 服务器

5.2.1 初始化 vsftpd 服务器运行环境

初始化 vsftpd 服务器运行环境如表 5-1 所示。

表 5-1　　　　　　　　　　vsftpd 服务器运行环境

操作系统	角色	IP 地址	主机名
CentOS 6.9 x86_64	vsftpd 服务器	192.168.2.106	www.ftp_server.com
CentOS 6.9 x86_64	vsftpd 客户端	192.168.2.107	www.ansible.com
Windows 10 x86_64	vsftpd 客户端	192.168.2.105	hanyw

1. 查看 vsftpd 服务器运行环境

(1) 确认基础环境信息。

登录 vsftpd 服务器，查看操作系统的版本、内核版本、主机名称、IP 地址、网络连通性等信息，代码如下：

```
[root@www.ftp_server.com ~]#cat /etc/redhat-release
CentOS release 6.9 (Final)
[root@www.ftp_server.com ~]#uname -r
2.6.32-696.el6.x86_64
[root@www.ftp_server.com ~]#ip ro |grep src |awk '{print $NF}'
192.168.2.106
[root@www.ftp_server.com ~]#uname -n
```

```
www.ftp_server.com
[root@www.ftp_server.com ~]#ping jd*.com -c 3
PING jd*.com (120.52.148.118) 56(84) bytes of data.
64 bytes from 120.52.148.118: icmp_seq=1 ttl=50 time=8.55 ms
64 bytes from 120.52.148.118: icmp_seq=2 ttl=50 time=7.58 ms
64 bytes from 120.52.148.118: icmp_seq=3 ttl=50 time=7.56 ms

--- jd*.com ping statistics ---
3 packets transmitted, 3 received, 0% packet loss, time 2010ms
rtt min/avg/max/mdev = 7.566/7.901/8.550/0.469 ms
```

上述代码可以确认 IP 地址等信息配置准确无误。

（2）确认防火墙和 SELinux 是否关闭。

登录 vsftpd 服务器，关闭防火墙和 SELinux，代码如下。

```
[root@www.ftp_server.com ~]# sestatus
SELINUX status:            disabled
[root@www.ftp_server.com ~]# /etc/init.d/iptables status
iptables: Firewall is not running.
[root@www.ftp_server.com ~]# iptables -L -n
Chain INPUT (policy ACCEPT)
target     prot opt source              destination

Chain FORWARD (policy ACCEPT)
target     prot opt source              destination

Chain OUTPUT (policy ACCEPT)
target     prot opt source              destination
```

从上述代码可以看到，防火墙和 SELinux 已经被关闭。

2. 查看 vsftpd 客户端运行环境

（1）确认基础环境信息。

登录 vsftpd 客户端确认基础环境信息，包括 IP 地址、主机名称、操作系统版本、内核版本、网络连通性等，代码如下。

```
[root@www.ansible.com ~]#cat /etc/redhat-release
CentOS release 6.9 (Final)
[root@www.ansible.com ~]#uname -r
2.6.32-696.el6.x86_64
[root@www.ansible.com ~]#uname -n
www.ansible.com
[root@www.ansible.com ~]#ip ro |grep src |awk '{print $NF}'
192.168.2.107
[root@www.ansible.com ~]#ping jd*.com -c 2
PING jd*.com (120.52.148.118) 56(84) bytes of data.
64 bytes from 120.52.148.118: icmp_seq=1 ttl=50 time=7.86 ms
64 bytes from 120.52.148.118: icmp_seq=2 ttl=50 time=8.25 ms

--- jd*.com ping statistics ---
2 packets transmitted, 2 received, 0% packet loss, time 1009ms
rtt min/avg/max/mdev = 7.868/8.061/8.255/0.213 ms
```

上述代码确认 IP 地址等信息配置准确无误。

（2）确认防火墙和 SELinux 是否关闭。

登录 vsftpd 客户端，确认防火墙和 SELinux 已经被关闭，代码如下。

```
[root@www.ansible.com ~]#sestatus
SELINUX status:            disabled
[root@www.ansible.com ~]#/etc/init.d/iptables status
iptables: Firewall is not running.
```

从上述代码可以看到，防火墙和 SELinux 已经被关闭。

5.2.2 安装 vsftpd 软件

1. 安装 vsftpd 软件

在 Linux 操作系统中，应用最广泛的 FTP 服务器软件是 vsftpd，vsftpd 软件提供了非常安全的 FTP 服务。vsftpd 是一个 UNIX 操作系统上运行的服务器的名字，它可以运行在诸如 Linux、BSD、Solaris、HP-UX 以及 IRIX 等主流操作系统上面，它支持很多其他的 FTP 服务器不支持的特征，如下所示。

- 更高级别的安全性需求。
- 精细的带宽限制。
- 虚拟用户定制化安全策略。
- 高速。

vsftpd 提供了一个快速的、稳定的且相当安全的 FTP 服务，CentOS 默认没有安装 vsftpd FTP 程序，读者可通过 yum -y install vsftpd 指令进行安装，代码如下。

```
[root@www.ftp_server.com ~]# yum -y install vsftpd
Loaded plugins: fastestmirror
Setting up Install Process
（中间代码略）

Installed:
  vsftpd.x86_64 0:2.2.2-24.el6

Complete!
```

vsftpd 软件已经安装完成，查看 vsftpd 服务的一些配置文件，其所有的配置文件都保存在 /etc/vsftpd/ 目录下，代码如下。

```
[root@www.ftp_server.com ~]# rpm -ql vsftpd
/etc/logrotate.d/vsftpd
/etc/pam.d/vsftpd
/etc/rc.d/init.d/vsftpd
/etc/vsftpd
/etc/vsftpd/ftpusers
/etc/vsftpd/user_list
/etc/vsftpd/vsftpd.conf
/etc/vsftpd/vsftpd_conf_migrate.sh
/usr/sbin/vsftpd
/usr/share/doc/vsftpd-2.2.2
/usr/share/doc/vsftpd-2.2.2/AUDIT
/usr/share/doc/vsftpd-2.2.2/BENCHMARKS
/usr/share/doc/vsftpd-2.2.2/BUGS
/usr/share/doc/vsftpd-2.2.2/COPYING
/usr/share/doc/vsftpd-2.2.2/Changelog
/usr/share/doc/vsftpd-2.2.2/EXAMPLE
/usr/share/doc/vsftpd-2.2.2/EXAMPLE/INTERNET_SITE
/usr/share/doc/vsftpd-2.2.2/EXAMPLE/INTERNET_SITE/README
/usr/share/doc/vsftpd-2.2.2/EXAMPLE/INTERNET_SITE/README.configuration
/usr/share/doc/vsftpd-2.2.2/EXAMPLE/INTERNET_SITE/vsftpd.conf
/usr/share/doc/vsftpd-2.2.2/EXAMPLE/INTERNET_SITE/vsftpd.xinetd
/usr/share/doc/vsftpd-2.2.2/EXAMPLE/INTERNET_SITE_NOINETD
/usr/share/doc/vsftpd-2.2.2/EXAMPLE/INTERNET_SITE_NOINETD/README
/usr/share/doc/vsftpd-2.2.2/EXAMPLE/INTERNET_SITE_NOINETD/README.configuration
/usr/share/doc/vsftpd-2.2.2/EXAMPLE/INTERNET_SITE_NOINETD/vsftpd.conf
/usr/share/doc/vsftpd-2.2.2/EXAMPLE/PER_IP_CONFIG
/usr/share/doc/vsftpd-2.2.2/EXAMPLE/PER_IP_CONFIG/README
/usr/share/doc/vsftpd-2.2.2/EXAMPLE/PER_IP_CONFIG/README.configuration
/usr/share/doc/vsftpd-2.2.2/EXAMPLE/PER_IP_CONFIG/hosts.allow
/usr/share/doc/vsftpd-2.2.2/EXAMPLE/README
```

```
/usr/share/doc/vsftpd-2.2.2/EXAMPLE/VIRTUAL_HOSTS
/usr/share/doc/vsftpd-2.2.2/EXAMPLE/VIRTUAL_HOSTS/README
/usr/share/doc/vsftpd-2.2.2/EXAMPLE/VIRTUAL_USERS
/usr/share/doc/vsftpd-2.2.2/EXAMPLE/VIRTUAL_USERS/README
/usr/share/doc/vsftpd-2.2.2/EXAMPLE/VIRTUAL_USERS/README.configuration
/usr/share/doc/vsftpd-2.2.2/EXAMPLE/VIRTUAL_USERS/logins.txt
/usr/share/doc/vsftpd-2.2.2/EXAMPLE/VIRTUAL_USERS/vsftpd.conf
/usr/share/doc/vsftpd-2.2.2/EXAMPLE/VIRTUAL_USERS/vsftpd.pam
/usr/share/doc/vsftpd-2.2.2/EXAMPLE/VIRTUAL_USERS_2
/usr/share/doc/vsftpd-2.2.2/EXAMPLE/VIRTUAL_USERS_2/README
/usr/share/doc/vsftpd-2.2.2/FAQ
/usr/share/doc/vsftpd-2.2.2/INSTALL
/usr/share/doc/vsftpd-2.2.2/LICENSE
/usr/share/doc/vsftpd-2.2.2/README
/usr/share/doc/vsftpd-2.2.2/README.security
/usr/share/doc/vsftpd-2.2.2/REWARD
/usr/share/doc/vsftpd-2.2.2/SECURITY
/usr/share/doc/vsftpd-2.2.2/SECURITY/DESIGN
/usr/share/doc/vsftpd-2.2.2/SECURITY/IMPLEMENTATION
/usr/share/doc/vsftpd-2.2.2/SECURITY/OVERVIEW
/usr/share/doc/vsftpd-2.2.2/SECURITY/TRUST
/usr/share/doc/vsftpd-2.2.2/SIZE
/usr/share/doc/vsftpd-2.2.2/SPEED
/usr/share/doc/vsftpd-2.2.2/TODO
/usr/share/doc/vsftpd-2.2.2/TUNING
/usr/share/doc/vsftpd-2.2.2/vsftpd.xinetd
/usr/share/man/man5/vsftpd.conf.5.gz
/usr/share/man/man8/vsftpd.8.gz
/var/ftp
/var/ftp/pub
[root@www.ftp_server.com ~]# cd /etc/vsftpd/
[root@www.ftp_server.com /etc/vsftpd]# ls -l
total 20
-rw-------  1 root root  125 Mar 22  2017 ftpusers
-rw-------  1 root root  361 Mar 22  2017 user_list
-rw-------  1 root root 4623 Dec 10 18:53 vsftpd.conf
-rwxr--r--  1 root root  338 Mar 22  2017 vsftpd_conf_migrate.sh
```

/etc/vsftpd/目录下一共有 4 个配置文件，如图 5-4 所示。

（1）ftpusers。

ftpusers 是 vsftpd 服务黑名单用户配置文件，通常系统用户还有根用户都是放在此配置文件中的。因为这些用户的权限很大，所以如果使用 FTP 服务可能会造成一些问题。

（2）user_list。

user_list 是 vsftpd 服务用户列表文件可以通过 user_list 在主配置文件里设置用户是黑名单用户还是白名单用户。

（3）vsftpd.conf。

vsftpd.conf 是 vsftpd 服务的主配置文件。

（4）vsftpd_conf_migrate.sh。

vsftpd_conf_migrate.sh 是 vsftpd 服务中的 FTP 服务迁移脚本文件。

vsftpd 软件安装好后，在/var 目录下有一个 ftp 文件夹，这个文件夹就是 vsftpd 服务默认共享目录，目录信息如图 5-5 所示。

目录信息的代码如下。

```
[root@www.ftp_server.com ~]# cd /var/ftp/
[root@www.ftp_server.com /var/ftp]# ls -l
total 4
drwxr-xr-x 2 root root 4096 Mar 22  2017 pub
```

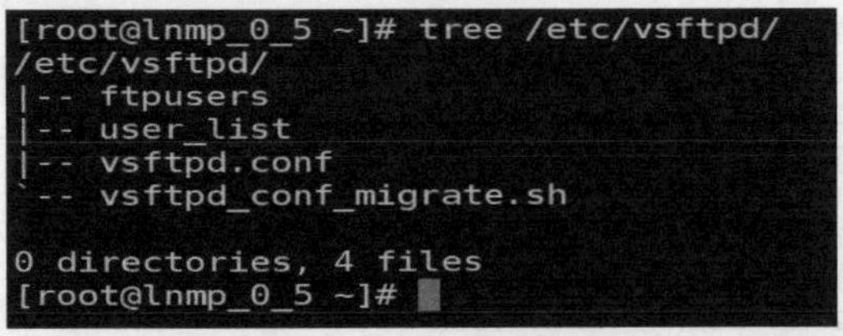

```
[root@lnmp_0_5 ~]# tree /etc/vsftpd/
/etc/vsftpd/
|-- ftpusers
|-- user_list
|-- vsftpd.conf
`-- vsftpd_conf_migrate.sh

0 directories, 4 files
[root@lnmp_0_5 ~]#
```

图 5-4　vsftpd 配置文件

```
[root@lnmp_0_5 ~]# tree /var/ftp/
/var/ftp/
`-- pub

1 directory, 0 files
[root@lnmp_0_5 ~]# ls -ld /var/ftp/pub/
drwxr-xr-x 2 root root 4096 Oct 31  2018 /var/ftp/pub/
[root@lnmp_0_5 ~]#
```

图 5-5　vsftpd 默认共享目录信息

上述代码中的 pub 目录，其他用户没有 w 权限。Linux 操作系统中，w 权限意味着该文件或目录的所有者有写入、删除、重命名等操作权限。

2．检测系统是否已安装 vsftpd

检测系统是否已安装 vsftpd 有两种方法，代码如下。

```
[root@lnmp_0_5 ~]# vsftpd -v
vsftpd: version 3.0.2
[root@lnmp_0_5 ~]# rpm -q vsftpd
vsftpd-3.0.2-25.el7.x86_64
```

3．查看 vsftpd 安装位置

查看 vsftpd 安装位置。

```
[root@lnmp_0_5 ~]# whereis vsftpd
vsftpd: /usr/sbin/vsftpd /etc/vsftpd /usr/share/man/man8/vsftpd.8.gz
```

从上述代码可以看到 vsftpd 安装位置位于/usr/sbin/vsftpd，配置文件目录为/etc/vsftpd，man 帮助文档位于/usr/share/man/。

5.2.3　访问 vsftpd 服务

1．重启 vsftpd 服务

CentOS 6 系列操作系统重启 vsftpd 服务的代码如下。

```
[root@www.ftp_server.com ~]#/etc/init.d/vsftpd restart
关闭 vsftpd:                                              [确定]
为 vsftpd 启动 vsftpd:                                    [确定]
```

CentOS 7 系列操作系统启动 vsftpd 服务的代码如下。

```
[root@lnmp_0_5 ~]# systemctl start vsftpd.service
[root@lnmp_0_5 ~]# systemctl status vsftpd.service
  vsftpd.service - Vsftpd ftp daemon
   Loaded: loaded (/usr/lib/systemd/system/vsftpd.service; disabled;
vendor preset: disabled)
   Active: active (running) since Wed 2019-09-25 08:00:06 CST; 10s ago
  Process: 8667 ExecStart=/usr/sbin/vsftpd /etc/vsftpd/vsftpd.conf (code=exited,
status=0/SUCCESS)
 Main PID: 8668 (vsftpd)
   CGroup: /system.slice/vsftpd.service
           └─8668 /usr/sbin/vsftpd /etc/vsftpd/vsftpd.conf

Sep 25 08:00:06 lnmp_0_5 systemd[1]: Starting Vsftpd ftp daemon...
Sep 25 08:00:06 lnmp_0_5 systemd[1]: Started Vsftpd ftp daemon.
```

从上述代码可看到，vsftpd 服务已经被成功启动（重启）。

2．查看端口和进程

```
[root@www.ftp_server.com ~]#netstat -ntpl |grep 21
tcp        0      0 0.0.0.0:21              0.0.0.0:*               LISTEN      1457/vsftpd
[root@www.ftp_server.com ~]#ps -ef |grep vsftpd
root      1457     1  0 19:07 ?        00:00:00 /usr/sbin/vsftpd /etc/vsftpd/vsftpd.conf
root      1462  1351  0 19:07 pts/0    00:00:00 grep vsftpd
```

```
[root@www.ftp_server.com ~]#lsof -i:21
COMMAND  PID  USER  FD   TYPE DEVICE SIZE/OFF NODE NAME
vsftpd  1457  root  3u   IPv4  11146       0t0  TCP *:ftp (LISTEN)
```

从上述代码可看到，vsftpd 服务进程已经存在，端口号已经被监听。

3. 用浏览器访问 vsftpd 服务

用浏览器访问 vsftpd 服务，如图 5-6 和图 5-7 所示。

图 5-6　浏览器访问 vsftpd 服务 1

图 5-7　浏览器访问 vsftpd 服务 2

在浏览器中输入 ftp://192.168.2.106，出现 upload 目录，说明 vsftpd 服务已经被成功搭建，进入 upload 目录，其目录内没有任何文件，这样就完成了最基本的 vsftpd 服务的搭建。

5.2.4　vsftpd iptables 设置

若通过浏览器无法访问 vsftpd 服务，请检查 iptables、firewalld、SELinux 安装策略是否放行，以及 vsftpd 服务是否启动。FTP 连接如下。

（1）一个控制连接。

该控制连接用于传递客户端的命令和服务器端对命令的响应，比如：登录使用的用户名与密码，变更目录指令 CWD、PUT、GET 文件。它使用 TCP 21 端口。

（2）多个数据连接。

这些连接用于传输文件和其他数据，比如：目录列表指令 list。使用端口依据 FTP 服务端工作模式决定。

vsftpd 主动与被动模式的区别在于 PORT 指令的发出方，或者说数据连接的主动发起方。

- 主动模式下，由客户端通过 PORT 告知服务端自己的监听端口，然后服务端通过自己定义的主动模式下的端口（默认为 20）发起到客户端宣告的端口的连接。
- 被动模式下，服务端在接到客户端的 PASV 指令后，通过 PORT 发送端口号给客户端，客户端连接这个端口进行数据传输。

1. 主动模式下 iptables 设置

这个模式下，因为客户端需要连接服务端的 21 端口，同时服务端的 20 端口主动外联客户端的端口，所以要确保 INPUT 方向的 21 端口允许访问，同时 OUTPUT 方向的 20 端口允许通过（通常 OUTPUT 默认为 ACCEPT，所以这个不用设置。如果为 DROP，则需要添加外出方向的 20 端口访问规则），以及 RELATED 与 ESTABLISHED 规则。

```
iptables -A INPUT -p tcp -m tcp --dport 21 -j ACCEPT
iptables -A INPUT -m state --state RELATED,ESTABLISHED -j ACCEPT
```

2. 被动模式下 iptables 设置

针对 vsftpd 的设置可以使用不同策略。

（1）vsftpd 未指定被动模式的端口范围。

在/etc/sysconfig/iptables-config 中添加 IPTABLES_MODULES="ip_conntrack_ftp"，加载 ip_conntrack_ftp 模块以过滤和传输与 FTP 控制连接相关的数据。修改设置后使用 service iptables restart 使新的模块加载，同时 iptables 需要被允许访问 21 端口。代码如下。

```
iptables -A INPUT -p tcp -m tcp --dport 21 -j ACCEPT
iptables -A INPUT -m state --state RELATED,ESTABLISHED -j ACCEPT
```

（2）vsftpd 指定被动模式的端口范围。

也可以使用（1）中的方案，或者可以在 iptables 的 INPUT 链中开放对指定范围端口的访问，在/etc/vsftpd/vsftpd.conf 文件中设置代码如下。

```
pasv_enable=YES
pasv_min_port=6666
pasv_max_port=8888
```

在 iptables 中开放 6666~8888 端口，代码如下。

```
iptables -A INPUT -p tcp --dport 21 -j ACCEPT
iptables -A INPUT -p tcp --dport 6666:8888 -j ACCEPT
iptables -A INPUT -m state --state ESTABLISHED,RELATED -j ACCEPT
```

3. iptables 设置 vsftpd 完整演示

使用 iptables 设置放行 vsftpd 服务，代码如下。

```
[root@lnmp_0_5 ~]# iptables -A INPUT -p tcp -m tcp --dport 21 -j ACCEPT
[root@lnmp_0_5 ~]# iptables -A INPUT -m state --state RELATED,ESTABLISHED -j ACCEPT
[root@lnmp_0_5 ~]# iptables -A INPUT -p tcp -m tcp --dport 21 -j ACCEPT
[root@lnmp_0_5 ~]# iptables -A INPUT -m state --state RELATED,ESTABLISHED -j ACCEPT
[root@lnmp_0_5 ~]# iptables -A INPUT -p tcp --dport 21 -j ACCEPT
[root@lnmp_0_5 ~]# iptables -A INPUT -p tcp --dport 6666:8888 -j ACCEPT
[root@lnmp_0_5 ~]# iptables -A INPUT -m state --state ESTABLISHED,RELATED -j ACCEPT
[root@lnmp_0_5 ~]# service iptables save
iptables: Saving firewall rules to /etc/sysconfig/iptables:[  OK  ]
[root@lnmp_0_5 ~]# iptables -L -n
Chain INPUT (policy ACCEPT)
target     prot opt source               destination
ACCEPT     tcp  --  0.0.0.0/0            0.0.0.0/0           tcp dpt:21
ACCEPT     all  --  0.0.0.0/0            0.0.0.0/0           state RELATED,ESTABLISHED
ACCEPT     tcp  --  0.0.0.0/0            0.0.0.0/0           tcp dpt:21
ACCEPT     all  --  0.0.0.0/0            0.0.0.0/0           state RELATED,ESTABLISHED
ACCEPT     tcp  --  0.0.0.0/0            0.0.0.0/0           tcp dpt:21
ACCEPT     tcp  --  0.0.0.0/0            0.0.0.0/0           tcp dpts:6666:8888
ACCEPT     all  --  0.0.0.0/0            0.0.0.0/0           state RELATED,ESTABLISHED
ACCEPT     tcp  --  0.0.0.0/0            0.0.0.0/0           tcp dpt:21
ACCEPT     all  --  0.0.0.0/0            0.0.0.0/0           state RELATED,ESTABLISHED
ACCEPT     tcp  --  0.0.0.0/0            0.0.0.0/0           tcp dpt:21
ACCEPT     tcp  --  0.0.0.0/0            0.0.0.0/0           tcp dpts:6666:8888

Chain FORWARD (policy ACCEPT)
target     prot opt source               destination

Chain OUTPUT (policy ACCEPT)
target     prot opt source               destination
```

上述代码放行 6666~8888 之间的端口，在/etc/vsftpd/vsftpd.conf 中设置端口范围，代码如下。

```
pasv_enable=YES
pasv_min_port=6666
pasv_max_port=8888
```

4. 查看完整 iptables 配置文件

查看完整 iptables 配置文件，代码如下。

```
[root@lnmp_0_5 ~]# cat /etc/sysconfig/iptables
# Generated by iptables-save v1.4.21 on Wed Sep 25 09:10:08 2019
*filter
:INPUT ACCEPT [18:824]
:FORWARD ACCEPT [0:0]
:OUTPUT ACCEPT [114:9841]
-A INPUT -p tcp -m tcp --dport 21 -j ACCEPT
-A INPUT -m state --state RELATED,ESTABLISHED -j ACCEPT
-A INPUT -p tcp -m tcp --dport 21 -j ACCEPT
-A INPUT -m state --state RELATED,ESTABLISHED -j ACCEPT
-A INPUT -p tcp -m tcp --dport 21 -j ACCEPT
-A INPUT -p tcp -m tcp --dport 6666:8888 -j ACCEPT
-A INPUT -m state --state RELATED,ESTABLISHED -j ACCEPT
-A INPUT -p tcp -m tcp --dport 21 -j ACCEPT
-A INPUT -m state --state RELATED,ESTABLISHED -j ACCEPT
-A INPUT -p tcp -m tcp --dport 21 -j ACCEPT
-A INPUT -m state --state RELATED,ESTABLISHED -j ACCEPT
-A INPUT -p tcp -m tcp --dport 21 -j ACCEPT
-A INPUT -p tcp -m tcp --dport 6666:8888 -j ACCEPT
COMMIT
# Completed on Wed Sep 25 09:10:08 2019
```

上述代码主要放行 FTP 服务和配置文件中设定的端口。

5.3 vsftpd 配置文件和日志配置

5.3.1 vsftpd 配置文件详解

1. 基本格式

vsftpd.conf 的内容非常简洁，每一行即为一项设定，若是空白行或是开头为 "#" 的一行，将会被忽略，内容格式如下。

```
option=value
```

要注意的是，"=" 两边不能有空格。

2. 允许匿名用户和本地用户登录

设置 vsftpd 允许匿名用户和本地用户登录，代码如下。

```
anonymous_enable=YES
local_enable=YES
```

- 匿名用户使用的登录名为 ftp 或 anonymous，口令为空，匿名用户不能离开匿名用户主目录/var/ftp，且只能下载不能上传。
- 本地用户的登录名为本地用户名，口令为此本地用户的口令，本地用户可以在主目录中进行读写操作，本地用户可以离开主目录切换至有权限访问的其他目录，并在权限允许的情况下进行上传或下载。

以下代码写在文件/etc/vsftpd.ftpusers 中，禁止本地用户登录。

```
write_enable=YES
```

3. 设置匿名用户

设置匿名用户常用参数如表 5-2 所示。

表 5-2　　　　　　　　　　　设置匿名用户常用参数

参数	说明
anonymous_enable	是否允许匿名用户登录，YES 为允许，NO 为不允许
write_enable	是否允许登录用户有写权限，属于全局设置
no_anon_password	启动这项功能，则使用匿名登录时，不会询问密码
ftp_username=ftp	定义匿名登录的用户名称
anon_root=/var/ftp	使用匿名登录时，所登录的目录
anon_upload_enable	当 write_enable=YES 时，此项才有效
anon_world_readable_only	允许匿名用户下载可阅读的档案
anon_mkdir_write_enable	允许匿名新增目录权限，当 write_enable=YES 时生效
anon_other_write_enable	允许匿名用户拥有上传或建立目录之外的权限
chown_uploads	是否改变匿名用户上传文件（非目录）的属主
chown_username	设置匿名用户上传文件（非目录）的属主名
anon_umask	设置匿名用户新增或上传档案时的 umask 值

4．系统用户设置

正常用户就是操作系统的系统用户，一般我们安装的各种服务都能通过系统用户登录来使用。

首先启动 vsftpd 服务，这里先通过 CentOS 提供的一个图形界面工具来使用系统用户登录 FTP，系统用户常用参数如表 5-3 所示。

表 5-3　　　　　　　　　　　vsftpd 系统用户常用参数

参数	说明
local_enable	控制是否允许本地用户登录
local_root	当本地用户登录时，进入指定目录，默认值为各用户的主目录
write_enable	是否允许登录用户有写权限，属于全局设置，默认值为 YES
local_umask	本地用户新增档案时的 umask 值，默认值为 077
file_open_mode	本地用户上传档案后的档案权限，与 chmod 所使用的数值相同

5．设置欢迎语

vsftpd 服务欢迎语设置，如表 5-4 所示。

表 5-4　　　　　　　　　　　vsftpd 服务欢迎语设置

参数	说明
dirmessage_enable	该档案会放置欢迎语，或是对该目录的说明
message_file	将要显示的信息写入该文件，默认值为 .message
banner_file	当使用者登录时，会显示此设定所在的档案内容
ftpd_banner	这里用来定义欢迎语的字符串

6．控制用户是否允许切换到上级目录

在默认配置下，本地用户登录 FTP 后可以使用 cd 指令切换到其他目录，这样会给系统带来安全隐患。可以通过以下 3 个配置参数来控制用户切换目录，如表 5-5 所示。

表 5-5　　　　　　　　　　　允许切换到上级目录

参数	说明
chroot_list_enable	设置是否启用 chroot_list_file 的用户列表文件
chroot_list_file	用户列表文件，控制用户切换到主目录的上级目录
chroot_local_user	指定用户列表文件中的用户是否允许切换到上级目录

将以上参数进行搭配能实现以下几种效果。

- 当 chroot_list_enable=YES、chroot_local_user=YES 时，在/etc/vsftpd.chroot_list 文件中列出的用户，可以切换到其他目录，未在文件中列出的用户，不能切换到其他目录。
- 当 chroot_list_enable=YES、chroot_local_user=NO 时，在/etc/vsftpd.chroot_list 文件中列出的用户，不能切换到其他目录，未在文件中列出的用户，可以切换到其他目录。
- 当 chroot_list_enable=NO、chroot_local_user=YES 时，所有的用户均不能切换到其他目录。
- 当 chroot_list_enable=NO、chroot_local_user=NO 时，所有的用户均可以切换到其他目录。

7. 数据传输模式设置

FTP 在传输数据时，可以使用二进制模式或 ASCII 模式来上传或下载数据。

设置是否启用 ASCII 模式上传数据。默认值为 NO，代码如下。

```
ascii_upload_enable=YES/NO（NO）
```

设置是否启用 ASCII 模式下载数据。默认值为 NO，代码如下。

```
ascii_download_enable=YES/NO（NO）
```

8. vsftpd 访问控制设置

两种控制方式：一种是控制主机访问，另一种是控制用户访问。

（1）控制主机访问，代码如下。

```
tcp_wrappers=YES/NO（YES）
```

设置 vsftpd 是否与 tcp wrapper 相结合来进行主机的访问控制，默认值为 YES。如果启用，则 vsftpd 服务器会检查/etc/hosts.allow 和/etc/hosts.deny 中的设置，来决定请求连接的主机，是否允许访问该 FTP 服务器。

这两个文件可以起到简易的防火墙功能。

比如，若要仅允许 192.168.0.1～192.168.0.254 的用户可以连接 FTP 服务器，则在/etc/hosts.allow 文件中添加以下内容。

```
vsftpd:192.168.0. :allow
all:all :deny
```

（2）控制用户访问。

对于用户的访问控制可以通过/etc 目录下的 vsftpd.user_list 和 ftpusers 文件来实现，代码如下。
控制用户访问 FTP 服务器的文件，里面写着用户名称，一个用户名称一行，代码如下。

```
userlist_file=/etc/vsftpd.user_list
```

是否启用 vsftpd.user_list 配置文件，代码如下。

```
userlist_enable=YES/NO（NO）
userlist_deny=YES/NO（YES）
```

上述配置决定 vsftpd.user_list 文件中的用户是否能够访问 FTP 服务器。若设置为 YES，则 vsftpd.user_list 文件中的用户不允许访问 FTP 服务器；若设置为 NO，则只有 vsftpd.user_list 文件中的用户才能访问 FTP 服务器。

/etc/vsftpd/ftpusers 文件专门用于定义不允许访问 FTP 服务器的用户列表（注意，如果 userlist_enable=YES、userlist_deny=NO，且在 vsftpd.user_list 和 ftpusers 中都有某个用户时，那么这个用户是不能够访问 FTP 服务器的，即 ftpusers 的优先级较高）。默认情况下 vsftpd.user_list 和 ftpusers，这两个文件已经设置了一些不允许访问 FTP 服务器的系统内部账户。如果系统没有这两个文件，那么新建这两个文件，将用户添加进去即可。

9. 访问速度设置

vsftpd 服务访问速度设置，代码如下。

```
anon_max_rate=0
```
设置匿名用户使用的最大传输速度,单位为 B/s,0 表示不限制速度。默认值为 0。
```
local_max_rate=0
```

10．超时时间设置

vsftpd 服务访问超时时间设置,代码如下。
```
accept_timeout=60
```
设置建立 FTP 连接的超时时间,单位为秒。默认值为 60。
```
connect_timeout=60
```
PORT 方式下建立数据连接的超时时间,单位为秒。默认值为 60。
```
data_connection_timeout=120
```
设置建立 FTP 数据连接的超时时间,单位为秒。默认值为 120。
```
idle_session_timeout=300
```
设置多长时间不对 FTP 服务器进行任何操作,则断开该 FTP 连接,单位为秒。默认值为 300。

11．日志文件设置

vsftpd 服务日志文件设置,代码如下。
```
xferlog_enable= YES/NO（YES）
```
是否启用上传或下载日志记录。如果启用,则上传与下载的信息将被完整记录在 xferlog_file 所定义的档案中。默认值为 YES。
```
xferlog_file=/var/log/vsftpd.log
```
设置日志文件名和路径,默认值为/var/log/vsftpd.log。

12．自定义用户设置

在 vsftpd 中,可以通过定义用户配置文件来实现不同的用户使用不同的配置。

设置用户配置文件所在的目录,代码如下。
```
user_config_dir=/etc/vsftpd/userconf
```
设置该配置项后,用户登录服务器后,系统就会到/etc/vsftpd/userconf 目录下,读取与当前用户名相同的文件,并根据文件中的配置命令,对当前用户进行更进一步的配置。

例如：定义 user_config_dir=/etc/vsftpd/userconf,且主机上有使用者 user1、user2,那么我们就在 user_config_dir 的目录中新增文件名为 user1 和 user2 两个文件。若是 user1 登录,则会读取 user_config_dir 下的 user1 这个档案内的设定。默认值为无。利用用户配置文件,可以实现对不同用户进行访问速度的控制,在各用户配置文件中定义 local_max_rate=XX 即可。

13．连接相关的设置

vsftpd 服务与连接相关的设置,代码如下。
```
listen=YES/NO（YES）
```
设置 vsftpd 服务器是否以 standalone 模式运行。以 standalone 模式运行是一种较好的方式,此时 listen 必须设置为 YES,此为默认值。建议不要更改,有很多与服务器运行相关的配置命令,需要在此模式下才有效。若设置为 NO,则 vsftpd 不是以独立的服务运行的,要受到 xinetd 服务的管控,功能上会受到限制。

设置允许客户端的最大连接数,代码如下。
```
max_clients=0
```
设置 vsftpd 允许的最大连接数,默认值为 0,表示不受限制。若设置为 100 时,则同时允许有 100 个连接,超出的将被拒绝。只有在 standalone 模式下运行才有效。

设置允许单 IP 地址的最大用户数,代码如下。
```
max_per_ip=0
```

设置每个 IP 地址允许与 FTP 服务器同时建立连接的数目。默认值为 0，表示不受限制。只有在 standalone 模式下运行才有效。

设置 FTP 服务器在指定的 IP 地址上侦听用户的 FTP 请求，代码如下。

```
listen_address=IP 地址
```

若不设置，则对服务器绑定的所有 IP 地址进行侦听。只有在 standalone 模式下运行才有效。

设置每个与 FTP 服务器的连接，是否以不同的进程表现出来，代码如下。

```
setproctitle_enable=YES/NO（NO）
```

默认值为 NO，若设置为 YES，则每个连接都会有一个 vsftpd 的进程。

14．其他设置

vsftpd 其他设置如下。

```
text_userdb_names= YES/NO（NO）
```

设置在执行 ls -la 之类的指令时，是显示 UID、GID 还是显示具体的用户名和组名。默认值为 NO，即以 UID 和 GID 方式显示。若希望显示用户名和组名，则设置为 YES。

```
ls_recurse_enable=YES/NO（NO）
```

若是启用此功能，则允许用户使用 ls -R（可以查看当前目录下子目录中的文件）指令。默认值为 NO。

```
hide_ids=YES/NO（NO）
```

如果启用此功能，所有档案的拥有者与群组都为 FTP，也就是用户登录使用 ls -al 之类的指令，所看到的档案拥有者和群组均为 FTP。默认值为 NO。

```
download_enable=YES/NO（YES）
```

如果设置为 NO，所有的文件都不能下载到本地，文件夹不受影响。默认值为 YES。

5.3.2 配置 vsftpd 日志

vsftpd 日志配置常见解决方案代码如下。

```
xferlog_enable=YES
xferlog_std_format=YES
xferlog_file=/var/log/xferlog
dual_log_enable=YES
vsftpd_log_file=/var/log/vsftpd.log
```

修改后 vsftpd.conf 配置文件代码如下。

```
[root@lnmp_0_5 ~]# grep -nE "xferlog_enable|xferlog_file|xferlog_std_format|
log_ftp_protocol|dual_log_enable|vsftpd_log_file" /etc/vsftpd/vsftpd.conf
40:xferlog_enable=YES
41:dual_log_enable=YES
42:vsftpd_log_file=/var/log/vsftpd.log
55:xferlog_file=/var/log/xferlog
59:xferlog_std_format=YES
60:log_ftp_protocol=YES
```

重启 vsftpd 服务，访问浏览器，查看日志输出，代码如下。

```
[root@lnmp_0_5 ~]# tail -fn100 /var/log/vsftpd.log
Wed Sep 25 12:32:24 2019 [pid 7420] CONNECT: Client "::ffff:118.74.58.73"
Wed Sep 25 12:32:24 2019 [pid 7419] [ftp] OK LOGIN: Client "::ffff:118.74.58.73",
anon password "chrome@example.com"
Wed Sep 25 12:32:26 2019 [pid 7424] CONNECT: Client "::ffff:118.74.58.73"
Wed Sep 25 12:32:26 2019 [pid 7423] [ftp] OK LOGIN: Client "::ffff:118.74.58.73",
anon password "chrome@example.com"
Wed Sep 25 12:32:28 2019 [pid 7429] CONNECT: Client "::ffff:118.74.58.73"
Wed Sep 25 12:32:28 2019 [pid 7428] [ftp] OK LOGIN: Client "::ffff:118.74.58.73",
anon password chrome@example.com
```

上述代码记录客户端的连接和登录信息。

5.4 vsftpd 匿名用户配置案例

匿名用户使用场景如下。

1. 下载网站

目前比较流行的下载网站会使用匿名用户进行登录，然后下载所需要的文件。

2. 公司共享文件

公司内部要共享一些文件，如公司制度和文化宣讲、公司形象宣传等资料。

5.4.1 vsftpd 服务匿名用户基础配置

1. vsftpd 服务匿名用户配置参数

vsftpd 服务匿名用户配置常用参数如表 5-6 所示。

表 5-6　　　　　　　　　vsftpd 服务匿名用户配置常用参数

参数	说明
write_enable	是否允许登录用户有写权限，全局设置默认 YES
anonymous_enable	是否允许匿名用户登录
anon_upload_enable	允许匿名用户上传文件，upload_enable=YES 有效
anon_mkdir_write_enable	允许匿名用户新增目录，write_enable=YES 有效

2. vsftpd 服务端配置匿名用户登录

（1）匿名用户基本配置代码如下。

```
[root@www.ftp_server.com vsftpd]#head -28  vsftpd.conf  |nl
     1  ####匿名登录设置####
     2  anonymous_enable=YES
     3  local_enable=YES
     4  write_enable=YES
     5  local_umask=022
     6  dirmessage_enable=YES
     7  xferlog_enable=YES
     8  connect_from_port_20=YES
     9  xferlog_file=/var/log/vsftpd.log
    10  xferlog_std_format=YES
    11  listen=YES
    12  userlist_enable=YES
    13  tcp_wrappers=YES
    14  max_per_ip=5
    15  max_clients=100

    16  # 日志中文
    17  syslog_enable=yes

    18  no_anon_password=YES
    19  idle_session_timeout=600
    20  banner_file=/etc/vsftpd/anon_welcome.txt

    21     ####匿名登录设置####
```

（2）匿名用户配置文件注解。

anonymous_enable=YES

开启匿名访问。
```
local_enable=YES
```
本地实体用户访问开启。
```
write_enable=YES
```
允许用户新增目录。
```
local_umask=022
```
创建新目录，控制文件的权限。
```
dirmessage_enable=YES
```
目录下有.message 文件则显示该文件的内容。
```
xferlog_enable=YES
```
日志文件记录于/var/log/vferlog。
```
connect_from_port_20=YES
```
支持主动式连接功能。
```
xferlog_std_format=YES
```
支持 WuFTP 的日志文件格式。

在 vsftpd.conf 配置文件后面添加的代码如下。
```
no_anon_password=YES
```
匿名登录时，不检验密码。
```
idle_session_timeout=600
```
匿名用户 10min 无操作则掉线。
```
banner_file=/etc/vsftpd/anon_welcome.txt
```
匿名用户登录后看到的欢迎信息。

(3) 查看欢迎文件信息，代码如下。
```
[root@www.ftp_server.com vsftpd]#cat /etc/vsftpd/anon_welcome.txt
************************Welcome**********************
```

(4) 先安装两个包，代码如下。
```
[root@www.ftp_server.com pub]#yum install finger ftp -y
已加载插件: fastestmirror
设置安装进程
Loading mirror speeds from cached hostfile
 * base: mirrors.163.com
 * epel: fedora.cs.nctu.edu.tw
 * extras: mirror.bit.edu.cn
 * updates: mirror.bit.edu.cn
包 finger-0.17-40.el6.x86_64 已安装并且是最新版本
包 ftp-0.17-54.el6.x86_64 已安装并且是最新版本
无须任何处理
```

(5) 匿名用户默认目录查看，代码如下。
```
[root@www.ftp_server.com pub]#finger ftp
Login: ftp                              Name: FTP User
Directory: /var/ftp                     Shell: /sbin/nologin
Never logged in.
No mail.
No Plan.
```

(6) 匿名用户登录，代码如下。
```
[root@www.ftp_server.com pub]#ftp localhost
Trying ::1...
ftp: connect to address ::1 拒绝连接
Trying 127.0.0.1...
Connected to localhost (127.0.0.1).
220-************************Welcome**********************
220
Name (localhost:root): anonymous
230 Login successful.
```

```
Remote system type is UNIX.
Using binary mode to transfer files.
ftp> exit
221 Goodbye.
[root@www.ftp_server.com pub]#ftp localhost
Trying ::1...
ftp: connect to address ::1 拒绝连接
Trying 127.0.0.1...
Connected to localhost (127.0.0.1).
220-*********************Welcome********************
220
Name (localhost:root): ftp
230 Login successful.
Remote system type is UNIX.
Using binary mode to transfer files.
ftp> ls
227 Entering Passive Mode (127,0,0,1,103,118).
150 Here comes the directory listing.
drwxr-xr-x    3 0        0            4096 May 14 12:11 pub
226 Directory send OK.
ftp> cd pub
250 Directory successfully changed.
ftp> ls
227 Entering Passive Mode (127,0,0,1,150,122).
150 Here comes the directory listing.
drwxr-xr-x    2 0        0            4096 May 14 12:11 abc
-rw-r--r--    1 0        0            1091 Apr 27 08:34 passwd
226 Directory send OK.
ftp> exit
221 Goodbye.
```

（7）匿名用户只需要输入 IP 地址即可登录，如图 5-8 所示。

图 5-8　匿名用户登录

（8）默认匿名用户只能下载文件和目录，不能创建文件和目录。

5.4.2　配置匿名用户上传、下载案例

1．开启匿名用户相关功能

备份原始配置文件，代码如下。

```
[root@laohan_vsftp ~]#cp -av /etc/vsftpd/vsftpd.conf{,_bak_$(date +%F-%T)}
'/etc/vsftpd/vsftpd.conf' -> '/etc/vsftpd/vsftpd.conf_bak_2019-09-25-17:02:03'
```

开启匿名用户，如图 5-9 所示。

开启匿名用户上传、下载文件权限，如图 5-10 所示。

2．基本权限设置

进入/var/目录，设置 ftp 目录权限，代码如下。

```
[root@laohan_vsftp var]#ls -ld ftp/
drwxr-xr-x 3 root root 4096 Sep 25 16:46 ftp/
```

5.4 vsftpd 匿名用户配置案例

图 5-9 开启匿名用户

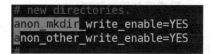

图 5-10 开启匿名用户上传、下载文件权限

设置 pub 目录权限为 777，添加该权限的目的是让匿名用户可以自由上传和下载，代码如下。

```
[root@laohan_vsftp var]#chmod 777 ftp/pub
[root@laohan_vsftp var]#cd /var/ftp/pub/
[root@laohan_vsftp pub]#ll
total 0
[root@laohan_vsftp pub]#cp -av /etc/passwd .
'/etc/passwd' -> './passwd'
```

3．测试

使用 ftp 用户登录，测试上传文件和目录权限（增加和删除），如图 5-11 所示。

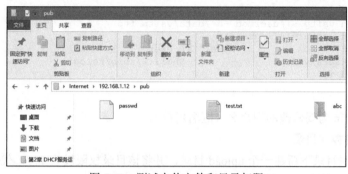

图 5-11 测试上传文件和目录权限

5.4.3 配置匿名用户仅有上传权限案例

目的：匿名用户可上传、不能删除、不能更名目录/文件。

方法：在匿名用户的主目录下新建一个 upload 目录，用来存放匿名用户上传的文件。

1．查看当前备份文件

```
[root@laohan_vsftp ~]#ll /etc/vsftpd/vsftpd.conf*
-rw------- 1 root root 4625 Sep 25 17:01 /etc/vsftpd/vsftpd.conf
-rw------- 1 root root 4599 Mar 22  2017 /etc/vsftpd/vsftpd.conf_bak_2019-09-25-16:46:50
-rw------- 1 root root 4625 Sep 25 17:01 /etc/vsftpd/vsftpd.conf_bak_2019-09-25-17:01:58
-rw------- 1 root root 4625 Sep 25 17:01 /etc/vsftpd/vsftpd.conf_bak_2019-09-25-17:02:01
-rw------- 1 root root 4625 Sep 25 17:01 /etc/vsftpd/vsftpd.conf_bak_2019-09-25-17:02:02
-rw------- 1 root root 4625 Sep 25 17:01 /etc/vsftpd/vsftpd.conf_bak_2019-09-25-17:02:03
```

上述代码备份 vsftpd.conf 到当前目录下，用以修改错误后及时回滚配置文件。

2．恢复备份文件到原始状态

恢复 vsftpd 配置文件到安装时的状态，代码如下。

```
[root@laohan_vsftp ~]#cp -av /etc/vsftpd/vsftpd.conf_bak_2019-09-25-17:02:03 /etc/vsftpd/vsftpd.conf
cp: overwrite '/etc/vsftpd/vsftpd.conf'? y
'/etc/vsftpd/vsftpd.conf_bak_2019-09-25-17:02:03' -> '/etc/vsftpd/vsftpd.conf'
```

3. 配置匿名用户权限

修改/etc/vsftpd/vsftpd.conf，配置匿名用户可上传文件，不能新建、删除目录和文件，代码如下。

```
anonymous_enable=YES              #启用匿名访问
ftp_username=ftp                  #指定匿名用户，默认为ftp
anon_root=/data/vsftp             #指定匿名用户登录后的主目录为/data/vsftp
write_enable=YES                  #配置登录的ftp用户写权限
anon_upload_enable=YES            #允许匿名用户上传文件
anon_mkdir_write_enable=NO        #不允许匿名用户创建目录
anon_other_write_enable=NO        #不允许匿名用户进行删除或者改名操作
```

4. 修改匿名用户主目录

修改匿名用户主目录，代码如下。

```
[root@laohan_vsftp ~]#finger ftp
Login: ftp                         Name: FTP User
Directory: /var/ftp                Shell: /sbin/nologin
Never logged in.
No mail.
No Plan.
[root@laohan_vsftp ~]#usermod -d /data/vsftp ftp
[root@laohan_vsftp ~]#finger ftp
Login: ftp                         Name: FTP User
Directory: /data/vsftp             Shell: /sbin/nologin
Never logged in.
No mail.
No Plan.
```

上述代码中 ftp 为要修改的用户名（匿名用户）。

5. 新建项目数据目录

在匿名用户主目录下新建一个 upload 目录，并将该目录权限设为 777，代码如下。

```
[root@laohan_vsftp ~]#mkdir -pv /data/vsftp/upload
mkdir: created directory '/data/vsftp'
mkdir: created directory '/data/vsftp/upload'
[root@laohan_vsftp ~]#chmod 777 /data/vsftp/upload
[root@laohan_vsftp ~]#ls -ld /data/vsftp/
drwxr-xr-x 3 root root 4096 Sep 25 17:31 /data/vsftp/
[root@laohan_vsftp ~]#ls -ld /data/vsftp/upload/
drwxrwxrwx 2 root root 4096 Sep 25 17:31 /data/vsftp/upload/
```

而此时主目录的属主应该是 root，代码如下。

```
[root@laohan_vsftp ~]#chown root.root /data/vsftp
[root@laohan_vsftp ~]#ls -ld /data/vsftp/
drwxr-xr-x 3 root root 4096 Sep 25 17:31 /data/vsftp/
```

这样匿名用户就对主目录有可读权限，而对 upload 有可读、可上传、非删除、非更名权限。如果在上面的条件下，要使匿名用户拥有 upload 目录下文件的删除、更名权限，则设置 anon_other_write_enable=YES 即可。

6. 测试匿名用户权限

重启服务，测试匿名用户权限，代码如下。

```
[root@laohan_vsftp ~]#/etc/init.d/vsftpd restart
Shutting down vsftpd:                                      [  OK  ]
Starting vsftpd for vsftpd:                                [  OK  ]
[root@laohan_vsftp ~]#netstat -ntpl
Active Internet connections (only servers)
Proto Recv-Q Send-Q Local Address           Foreign Address         State       PID/Program name
tcp        0      0 0.0.0.0:21              0.0.0.0:*               LISTEN      2051/vsftpd
tcp        0      0 0.0.0.0:22              0.0.0.0:*               LISTEN      1258/sshd
```

```
tcp        0      0 127.0.0.1:25            0.0.0.0:*               LISTEN      1337/master
tcp        0      0 :::22                   :::*                    LISTEN      1258/sshd
tcp        0      0 ::1:25                  :::*                    LISTEN      1337/master
[root@laohan_vsftp ~]#ps -ef |grep vsftpd
root      2051     1  0 17:33 ?        00:00:00 /usr/sbin/vsftpd /etc/vsftpd/vsftpd.conf
root      2055  1996  0 17:33 pts/1    00:00:00 grep vsftpd
```

测试匿名用户上传文件、新建和删除目录权限，如图 5-12 所示。

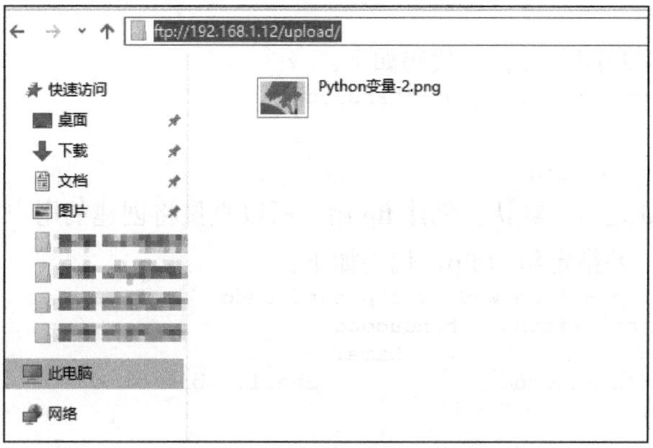

图 5-12　测试匿名用户权限

5.5　vsftpd 本地用户

本地用户多用于简单的企业场景，如企业对安全性和权限控制有要求，可以使用本地用户，建议在企业内部特定部门使用。一般情况下不建议使用本地用户，基本满足需求的前提下可以使用匿名用户，并做好权限控制，若对权限管理需求较高，建议使用 vsftpd 中高安全级别的虚拟用户。

5.5.1　本地用户案例

1．配置本地用户上传文件

修改 /etc/vsftpd/vsftpd.conf。

（1）禁用匿名登录，代码如下。

```
anonymous_enable=NO
local_enable=YES
write_enable=YES
```

（2）设置本地用户上传的文件权限，代码如下。

```
local_umask=022
dirmessage_enable=YES
xferlog_enable=YES
connect_from_port_20=YES
xferlog_std_format=YES
listen=YES
pam_service_name=vsftpd
userlist_enable=YES
tcp_wrappers=YES
```

(3) 限制用户在主目录之内,代码如下。
```
chroot_local_user=YES
```
(4) 针对不同的用户做不同的配置,这个目录下都是以用户名作为文件名,代码如下。
```
user_config_dir=/etc/vsftpd/vsftpd_user_conf
```

2. 创建用户配置目录

```
mkdir -pv /etc/vsftpd/vsftpd_user_conf
```

代码输出如下。

```
[root@laohan_vsftp ~]#mkdir -pv /etc/vsftpd/vsftpd_user_conf
mkdir: created directory '/etc/vsftpd/vsftpd_user_conf'
```

3. 配置本地用户

编辑 handuoduo 用户配置文件,代码如下。

```
vim /etc/vsftpd/vsftpd_user_conf/handuoduo
```

内容如下。

```
local_root=/data/handuoduo
```

因为安装好 vsftpd 之后,默认会创建 ftp 组,所以直接将创建的用户加入 ftp 组即可。然后创建 handuoduo 用户,并指定组为 ftp,代码如下。

```
[root@laohan_vsftp ~]#useradd -g ftp handuoduo
[root@laohan_vsftp ~]#finger handuoduo
Login: handuoduo                    Name:
Directory: /home/handuoduo          Shell: /bin/bash
Never logged in.
No mail.
No Plan.
```

设置用户密码,代码如下。

```
echo 1 |passwd --stdin handuoduo
Changing password for user handuoduo.
passwd: all authentication tokens updated successfully.
```

创建目录。

```
[root@laohan_vsftp ~]#mkdir -pv /data/handuoduo
mkdir: created directory '/data/handuoduo'
```

设置目录相关权限,设置目录用户为 handuoduo,代码如下。

```
chown -R handuoduo /data/handuoduo
```

代码输出如下。

```
[root@laohan_vsftp ~]#chown -R handuoduo /data/handuoduo
[root@laohan_vsftp ~]#ls -ld /data/handuoduo
drwxr-xr-x 2 handuoduo root 4096 Sep 25 18:10 /data/handuoduo
```

设置组有写入权限,代码如下。

```
chmod 755 -R /data/handuoduo
[root@laohan_vsftp ~]#chmod 755 -R /data/handuoduo
[root@laohan_vsftp ~]#ls -ld /data/handuoduo
drwxr-xr-x 2 handuoduo root 4096 Sep 25 18:10 /data/handuoduo
```

4. 测试本地用户

重启 vsftpd,代码如下。

```
/etc/init.d/vsftpd start
[root@laohan_vsftp ~]#/etc/init.d/vsftpd restart
Shutting down vsftpd:                              [  OK  ]
Starting vsftpd for vsftpd:                        [  OK  ]
```

访问 FTP 服务,代码如下。

```
ftp://192.168.1.12/
```

输入用户名和密码,成功之后,进入一个目录,测试上传文件和目录权限,效果如图 5-13 和图 5-14 所示。

图 5-13 本地用户登录

图 5-14 本地用户上传文件

5. 优化本地用户

优化本地用户，代码如下。

```
[root@laohan_vsftp ~]# adduser -g ftp -s /sbin/nologin hanmingze
[root@laohan_vsftp ~]# passwd beinan
Changing password for user beinan.
New password:
Retype new password:
passwd: all authentication tokens updated successfully.
```

上述命令中，新建用户系统会在/home 目录下建立一个与用户同名的文件夹。

ftp 用户登录系统后进入特定的目录可使用如下命令。

```
[root@laohan_vsftp ~]# adduser -d /opt/hanmingze -g ftp -s /sbin/nologin hanmingze
[root@laohan_vsftp ~]# passwd hanmingze
Changing password for user beinan.
New password:
Retype new password:
passwd: all authentication tokens updated successfully.
```

上述命令在系统中创建了名为 hanmingze 的用户，用户的主目录指向/opt/hanmingze，实际应用中，建议读者使用此种方法创建 ftp 用户（SSH 不可登录），增强 vsftpd 服务的健壮性。

5.5.2 配置本地用户经验谈

vsftpd 软件是可以以系统的普通用户来登录的。比如创建一个系统的普通用户 handuoduo，并设置一个密码，我们可以用 handuoduo 这个用户名和它的密码来登录 vsftpd，只不过这个登录不是 SSH 方式，而是 FTP 方式。登录后会进入 handuoduo 这个用户的主目录下，但是这样是不安全的，毕竟给这个用户设置了密码，它就能够登录操作系统。这时候，我们可以采用给 FTP 设置虚拟用户的方式解决这个问题，虚拟用户映射成系统的一个普通用户，或者也可以映射成多个普通用户，这样就算是有密码，也不能通过 SSH 去登录这台服务器，这样相对来说安全很多。具体代码如下。

```
[root@laohan_vsftp ~]#ssh  handuoduo@localhost
The authenticity of host 'localhost (::1)' can't be established.
RSA key fingerprint is 55:df:11:1c:33:e2:ae:44:19:18:77:fe:39:7b:b4:fa.
Are you sure you want to continue connecting (yes/no)? yes
Warning: Permanently added 'localhost' (RSA) to the list of known hosts.
handuoduo@localhost's password:
[handuoduo@laohan_vsftp ~]$whoami
handuoduo
```

登录系统后查看本地用户名和密码，代码如下。

```
[handuoduo@laohan_vsftp ~]$cat /etc/passwd
root:x:0:0:root:/root:/bin/bash
bin:x:1:1:bin:/bin:/sbin/nologin
```

```
daemon:x:2:2:daemon:/sbin:/sbin/nologin
adm:x:3:4:adm:/var/adm:/sbin/nologin
lp:x:4:7:lp:/var/spool/lpd:/sbin/nologin
sync:x:5:0:sync:/sbin:/bin/sync
shutdown:x:6:0:shutdown:/sbin:/sbin/shutdown
halt:x:7:0:halt:/sbin:/sbin/halt
mail:x:8:12:mail:/var/spool/mail:/sbin/nologin
uucp:x:10:14:uucp:/var/spool/uucp:/sbin/nologin
operator:x:11:0:operator:/root:/sbin/nologin
games:x:12:100:games:/usr/games:/sbin/nologin
gopher:x:13:30:gopher:/var/gopher:/sbin/nologin
ftp:x:14:50:FTP User:/data/vsftp:/sbin/nologin
nobody:x:99:99:Nobody:/:/sbin/nologin
vcsa:x:69:69:virtual console memory owner:/dev:/sbin/nologin
saslauth:x:499:76:Saslauthd user:/var/empty/saslauth:/sbin/nologin
postfix:x:89:89::/var/spool/postfix:/sbin/nologin
sshd:x:74:74:Privilege-separated SSH:/var/empty/sshd:/sbin/nologin
ntp:x:38:38::/etc/ntp:/sbin/nologin
nginx:x:498:498:Nginx web server:/var/lib/nginx:/sbin/nologin
dbus:x:81:81:System message bus:/:/sbin/nologin
apache:x:48:48:Apache:/var/www:/sbin/nologin
handuoduo:x:500:50::/data/handuoduo:/bin/bash
```

从上述代码可以看到，vsftpd 服务中的本地用户非常不安全，容易造成操作系统信息泄露，并带来安全隐患。

5.6 vsftpd 虚拟用户配置案例

如果以 vsftpd 系统用户访问 FTP 服务器，系统用户越多越不利于管理，而且不利于系统安全。为了能更加安全地使用 vsftpd，需使用 vsftpd 虚拟用户方式。

vsftpd 虚拟用户原理：虚拟用户就是没有创建真实系统用户，而是通过映射到其中一个真实用户和设置相应的权限来实现访问验证。虚拟用户不能登录 Linux 操作系统，从而让系统更加安全可靠。

5.6.1 配置 vsftpd 虚拟用户

1. 安装相关软件包

安装相关软件包，代码如下。

```
[root@www.ftp_server.com ~]#yum -y install pam vsftpd db4 db4-utils
[root@www.ftp_server.com ~]#rpm -q pam vsftpd db4 db4-utils
pam-1.1.1-24.el6.x86_64
vsftpd-2.2.2-24.el6.x86_64
db4-4.7.25-22.el6.x86_64
db4-utils-4.7.25-22.el6.x86_64
```

- pam 是用来提供身份验证的。
- vsftpd 是 FTP 服务的主程序。
- db4 支持文件数据库。
- db4 的软件包。

2. 创建虚拟用户

创建一个不能登录的用户，用作 FTP 服务的虚拟用户。由于这个 FTP 用户将来都是用来向网站上传文件的，因此创建用户的时候将这个用户的主目录设置在 Web 服务器的根目录，代码如下。
```
[root@www.ftp_server.com ~]# useradd -d /data/web/ -s /sbin/nologin vuser_ftp
[root@www.ftp_server.com ~]# id vuser_ftp
uid=891(vuser_ftp) gid=891(vuser_ftp) groups=891(vuser_ftp)
[root@www.ftp_server.com ~]# ls -ld /data/web/
drwx------ 2 vuser_ftp vuser_ftp 4096 Dec 11 22:16 /data/web/
```
- -d /data/web/ 表示指定用户的主目录。
- /data/web/ 指定 Web 服务器的根目录。
- -s /sbin/nologin 指令指定用户的 Shell，/sbin/nologin 表示此用户不能登录当前系统。
- 使用 vuser_ftp 作为虚拟用户的映射对象。

3．新建虚拟用户配置文件

创建一个记录 FTP 虚拟用户的用户名和密码的文件（文件名可随便指定，这里用 login.txt），此文件的格式如下。

- 一行用户名，换一行密码。
- 多个用户分行写入。

如创建/etc/vsftpd/login.txt 文件，并设定用户名和密码，代码如下。
```
cat >/etc/vsftpd/login.txt<<EOF
user1
user1passwd
user2
user2passwd
EOF
```

/etc/vsftpd/login.txt 文件内容如下。
```
[root@www.ftp_server.com ~]# cat /etc/vsftpd/login.txt
user1
user1passwd
user2
user2passwd
```

注意：用户名和密码要分开书写。

4．生成虚拟用户认证文件

使用 db_load 指令生成虚拟用户认证文件，代码如下。
```
[root@www.ftp_server.com ~]#db_load -T -t hash -f /etc/vsftpd/login.txt /etc/vsftpd/vsftpd_login.db
```

查看/etc/vsftpd/vsftpd_login.db 密码认证文件权限，代码如下。
```
[root@www.ftp_server.com ~]# ls -l /etc/vsftpd/vsftpd_login.db
```

把/etc/vsftpd/vsftpd_login.db 文件权限修改为 600。
```
[root@www.ftp_server.com ~]#ls -l /etc/vsftpd/vsftpd_login.db
-rw------- 1 root root 12288 Apr 27 12:25 /etc/vsftpd/vsftpd_login.db
[root@www.ftp_server.com ~]# chmod 600 /etc/vsftpd/vsftpd_login.db
[root@www.ftp_server.com ~]# ls -l /etc/vsftpd/vsftpd_login.db
-rw------- 1 root root 12288 Dec 11 22:18 /etc/vsftpd/vsftpd_login.db
```

- login.txt 文件是新建的用户名和密码文件。
- vsftpd_login.db 文件是 db_load 指令生成的虚拟用户认证文件，这个文件用 Vi 打开是看不到的，而且这个文件的权限为 600。
- db_load 指令是由 db4-utils 软件包提供的，如果没有这个指令就需要安装 db4-utils 软件包。

注意以下几点。

- 目前 login.txt 内容是实验环境，如果需要增加用户，就把用户名和密码按照规定的格式写入 login.txt 文件。

- 每次增加虚拟用户之后都使用 db_load 指令来更新 vsftpd_login.db 这个虚拟用户认证文件。
- 虚拟用户名单文件中用户的用户名和密码信息格式为"奇数行用户名,偶数行密码"。

5. 修改主配置文件

修改 FTP 服务主配置文件/etc/vsftpd/vsftpd.conf,可以把/etc/vsftpd/vsftpd.conf 备份,然后修改 vsftpd.conf,或者备份之后重写一个 vsftpd.conf 文件。

本次实验的 vsftpd.conf 内容如下。

```
[root@www.ftp_server.com ~]# cd /etc/vsftpd/
[root@www.ftp_server.com /etc/vsftpd]# cp -av vsftpd.conf vsftpd.conf_bak
'vsftpd.conf' -> 'vsftpd.conf_bak'
cat >/etc/vsftpd/vsftpd.conf<<EOF
anonymous_enable=NO
local_enable=YES
write_enable=YES
local_umask=022
dirmessage_enable=YES
xferlog_enable=YES
connect_from_port_20=YES
xferlog_file=/var/log/vsftpd.log
xferlog_std_format=YES
listen=YES
userlist_enable=YES
tcp_wrappers=YES
max_per_ip=5
max_clients=100
###### 下面是关于虚拟用户的配置-begin ######
#打开虚拟用户功能
guest_enable=YES
#将所有虚拟用户映射成 vuer_ftp 这个本地用户,此用户是之前新建的用户
guest_username=vuser_ftp
#ftp 用户的 pam 验证方式,默认是 vsftpd,必须改掉
pam_service_name=ftp.vu
#这里放置每个虚拟用户的配置文件
user_config_dir=/etc/vsftpd/vsftpd_user_conf
#####特别注意:vsftpd.conf 这个配置文件中每行的两端都不能有空格######
###### 下面是关于虚拟用户的配置-end ######
EOF
```

上述代码使用重定向的方式写入配置文件。

6. 创建 vsftpd 服务验证配置文件

(1)使用 rpm -ql vsftpd 指令查找验证模块信息。

```
[root@www.ftp_server.com /etc/vsftpd]# rpm -ql vsftpd  |grep vsftpd.pam
/usr/share/doc/vsftpd-2.2.2/EXAMPLE/VIRTUAL_USERS/vsftpd.pam
```

从上述代码可以看到,vsftpd.pam 文件已经存在。

(2)复制模板文件并改变文件名。

pam_service_name=ftp.vu 中 ftp.vu 表示使用相对路径,其绝对路径为/etc/pam.d/ftp.vu。

```
[root@www.ftp_server.com /etc/vsftpd]# pam_config="/usr/share/doc/vsftpd-2.2.2
/EXAMPLE/VIRTUAL_USERS/vsftpd.pam"
[root@www.ftp_server.com /etc/vsftpd]# echo $pam_config
/usr/share/doc/vsftpd-2.2.2/EXAMPLE/VIRTUAL_USERS/vsftpd.pam
/bin/cp -av ${pam_config} /etc/pam.d/ftp.vu
'/usr/share/doc/vsftpd-2.2.2/EXAMPLE/VIRTUAL_USERS/vsftpd.pam' -> '/etc/pam.d
/ftp.vu'
```

上述指令把 vsftpd 程序自带的关于 pam 认证的模板文件复制到 pam.d 这个服务的工作目录,同时修改文件名为 ftp.vu。/etc/pam.d/目录下已经有一个 vsftpd.pam 文件。现在要做的是让 vsftpd

虚拟用户的这个功能得到一个特殊的 pam 认证，还要修改 ftp.vu 文件，原文件内容如下。

```
[root@www.ftp_server.com ~]# rpm -ql vsftpd |grep vsftpd.pam
/usr/share/doc/vsftpd-2.2.2/EXAMPLE/VIRTUAL_USERS/vsftpd.pam
[root@www.ftp_server.com ~]# cat /usr/share/doc/vsftpd-2.2.2/EXAMPLE
/VIRTUAL_USERS/vsftpd.pam
auth required /lib/security/pam_userdb.so db=/etc/vsftpd_login
account required /lib/security/pam_userdb.so db=/etc/vsftpd_login
```

本次实验修改后的内容如下。

```
cat  >/etc/pam.d/ftp.vu<<EOF
auth required /lib64/security/pam_userdb.so db=/etc/vsftpd/vsftpd_login
account required /lib64/security/pam_userdb.so db=/etc/vsftpd/vsftpd_login
EOF
[root@www.ftp_server.com ~]# cat /etc/pam.d/ftp.vu
auth required /lib64/security/pam_userdb.so db=/etc/vsftpd/vsftpd_login
account required /lib64/security/pam_userdb.so db=/etc/vsftpd/vsftpd_login
```

有两个需要注意的地方。

（1）db=/etc/vsftpd/login 修改成 db=/etc/vsftpd/vsftpd_login，db=定义的是验证数据文件存放的位置，这个文件是以.db 结尾的。但是在/etc/pam.d/ftp.vu 中配置的时候不要加上.db，如图 5-15 所示。

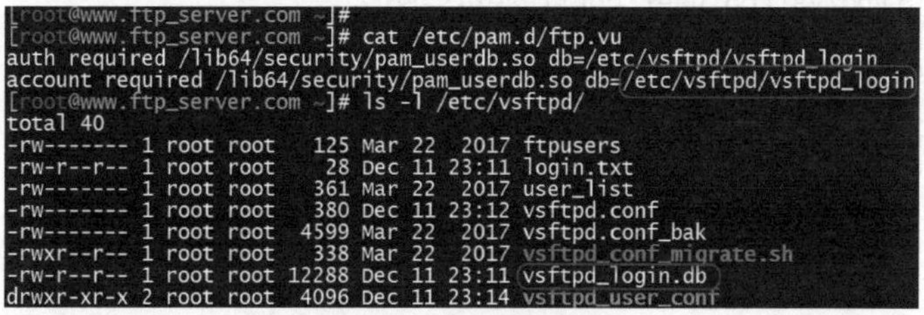

```
[root@www.ftp_server.com ~]#
[root@www.ftp_server.com ~]# cat /etc/pam.d/ftp.vu
auth required /lib64/security/pam_userdb.so db=/etc/vsftpd/vsftpd_login
account required /lib64/security/pam_userdb.so db=/etc/vsftpd/vsftpd_login
[root@www.ftp_server.com ~]# ls -l /etc/vsftpd/
total 40
-rw------- 1 root root   125 Mar 22  2017 ftpusers
-rw-r--r-- 1 root root    28 Dec 11 23:11 login.txt
-rw------- 1 root root   361 Mar 22  2017 user_list
-rw------- 1 root root   380 Dec 11 23:12 vsftpd.conf
-rw------- 1 root root  4599 Mar 22  2017 vsftpd.conf_bak
-rwxr--r-- 1 root root   338 Mar 22  2017 vsftpd_conf_migrate.sh
-rw-r--r-- 1 root root 12288 Dec 11 23:11 vsftpd_login.db
drwxr-xr-x 2 root root  4096 Dec 11 23:14 vsftpd_user_conf
```

图 5-15　配置虚拟用户

（2）注意配置文件中间的 /lib/security/pam_userdb.so，如果你的系统是 64 位的，那么相应的路径应是/lib64/security/pam-userdb.so，不然会出错。可使用下面的指令查看当前系统是否为 64 位，代码如下。

```
[root@www.ftp_server.com ~]# getconf LONG_BIT
64
```

上述代码返回结果中的 64 表示当前操作系统为 64 位。

5.6.2 创建虚拟用户目录

1. 创建虚拟用户配置文件存放目录

创建 vsftpd.conf 中提到的虚拟用户配置目录。

```
user_config_dir=/etc/vsftpd/vsftpd_user_conf
```

创建每个用户的权限配置文件，创建/etc/vsftpd/vsftpd_user_conf 目录，代码如下。

```
[root@www.ftp_server.com ~]# cd /etc/vsftpd/
[root@www.ftp_server.com /etc/vsftpd]# ll
total 40
-rw------- 1 root root   125 Mar 22  2017 ftpusers
-rw-r--r-- 1 root root    36 Dec 12 21:18 login.txt
-rw------- 1 root root   361 Mar 22  2017 user_list
-rw------- 1 root root   812 Dec 12 21:19 vsftpd.conf
-rw------- 1 root root  4599 Mar 22  2017 vsftpd.conf_bak
-rwxr--r-- 1 root root   338 Mar 22  2017 vsftpd_conf_migrate.sh
-rw------- 1 root root 12288 Dec 12 21:18 vsftpd_login.db
[root@www.ftp_server.com ~]# mkdir -pv /etc/vsftpd/vsftpd_user_conf
mkdir: created directory '/etc/vsftpd/vsftpd_user_conf'
```

```
[root@www.ftp_server.com /etc/vsftpd]# ll
total 44
-rw-------  1 root root   125 Mar 22  2017 ftpusers
-rw-r--r--  1 root root    36 Dec 12 21:18 login.txt
-rw-------  1 root root   361 Mar 22  2017 user_list
-rw-------  1 root root   812 Dec 12 21:19 vsftpd.conf
-rw-------  1 root root  4599 Mar 22  2017 vsftpd.conf_bak
-rwxr--r--  1 root root   338 Mar 22  2017 vsftpd_conf_migrate.sh
-rw-------  1 root root 12288 Dec 12 21:18 vsftpd_login.db
drwxr-xr-x  2 root root  4096 Dec 12 21:22 vsftpd_user_conf
[root@www.ftp_server.com /etc/vsftpd]# pwd
/etc/vsftpd
```

2．创建认证文件

在/etc/vsftpd/vsftpd_user_conf 目录下面分别创建在 login.txt 虚拟用户名和密码文件中提到的两个虚拟用户的权限配置文件 user1 和 user2，可以使用 touch user1 指令先创建 user1 文件，然后再用 vi 指令修改 user1 文件，代码如下。

```
#vi /etc/vsftpd/vsftpd_user_conf/user1
```

3．配置文件内容

虚拟用户配置文件内容如下。

```
cat >/etc/vsftpd/vsftpd_user_conf/user1<<EOF
anon_world_readable_only=no
write_enable=yes
anon_upload_enable=yes
anon_mkdir_write_enable=yes
anon_other_write_enable=yes
local_root=/data/web/html
EOF
[root@www.ftp_server.com /etc/vsftpd]# ll vsftpd_user_conf/
total 4
-rw-r--r-- 1 root root 150 Dec 12 21:23 user1
[root@www.ftp_server.com /etc/vsftpd]# cat /etc/vsftpd/vsftpd_user_conf/user1
anon_world_readable_only=no
write_enable=yes
anon_upload_enable=yes
anon_mkdir_write_enable=yes
anon_other_write_enable=yes
local_root=/data/web/html
```

上述代码中配置语句的两端都不能有空格。

4．创建权限控制文件

创建权限控制文件 user2，代码如下。

```
[root@www.ftp_server.com ~]# cp -av /etc/vsftpd/vsftpd_user_conf/user1 /etc/vsftpd/vsftpd_user_conf/user2
'/etc/vsftpd/vsftpd_user_conf/user1' -> '/etc/vsftpd/vsftpd_user_conf/user2'
[root@www.ftp_server.com ~]# cat >/etc/vsftpd/vsftpd_user_conf/user2<<EOF
anon_world_readable_only=no
write_enable=yes
anon_upload_enable=yes
anon_mkdir_write_enable=yes
anon_other_write_enable=yes
local_root=/data/web/vsftp_data
EOF
[root@www.ftp_server.com ~]# tree /etc/vsftpd/vsftpd_user_conf/
/etc/vsftpd/vsftpd_user_conf/
├── user1
└── user2

0 directories, 2 files
[root@www.ftp_server.com ~]# cat /etc/vsftpd/vsftpd_user_conf/user1
anon_world_readable_only=no
```

```
write_enable=yes
anon_upload_enable=yes
anon_mkdir_write_enable=yes
anon_other_write_enable=yes
local_root=/data/web/html
[root@www.ftp_server.com ~]# cat /etc/vsftpd/vsftpd_user_conf/user2
anon_world_readable_only=no
write_enable=yes
anon_upload_enable=yes
anon_mkdir_write_enable=yes
anon_other_write_enable=yes
local_root=/data/web/vsftp_data
```

在 user2 文件中修改 user2 文件的主目录即可，虚拟用户权限控制常用选项如表 5-7 所示。

表 5-7　　　　　　　　　　　　虚拟用户权限控制常用选项

选项	说明
anon_world_readable_only=no	用户可以浏览和下载文件
write_enable=yes	用户可以创建文件
anon_upload_enable=yes	用户可以上传文件
anon_mkdir_write_enable=yes	用户具有创建和删除目录的权限
anon_other_write_enable=yes	用户具有文件改名和删除文件的权限
local_root=/data/web//html	指定这个虚拟 ftp 用户的主目录

5. 创建虚拟用户主目录

创建虚拟用户主目录，代码如下。

```
[root@www.ftp_server.com ~]# mkdir -pv /data/web/{html,vsftp_data}
mkdir: created directory '/data/web/html'
mkdir: created directory '/data/web/vsftp_data'
[root@www.ftp_server.com ~]# tree /data/web/
/data/web/
├── html
└── vsftp_data

2 directories, 0 files
```

6. 修改目录权限

修改目录权限，代码如下。

```
[root@www.ftp_server.com ~]# cd /data/web/
[root@www.ftp_server.com /data/web]#
[root@www.ftp_server.com /data/web]# ll
total 8
drwxr-xr-x 2 root root 4096 Dec 12 21:25 html
drwxr-xr-x 2 root root 4096 Dec 12 21:25 vsftp_data
[root@www.ftp_server.com /data/web]# chown -R vuser_ftp. *
[root@www.ftp_server.com /data/web]#
[root@www.ftp_server.com /data/web]#
[root@www.ftp_server.com /data/web]# ll
total 8
drwxr-xr-x 2 vuser_ftp vuser_ftp 4096 Dec 12 21:25 html
drwxr-xr-x 2 vuser_ftp vuser_ftp 4096 Dec 12 21:25 vsftp_data
```

上述代码表示修改 vsftp_data 权限成功。

5.6.3 验证 vsftpd 服务

1. 重启 vsftpd 服务

虚拟用户配置完毕，重启 vsftpd 服务就可以使用虚拟用户登录 FTP。

vuser_ftp 这个用户是被映射的，因此该用户不能登录 FTP。

```
[root@www.ftp_server.com ~]# /etc/init.d/vsftpd start
Starting vsftpd for vsftpd:                           [  OK  ]
```
查看 vsftpd 服务。
```
[root@www.ftp_server.com ~]# netstat -ntpl |grep 21
tcp        0      0 0.0.0.0:21              0.0.0.0:*               LISTEN      1548/vsftpd
[root@www.ftp_server.com ~]# lsof -i tcp:21
COMMAND  PID USER   FD   TYPE DEVICE SIZE/OFF NODE NAME
vsftpd  1548 root    3u  IPv4  11262      0t0  TCP *:ftp (LISTEN)
[root@www.ftp_server.com ~]# ps aux |grep vsftpd
root      1548  0.0  0.0  52132   808 ?        Ss   22:50   0:00 /usr/sbin/vsftpd /etc/vsftpd/vsftpd.conf
root      1572  0.0  0.0 103328   880 pts/0    S+   22:53   0:00 grep vsftpd
```
从上述代码可以看到，21 端口已经被监听。

2．遇到的问题和解决方法

正确配置 vsftpd 的虚拟用户，使用 FTP 客户端时提示登录不了，或者可以登录但没有操作这个目录的权限，最有可能的"罪魁祸首"是 SELinux 和防火墙。

（1）配置防火墙。

防火墙如果开启而 FTP 端口没有打开，就不能连接到 FTP，关闭防火墙，代码如下。
```
[root@www.ftp_server.com ~]#/etc/init.d/iptables stop
```
（2）查看 SELinux。

SELinux 如果没有关闭，它内置的一些规则就会让虚拟 FTP 用户服务器无法登录或者获得相应的权限，临时关闭 SELinux，代码如下。
```
[root@www.ftp_server.com ~]#setenforce 0
setenforce: SELINUX is disabled
```
如果关闭 SELinux，依然无法访问 vsftpd，则应查看对应的配置文件是否正确。

3．vsftpd 运维经验分享

（1）设置权限注意事项。

如果虚拟用户的根目录在虚拟用户的配置文件中没有定义（local_root=），那么默认的根目录，就会被映射为本地用户的根目录（之前建的 vuser_ftp）。因此虚拟用户要在这个目录中有权限的话，必须先使被映射的这个本地用户对这个虚拟用户指定的根目录有相应的权限。

（2）映射用户注意事项。

这里实验用的本地用户是 vuser_ftp，在实际的配置中可以用 httpd 这个服务用户来作为映射的用户，这样一来，整个网站的目录只需要一个用户，很容易实现权限控制。

4．用户认证是核心

新增用户步骤如下。

（1）在/etc/vsftpd/login.txt 文件里面添加用户名和密码，使用如下指令重新生成认证文件。用户信息配置文件如图 5-16 所示。
```
db_load -T -t hash -f /etc/vsftpd/login.txt /etc/vsftpd/vsftpd_login.db
```

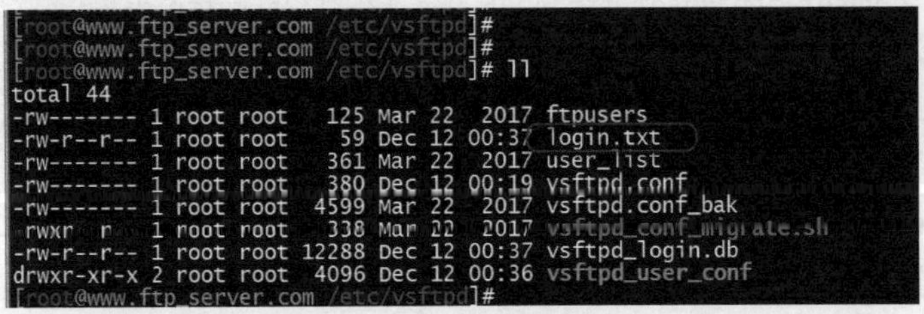

```
[root@www.ftp_server.com /etc/vsftpd]#
[root@www.ftp_server.com /etc/vsftpd]#
[root@www.ftp_server.com /etc/vsftpd]# ll
total 44
-rw-------  1 root root   125 Mar 22  2017 ftpusers
-rw-r--r--  1 root root    59 Dec 12 00:37 login.txt
-rw-r--r--  1 root root   361 Mar 22  2017 user_list
-rw-------  1 root root   380 Dec 12 00:19 vsftpd.conf
-rw-r--r--  1 root root  4599 Mar 22  2017 vsftpd.conf_bak
-rwxr--r--  1 root root   338 Mar 22  2017 vsftpd_conf_migrate.sh
-rw-r--r--  1 root root 12288 Dec 12 00:37 vsftpd_login.db
drwxr-xr-x  2 root root  4096 Dec 12 00:36 vsftpd_user_conf
[root@www.ftp_server.com /etc/vsftpd]#
```

图 5-16 用户信息配置文件

（2）多用户配置。

复制用户认证文件，修改不同目录即可。

```
[root@www.ftp_server.com /etc/vsftpd/vsftpd_user_conf]# pwd
/etc/vsftpd/vsftpd_user_conf
[root@www.ftp_server.com /etc/vsftpd/vsftpd_user_conf]# ls
hanmeimei    test1
[root@www.ftp_server.com /etc/vsftpd/vsftpd_user_conf]# cat test1
anon_world_readable_only=no
write_enable=yes
anon_upload_enable=yes
anon_mkdir_write_enable=yes
anon_other_write_enable=yes
local_root=/home/web/html
[root@www.ftp_server.com /etc/vsftpd/vsftpd_user_conf]# cat hanmeimei
anon_world_readable_only=no
write_enable=yes
anon_upload_enable=yes
anon_mkdir_write_enable=yes
anon_other_write_enable=yes
local_root=/home/web/hanmeimei
```

上述代码配置多用户。

第6章 rsync 服务

6.1 rsync 基础知识

6.1.1 rsync 快速入门

1. Rsync 介绍

rsync 是类 UNIX 操作系统的一款应用软件，它能同步更新单台或多台服务器的文件与目录。

rsync 用"rsync 算法"来使本地和远程两台主机之间的文件达到同步，这个算法只传送两个文件的不同部分，因此速度相当快，通常可以作为**备份工具**来使用。运行 rsync 服务器的计算机也叫 backup 服务器，一个 rsync 服务器可以同时备份多个 rsync 客户端的数据，多个 rsync 服务器也可以备份一个 rsync 客户端的数据。

rsync 除了使用 SSH 作为底层传输通道传输数据外，还可以使用 daemon 模式。rsync 服务器会打开 873 端口，等待对方的 rsync 连接。

连接 rsync 时，rsync 服务器会检查口令是否相符。若通过口令校验，则可以开始进行文件传输。第一次连接完成时，会把整份文件传输一次，下一次再进行同步时就只传送两个文件中不同的部分。

SSH 模式下，rsync 客户端程序必须同时在本地和远程的机器提前部署和安装。

2. rsync 基本特点

rsync 同步和备份工具的基本特点如下。
- 可以镜像保存整个目录树和文件系统。
- 可以很容易做到保持原来文件的权限、时间、软/硬链接等。
- 无须特殊权限即可安装。
- 可优化流程，文件传输效率高。
- 可以使用 RCP、SSH 等方式来传输文件，当然也可以通过直接的 socket 连接。
- 支持匿名传输。

3. rsync 指令基本格式

rsync 指令的基本格式有以下 6 种。
```
rsync [OPTION]… SRC DEST
rsync [OPTION]… SRC [USER@]HOST:DEST
rsync [OPTION]… [USER@]HOST:SRC DEST
rsync [OPTION]… [USER@]HOST::SRC DEST
rsync [OPTION]… SRC [USER@]HOST::DEST
rsync [OPTION]… rsync://[USER@]HOST[:PORT]/SRC [DEST]
```

4. rsync 的 6 种工作模式

了解并掌握 rsync 的工作模式，有助于用户在不同的工作场景下选择和采用最优解决方案，rsync 的 6 种工作模式讲解如下。

（1）复制本地文件。

当 SRC 和 DES 路径信息都不包含单个":"分隔符时，就启动这种工作模式，代码如下。
```
1   [root@lnmp_0_5 ~]# echo "跟老韩学 Shell 编程" > laohan/laohan.txt
2   [root@lnmp_0_5 ~]# cat laohan/laohan.txt
3   跟老韩学 Shell 编程
4   [root@lnmp_0_5 ~]# echo "跟老韩学 rsync" > rsync/rsync.txt
```

```
5   [root@lnmp_0_5 ~]# cat rsync/rsync.txt
6   跟老韩学 rsync
7   [root@lnmp_0_5 ~]# rsync -a rsync/ laohan/
8   [root@lnmp_0_5 ~]# ll laohan/
9   total 8
10  -rw-r--r-- 1 root root 24 Apr  3 08:31 laohan.txt
11  -rw-r--r-- 1 root root 18 Apr  3 08:31 rsync.txt
12  [root@lnmp_0_5 ~]# rsync -a rsync laohan/
13  [root@lnmp_0_5 ~]# ls -lhrt laohan/
14  total 12K
15  -rw-r--r-- 1 root root    24 Apr  3 08:31 laohan.txt
16  -rw-r--r-- 1 root root    18 Apr  3 08:31 rsync.txt
17  drwxr-xr-x 2 root root  4.0K Apr  3 08:31 rsync
18  [root@lnmp_0_5 ~]# tree laohan/
19  laohan/
20  |-- laohan.txt
21  |-- rsync
22  |   '-- rsync.txt
23  '-- rsync.txt
24
25  1 directory, 3 files
```

上述代码中第1～6行为测试文件和内容，第7行将rsync目录下的文件同步到laohan目录下，第8～11行为laohan目录下的文件列表信息。

第12行同步rsync目录和目录下的所有文件到laohan目录，第13～25行为laohan目录下文件列表信息和目录层次结构。

（2）复制本地文件到远程服务器。

使用远程Shell程序实现将本地文件复制到远程服务器。当DST路径信息包含单个":"分隔符时启动该工作模式，代码如下。

```
1   [root@lnmp_0_5 ~]# rsync -avz -e "ssh -p51518 " laohan 172.16.0.16:/root/
2   The authenticity of host '[172.16.0.16]:51518 ([172.16.0.16]:51518)' can't be
    established.
3   ECDSA key fingerprint is SHA256:/F3TPcw718AtyI1MZ9l/zHdl+p9GarnuIVIIyEoeXSs.
4   ECDSA key fingerprint is MD5:7d:c0:3e:6b:99:9e:9f:70:bd:e2:18:23:7f:e1:86:01.
5   Are you sure you want to continue connecting (yes/no)? yes
6   Warning: Permanently added '[172.16.0.16]:51518' (ECDSA) to the list of
    known hosts.
7   root@172.16.0.16's password:
8   sending incremental file list
9   laohan/
10  laohan/laohan.txt
11  laohan/rsync.txt
12  laohan/rsync/
13  laohan/rsync/rsync.txt
14
15  sent 362 bytes  received 89 bytes  12.70 bytes/sec
16  total size is 60  speedup is 0.13
17  [root@lnmp_0_16 ~]# tree /root/laohan
18  /root/laohan
19  |-- laohan.txt
20  |-- rsync
21  |   '-- rsync.txt
22  '-- rsync.txt
23
24  1 directory, 3 files
```

上述代码中第1～16行为本地服务器操作指令和返回结果，第17～24行为目标服务器同步的文件列表信息。

（3）复制远程服务器文件到本地服务器。

使用远程Shell程序实现将远程服务器的文件复制到本地服务器。当SRC路径信息包含单

个 ":" 分隔符时启动该工作模式，代码如下。

```
1   [root@lnmp_0_5 ~]# ls -lhrt
2   total 1.1M
3   -rw-r--r-- 1 root root 1009K Aug 14  2019 nginx-1.16.1.tar.gz
4   -rw-r--r-- 1 root root   148 Oct 13 18:07 a.sh
5   -rw-r--r-- 1 root root   315 Feb 27 21:20 a.py
6   drwxr-xr-x 2 root root  4.0K Apr  3 08:31 rsync
7   drwxr-xr-x 3 root root  4.0K Apr  3 08:32 laohan
8   [root@lnmp_0_5 ~]# rm -rf laohan/
9   [root@lnmp_0_5 ~]# rsync -avz  -e "ssh -p51518 " 172.16.0.16:/root/laohan .
10  root@172.16.0.16's password:
11  receiving incremental file list
12  laohan/
13  laohan/laohan.txt
14  laohan/rsync.txt
15  laohan/rsync/
16  laohan/rsync/rsync.txt
17
18  sent 97 bytes   received 358 bytes    70.00 bytes/sec
19  total size is 60  speedup is 0.13
20  [root@lnmp_0_5 ~]# tree /root/laohan/
21  /root/laohan/
22  |-- laohan.txt
23  |-- rsync
24  |   '-- rsync.txt
25  '-- rsync.txt
26
27  1 directory, 3 files
```

上述代码中第 1~8 行为本地服务器文件操作，即删除当前用户主目录下的 laohan 文件夹。第 9 行使用 rsync 指令将远程服务器 root 用户主目录下的 laohan 文件夹同步到本地的当前路径下。第 10~19 行为同步过程和结果。

（4）从远程 rsync 服务器中复制文件到本地服务器。

当 SRC 路径信息包含 "::" 分隔符时启动该工作模式，代码如下。

```
rsync -av root@172.16.78.192::www /databack
```

上述代码中远程服务器启动 rsync 服务（**rsync 守护进程**），并设置 www 同步模块，将 www 模块下的目录同步到本地的 /databack 文件夹中。

（5）从本地服务器复制文件到远程 rsync 服务器中。

当 DST 路径信息包含 "::" 分隔符时启动该工作模式，代码如下。

```
rsync -av /databack root@172.16.78.192::www
```

上述代码中远程服务器 rsync 作为后台运行进程，并设置 www 模块，将本地的 /databack 文件夹内容同步到远程的 www 模块的目录中。

（6）列出远程服务器的文件信息。

列出远程服务器的文件信息代码如下。

```
1   [root@lnmp_0_5 ~]# rsync -v  -e "ssh -p51518 " 172.16.0.16:/root/laohan
2   root@172.16.0.16's password:
3   receiving file list ... done
4   drwxr-xr-x         4,096 2020/04/03 08:32:14 laohan
5
6   sent 20 bytes   received 40 bytes   13.33 bytes/sec
7   total size is 0  speedup is 0.00
8   [root@lnmp_0_5 ~]# rsync -v  -e "ssh -p51518 " 172.16.0.16:/root/laohan/
9   root@172.16.0.16's password:
10  receiving file list ... done
11  drwxr-xr-x         4,096 2020/04/03 08:32:14 .
12  -rw-r--r--            24 2020/04/03 08:31:22 laohan.txt
13  -rw-r--r--            18 2020/04/03 08:31:40 rsync.txt
```

```
14 drwxr-xr-x              4,096 2020/04/03 08:31:40 rsync
15
16 sent 20 bytes  received 108 bytes   12.19 bytes/sec
17 total size is 42   speedup is 0.33
```

上述代码中第 1～7 行列出远程服务器目录的信息，第 8～17 行列出远程服务器目录下的所有文件信息。

5. rsync 工作模式总结

rsync 不能实现直接控制远程服务器和远程服务器之间的复制，对应于以上 6 种指令格式，可以总结为 rsync 有两种不同的模式。

（1）Shell 模式。

使用远程 Shell 程序进行连接，当源路径或目的路径的主机名后面包含一个 "：" 分隔符时使用这种模式，rsync 安装完成后就可以直接使用该模式。

（2）daemon 模式。

当源路径或目的路径的主机名后面包含 "::" 时，必须在一台服务器上启动 rsync 守护进程，通过 rsync --daemon 独立进程或者 xinetd 超级进程来管理 rsync 后台进程。

当 rsync 作为守护进程运行时，它需要一个运行该进程的用户。该用户必须对相应的模块和目录拥有读写数据、日志的权限。当 rsync 以 daemon 模式运行时，还需要 rsyncd.conf 配置文件。

6. 安装 rsync 软件

rsync 软件在 CentOS 6 上默认已经安装，也可使用 yum install rsync –y 指令安装。rsync 服务器和客户端使用同一个软件包，可使用如下指令查询 rsync 软件包的详细信息。

```
1  [root@lnmp_0_5 ~]# yum search rsync
2  Loaded plugins: fastestmirror, langpacks
3  Repository updates is listed more than once in the configuration
4  Repository extras is listed more than once in the configuration
5  Loading mirror speeds from cached hostfile
6   * base: mirrors.aliyun.com
7  ============================================================== N/S matched: rsync ==============================================================
8  grsync.x86_64 : A Gtk+ GUI for rsync
9  libguestfs-rsync.x86_64 : rsync support for libguestfs
10 librsync.x86_64 : Rsync remote-delta algorithm library
11 librsync-devel.x86_64 : Headers and development libraries for librsync
12 librsync-doc.noarch : Documentation files for librsync
13 perl-File-RsyncP.x86_64 : A perl implementation of an Rsync client
14 rsync-bpc.x86_64 : A customized version of rsync that is used as part of BackupPC
15 duplicity.x86_64 : Encrypted bandwidth-efficient backup using rsync algorithm
16 rclone.x86_64 : Rsync for cloud storage
17 rsync.x86_64 : A program for synchronizing files over a network
18
19    Name and summary matches only, use "search all" for everything.
20 [root@lnmp_0_5 ~]# yum info rsync
21 Loaded plugins: fastestmirror, langpacks
22 Repository updates is listed more than once in the configuration
23 Repository extras is listed more than once in the configuration
24 Loading mirror speeds from cached hostfile
25  * base: mirrors.aliyun.com
26 Installed Packages
27 Name        : rsync
28 Arch        : x86_64
29 Version     : 3.1.2
30 Release     : 6.el7_6.1
31 Size        : 815 k
32 Repo        : installed
```

```
33  From repo   : updates
34  Summary     : A program for synchronizing files over a network
35  URL         : http://rsync.samba.org/
36  License     : GPLv3+
37  Description : Rsync uses a reliable algorithm to bring remote and host files into
38              : sync very quickly. Rsync is fast because it just sends the
                  differences
39              : in the files over the network instead of sending the complete
40              : files. Rsync is often used as a very powerful mirroring process or
41              : just as a more capable replacement for the rcp command. A technical
42              : report which describes the rsync algorithm is included in this
43              : package.
```

上述代码中第 1～19 行为用 yum search 指令查看 rsync 软件包的输出，第 20～43 行为用 yum info 指令查看 rsync 软件包的输出。

6.1.2 rsync 特性和核心算法

1. rsync 的特性

- 支持复制特殊文件，如链接和设备文件等。
- 可以排除指定文件或同步目录，相当于打包指令 tar。
- 可以保持原文件或目录的权限、时间戳和软/硬链接等属性均不改变。
- 可实现增量同步，即只同步已发生变化的数据，因此数据传输效率更高。
- 可以使用 RCP、RSH、SSH 等方式来配合传输文件，也可以通过直接的 socket 连接进行操作。
- 支持匿名的或认证的进程模式传输，方便进行数据备份。

2. rsync 核心算法介绍

假定在名为 Nginx_11 和 Nginx_12 的两台计算机之间同步相似的文件 A 与 B，其中 Nginx_11 对文件 A 拥有访问权，Nginx_12 对文件 B 拥有访问权，并且假定主机 Nginx_11 与 Nginx_12 之间的网络带宽很小，那么 rsync 算法将通过下面的 5 个步骤来完成。

（1）Nginx_12 将文件 B 分割成一组不重叠的、固定大小为 n B 的数据块。最后一块可能会比 n B 小。

（2）Nginx_12 对每一个分割好的数据块执行两种校验：一种是 32bit 的滚动弱校验，另一种是 128bit 的 MD4 强校验。

（3）Nginx_12 将这些校验结果发给 Nginx_11。

（4）Nginx_11 通过搜索文件 A 的所有大小为 n B 的数据块（偏移量可以任选，不一定非要是 n 的倍数），来寻找与文件 B 的某一块有着相同的弱校验码和强校验码的数据块。这项工作可以借助滚动校验的特性很快完成。

（5）Nginx_11 发给 Nginx_12 一串指令来生成文件 A 在 Nginx_12 上的备份。这里的每一条指令要么是对文件 B 拥有某一个数据块而不须重传的证明，要么是一个数据块，这个数据块肯定是没有与文件 B 的任何一个数据块匹配上。

3. rsync 的优点与缺点

（1）rsync 的优点。

- 速度快：除首次全复制外，其他时候实现增量复制，传输速度快。
- 更安全：传输数据时可用 SSH 加密传输。

- 带宽占用更少：rsync 可对数据进行分块压缩传输，相比其他文件传输工具占用更少带宽。
- 权限限制：非 root 用户也可安装和执行 rsync 指令。
- 支持目录层级递归复制：rsync 可以镜像保存整个目录树和文件系统。
- 支持限速：传输文件时使用传输限速功能，避免服务器网卡流量写满，导致服务不可用。
- 支持断点续传：同步过程中由于其他原因导致同步工作未完成，下次执行该同步操作时会继续从断点处执行同步。
- 支持 128bit MD4 校验（3.0 以后的版本使用 MD5 加密）。
- 可以很容易做到保持原来文件的权限、时间、软/硬链接等文件属性信息。
- 与传统的 cp、tar 备份方式相比，rsync 具有安全性高、备份迅速、支持增量备份等优点。rsync 可以满足对实时性要求不高的数据备份需求，例如，定期地备份服务器文件数据到远程服务器、定期对本地磁盘进行数据镜像等。

（2）rsync 的缺点。

随着系统规模的不断扩大，rsync 的缺点逐渐暴露出来。

- rsync 数据同步时，需要先扫描所有文件，然后进行对比，最后进行差量传输。如果文件数量很多或文件体积很大，rsync 扫描文件时非常耗时，而且发生变化的文件往往是很少一部分，因此 rsync 扫描海量文件是非常低效的。
- rsync 不能实时监测、同步数据，虽然它可以通过 Linux 守护进程的方式触发同步，但是两次触发动作一定会有时间差，可能导致服务器和客户端数据出现不一致。

4．rsync 同步过程

rsync 文件同步涉及**源文件**和**目标文件**的概念，还涉及以哪个文件为**同步基准**。

rsync 同步过程由两种模式组成：决定哪些文件需要同步的检查模式和文件同步时的同步模式。

（1）rsync 检查模式。

检查模式是指按照指定规则来检查哪些文件需要被同步，哪些文件是明确被排除不传输的。默认情况下，rsync 使用 quick check（快速检查）算法快速检查源文件和目标文件的大小、修改时间是否一致，如果不一致则需要传输。我们也可以通过在 rsync 指令行中指定某些选项来改变 quick check 的检查模式，比如--size-only 选项表示 quick check 将仅检查文件大小不同的文件。

（2）rsync 同步模式。

同步模式是指在文件确定同步后，在同步过程发生之前要做哪些额外工作。如是否要先删除源主机上没有，但目标主机上有的文件，是否要先备份已存在的目标文件，是否要追踪链接文件等。

rsync 提供非常多的选项使得同步模式变得更具灵活性。

rsync 手动指定同步模式的选项更为常用，只有在有特殊需求时才指定检查模式，因为大多数检查模式选项都可能会影响 rsync 的性能。

6.1.3 rsync 基础运维实例

本小节采用实例的方式讲解 rsync 在实际企业环境中的具体使用，基本操作环境如表 6-1 所示。

6.1 rsync 基础知识

表 6-1 rsync 基本操作环境

角色	操作系统	IP 地址	主机名
服务器	CentOS 6.9 x84_64	192.168.2.156	rsync_server_156

1. 单台服务器内同步

（1）服务器安装 rsync 软件，代码如下。

```
[root@rsync_server_156 ~]# yum install epel-release -y
已加载插件：fastestmirror
设置安装进程
Determining fastest mirrors
 * base: mirror.bit.edu.cn
 * extras: mirrors.huaweicloud.com
 * updates: mirrors.huaweicloud.com
base
（中间代码略）

已安装:
  epel-release.noarch 0:6-8

完毕!
[root@rsync_server_156 ~]# ll /etc/yum.repos.d/ |grep epel
-rw-r--r--. 1 root root  957 11月  5 2012 epel.repo
-rw-r--r--. 1 root root 1056 11月  5 2012 epel-testing.repo
```

上述代码表示 epel 源已经安装完成。

```
[root@rsync_server_156 ~]# yum install rsync -y
已加载插件：fastestmirror
设置安装进程
Loading mirror speeds from cached hostfile
epel/metalink
| 7.6 kB     00:00
 * base: mirror.bit.edu.cn
 * epel: mirrors.tuna.tsinghua.edu.cn
 * extras: mirrors.huaweicloud.com
 * updates: mirrors.huaweicloud.com
（中间代码略）

已安装:
  rsync.x86_64 0:3.0.6-12.el6

完毕!
[root@rsync_server_156 ~]# rsync --version
rsync  version 3.0.6  protocol version 30
Copyright (C) 1996-2009 by Andrew Tridgell, Wayne Davison, and others.
Web site: http://rsync.samba.org/
Capabilities:
    64-bit files, 64-bit inums, 64-bit timestamps, 64-bit long ints,
    socketpairs, hardlinks, symlinks, IPv6, batchfiles, inplace,
    append, ACLs, xattrs, iconv, symtimes

rsync comes with ABSOLUTELY NO WARRANTY.  This is free software, and you
are welcome to redistribute it under certain conditions.  See the GNU
General Public Licence for details.
```

上述代码表示 rsync 软件已经安装成功。

（2）查看防火墙和 SELinux 是否关闭。

```
[root@rsync_server_156 ~]# sestatus
SELINUX status:                 disabled
[root@rsync_server_156 ~]# /etc/init.d/iptables status
iptables: 未运行防火墙。
```

从上述代码可以看到，防火墙和 SELinux 均已经处于关闭状态。

(3) 单台服务器同步。

创建源测试目录，代码如下。

```
[root@rsync_server_156 ~]# mkdir -pv  /src/server_156/
mkdir: 已创建目录 "/src"
mkdir: 已创建目录 "/src/server_156/"
[root@rsync_server_156 ~]# touch  /src/server_156/{1,2,3,4,5,6,7}
[root@rsync_server_156 ~]# ll /src/server_156/
总用量 0
-rw-r--r--. 1 root root 0 6月   4 23:11 1
-rw-r--r--. 1 root root 0 6月   4 23:11 2
-rw-r--r--. 1 root root 0 6月   4 23:11 3
-rw-r--r--. 1 root root 0 6月   4 23:11 4
-rw-r--r--. 1 root root 0 6月   4 23:11 5
-rw-r--r--. 1 root root 0 6月   4 23:11 6
-rw-r--r--. 1 root root 0 6月   4 23:11 7
[root@rsync_server_156 ~]# tree /src/server_156/
/src/server_156/
├── 1
├── 2
├── 3
├── 4
├── 5
├── 6
└── 7

0 directories, 7 files
```

上述代码创建源测试目录。

```
[root@rsync_server_156 ~]# mkdir -pv  /dest/server_156/
mkdir: 已创建目录 "/dest"
mkdir: 已创建目录 "/dest/server_156/"
[root@rsync_server_156 ~]# tree /dest/server_156/
/dest/server_156/

0 directories, 0 files
```

上述代码创建目标测试目录。

【实例 6-1】同步/src/server_156/目录内所有文件到/dest/server_156/目录

将/src/server_156/目录里的所有的文件同步至/dest/server_156/目录（不包含/src/server_156/目录本身），代码如下。

```
[root@rsync_server_156 ~]# rsync -av /src/server_156/ /dest/server_156/
sending incremental file list
./
1
2
3
4
5
6
7

sent 348 bytes  received 148 bytes  992.00 bytes/sec
total size is 0  speedup is 0.00
```

上述代码将/dest/server_156/目录内的所有文件同步到/src/server_156/目录内。

【实例 6-2】同步目录内所有文件和目录（包括目录自身）到目标目录

将/src/server_156/目录里的所有文件同步至/dest/server_156/目录（包含/src/server_156/目录本身），代码如下。

```
[root@rsync_server_156 ~]# rsync -av /src/server_156 /dest/server_156/
sending incremental file list
server_156/
```

```
server_156/1
server_156/2
server_156/3
server_156/4
server_156/5
server_156/6
server_156/7

sent 373 bytes  received 149 bytes   1044.00 bytes/sec
total size is 0  speedup is 0.00
```

下面代码与上述代码作用一致。

```
[root@rsync_server_156 ~]# rsync -avR /src/server_156 /dest/server_156/
sending incremental file list
/src/
/src/server_156/
/src/server_156/1
/src/server_156/2
/src/server_156/3
/src/server_156/4
/src/server_156/5
/src/server_156/6
/src/server_156/7

sent 396 bytes  received 153 bytes   1098.00 bytes/sec
total size is 0  speedup is 0.00
```

将/src/server_156/目录整个同步至/dest/server_156/目录。

2. 局域网同步

局域网内各服务器之间数据同步，rsync 服务器环境信息如表 6-2 所示。

表 6-2　　　　　　　　　　　rsync 服务器环境信息

角色	操作系统	IP 地址	主机名
本地服务器	CentOS 6.9 x84_64	192.168.2.156	rsync_server_156
客户端	CentOS 6.9 x84_64	192.168.2.157	rsync_client_157

（1）确认上述计算机关闭了防火墙和 SELinux。

```
[root@rsync_server_156 ~]# sestatus
SELINUX status:                 disabled
[root@rsync_server_156 ~]# /etc/init.d/iptables status
iptables: 未运行防火墙。
[root@rsync_client_157 ~]# /etc/init.d/iptables status
iptables: 未运行防火墙。
[root@rsync_client_157 ~]# sestatus
SELINUX status:                 disabled
```

从上述代码可以看到，防火墙和 SELinux 均已经处于关闭状态。

（2）实现密钥认证。

```
[root@rsync_server_156 ~]# ssh-keygen
Generating public/private rsa key pair.
Enter file in which to save the key (/root/.ssh/id_rsa):
Created directory '/root/.ssh'.
Enter passphrase (empty for no passphrase):
Enter same passphrase again:
Your identification has been saved in /root/.ssh/id_rsa.
Your public key has been saved in /root/.ssh/id_rsa.pub.
The key fingerprint is:
bf:cc:26:9d:aa:4e:1c:71:e1:0b:f7:83:78:1e:af:f5 root@rsync_server_156
The key's randomart image is:
+--[ RSA 2048]----+
|         .       |
|        . .      |
```

```
|        o +          |
|         * +         |
|        o S o        |
|       . + + .       |
|        o ..+.       |
|       .  .=+o       |
|       ..o.o++ E     |
+-----------------+
```

上述代码使用 ssh-keygen 指令生成密钥对。

```
[root@rsync_server_156 ~]# ssh-copy-id -i /root/.ssh/id_rsa.pub  192.168.2.157
The authenticity of host '192.168.2.157 (192.168.2.157)' can't be established.
RSA key fingerprint is 82:e8:a0:65:a9:90:8e:f4:60:4d:6e:0e:7d:34:e0:75.
Are you sure you want to continue connecting (yes/no)? yes
Warning: Permanently added '192.168.2.157' (RSA) to the list of known hosts.
root@192.168.2.157's password:
Now try logging into the machine, with "ssh '192.168.2.157'", and check in:

  .ssh/authorized_keys

to make sure we haven't added extra keys that you weren't expecting.
```

上述代码传递公钥文件到 192.168.2.157。

```
[root@rsync_server_156 ~]# ssh 192.168.2.157
Last login: Sat Jun  8 10:29:13 2019 from 192.168.2.106
[root@rsync_client_157 ~]# logout
Connection to 192.168.2.157 closed.
```

上述代码用于测试密钥文件登录。

```
[root@rsync_client_157 ~]# ssh-keygen
Generating public/private rsa key pair.
Enter file in which to save the key (/root/.ssh/id_rsa):
Enter passphrase (empty for no passphrase):
Enter same passphrase again:
Your identification has been saved in /root/.ssh/id_rsa.
Your public key has been saved in /root/.ssh/id_rsa.pub.
The key fingerprint is:
a9:ce:45:7d:c7:56:17:c7:59:0e:66:33:00:5b:88:8e root@rsync_client_157
The key's randomart image is:
+--[ RSA 2048]----+
|         ..oo.*o=|
|          . ..o o *+|
|         o .    +|
|        E .o . ..|
|         S . . + |
|          o . o  |
|         . .     |
|        o .      |
|         o       |
+-----------------+
[root@rsync_client_157 ~]# ssh-copy-id -i /root/.ssh/id_rsa.pub  192.168.2.156
The authenticity of host '192.168.2.156 (192.168.2.156)' can't be established.
RSA key fingerprint is 82:e8:a0:65:a9:90:8e:f4:60:4d:6e:0e:7d:34:e0:75.
Are you sure you want to continue connecting (yes/no)? yes
Warning: Permanently added '192.168.2.156' (RSA) to the list of known hosts.
root@192.168.2.156's password:
Now try logging into the machine, with "ssh '192.168.2.156'", and check in:

  .ssh/authorized_keys

to make sure we haven't added extra keys that you weren't expecting.

[root@rsync_client_157 ~]# ssh 192.168.2.156
Last login: Sat Jun  8 10:29:07 2019 from 192.168.2.106
[root@rsync_server_156 ~]# exit
logout
```

```
Connection to 192.168.2.156 closed.
```

上述代码首先在本地服务器生成密钥对，然后传输公钥文件到远程服务器，最后进行密钥文件登录验证。

【实例 6-3】 同步远程服务器文件到本机

远程服务器创建测试文件，代码如下。

```
[root@rsync_client_157 ~]# echo "This is a test file">client_157.log
[root@rsync_client_157 ~]# cat client_157.log
This is a test file
```

开始执行同步，代码如下。

```
[root@rsync_server_156 ~]# mkdir -pv  rsync_bak
mkdir: 已创建目录 "rsync_bak"
```

上述代码创建存放文件目录。

同步远程服务器 192.168.2.157 上的/client_157.log 文件到本机 rsync_bak/目录，代码如下。

```
[root@rsync_server_156 ~]# rsync -av 192.168.2.157:/root/client_157.log rsync_bak/
bash: rsync: command not found
rsync: connection unexpectedly closed (0 bytes received so far) [receiver]
rsync error: error in rsync protocol data stream (code 12) at io.c(600) [receiver=3.0.6]
[root@rsync_server_156 ~]# rsync -av 192.168.2.157:/root/client_157.log rsync_bak/
receiving incremental file list
client_157.log

sent 30 bytes  received 107 bytes  91.33 bytes/sec
total size is 20  speedup is 0.15
[root@rsync_server_156 ~]# tree rsync_bak/
rsync_bak/
└── client_157.log

0 directories, 1 file
```

查看文件内容是否正确，代码如下。

```
[root@rsync_server_156 ~]# cat rsync_bak/client_157.log
This is a test file
```

上述代码中，远程服务器 rsync 软件未安装，可以使用如下代码进行安装。

```
[root@rsync_client_157 ~]# yum install rsync -y
已加载插件：fastestmirror
设置安装进程
Determining fastest mirrors
 * base: mirrors.tuna.tsinghua.edu.cn
 * extras: mirror.bit.edu.cn
 * updates: mirror.bit.edu.cn
（中间代码略）

已安装:
  rsync.x86_64 0:3.0.6-12.el6

完毕!
[root@rsync_client_157 ~]# rsync  --version
rsync  version 3.0.6  protocol version 30
Copyright (C) 1996-2009 by Andrew Tridgell, Wayne Davison, and others.
Web site: http://rsync.samba.org/
Capabilities:
   64-bit files, 64-bit inums, 64-bit timestamps, 64-bit long ints,
   socketpairs, hardlinks, symlinks, IPv6, batchfiles, inplace,
   append, ACLs, xattrs, iconv, symtimes

rsync comes with ABSOLUTELY NO WARRANTY.  This is free software, and you
are welcome to redistribute it under certain conditions.  See the GNU
General Public Licence for details.
```

上述代码表示，rsync 软件已经被安装成功。

【实例 6-4】 同步本地文件到远程服务器

同步本地文件到远程服务器，代码如下。

```
[root@rsync_server_156 ~]# echo 'This is a test file' >rsync_server_156.log
[root@rsync_server_156 ~]# cat rsync_server_156.log
This is a test file
[root@rsync_server_156 ~]# rsync  -av rsync_server_156.log 192.168.2.157:/root/
sending incremental file list
rsync_server_156.log

sent 108 bytes   received 31 bytes   278.00 bytes/sec
total size is 20  speedup is 0.14
[root@rsync_client_157 ~]# cat rsync_server_156.log
This is a test file
```

从上述代码可以看到，文件内容无误，说明同步成功。

3．局域网指定用户同步

【实例 6-5】 局域网同步本地文件到远程服务器

同步本地文件到远程服务器，并指定端口和用户，代码如下。

```
[root@rsync_server_156 ~]#  rsync -avz  -e "ssh -p 22 -o StrictHostKeyChecking=no -l root" /root/rsync_server_156.log  192.168.2.157:/tmp
sending incremental file list
rsync_server_156.log

sent 82 bytes   received 37 bytes   238.00 bytes/sec
total size is 20  speedup is 0.17
[root@rsync_server_156 ~]# ssh 192.168.2.157 cat /tmp/rsync_server_156.log
This is a test file
```

rsync 程序的**-e** 选项表示使用**远程服务器**的 SSH 连接相关信息，代码如下。

```
[root@rsync_server_156 ~]# rsync  --help |grep -A5 -w '\-e'
 -e, --rsh=COMMAND           specify the remote shell to use
     --rsync-path=PROGRAM    specify the rsync to run on the remote machine
     --existing              skip creating new files on receiver
     --ignore-existing       skip updating files that already exist on receiver
     --remove-source-files   sender removes synchronized files (non-dirs)
     --del                   an alias for --delete-during
```

4．rsync 其他常用方法介绍

【实例 6-6】 只更新客户端已存在的文件

（1）创建服务器测试文件，代码如下。

```
[root@rsync_server_156 ~]# mkdir -pv  156_test/{a,b,c}
mkdir: 已创建目录 "156_test"
mkdir: 已创建目录 "156_test/a"
mkdir: 已创建目录 "156_test/b"
mkdir: 已创建目录 "156_test/c"
[root@rsync_server_156 ~]# for dir in $(echo  156_test/{a..c} |tr ' ' '\n');
do cp -av rsync_server_156.log $dir;done
"rsync_server_156.log" -> "156_test/a/rsync_server_156.log"
"rsync_server_156.log" -> "156_test/b/rsync_server_156.log"
"rsync_server_156.log" -> "156_test/c/rsync_server_156.log"
```

上述代码表示创建服务器测试文件。

（2）客户端创建测试目录，代码如下。

```
[root@rsync_client_157 ~]# mkdir test_157
```

上述代码表示创建测试目录。

（3）同步文件到客户端，代码如下。

```
[root@rsync_server_156 ~]# rsync -avz 156_test 192.168.2.157:/root/test_157
sending incremental file list
```

```
156_test/
156_test/a/
156_test/a/rsync_server_156.log
156_test/b/
156_test/b/rsync_server_156.log
156_test/c/
156_test/c/rsync_server_156.log

sent 360 bytes  received 85 bytes  890.00 bytes/sec
total size is 60  speedup is 0.13
```
上述代码表示同步文件到客户端。

（4）客户端删除单个目录，代码如下。
```
[root@rsync_client_157 test_157]# ll
总用量 4
drwxr-xr-x 5 root root 4096 6月  8 11:21 156_test
[root@rsync_client_157 test_157]# cd 156_test/
[root@rsync_client_157 156_test]# ll
总用量 12
drwxr-xr-x 2 root root 4096 6月  8 11:22 a
drwxr-xr-x 2 root root 4096 6月  8 11:22 b
drwxr-xr-x 2 root root 4096 6月  8 11:22 c
[root@rsync_client_157 156_test]# rm -rfv c/
已删除"c/rsync_server_156.log"
已删除目录:"c"
```
上述代码表示在客户端删除单个目录。

（5）开始同步已经存在的文件，代码如下。
```
[root@rsync_server_156 ~]# rsync -r -v --existing 156_test/a 192.168.2.157:/root/test_157
sending incremental file list

sent 67 bytes  received 13 bytes  53.33 bytes/sec
total size is 20  speedup is 0.25
```
上述代码表示只更新客户端已经存在的文件。

【实例6-7】rsync文件同步测试

（1）创建156.log测试文件，写入"line 1"数据进行测试，代码如下。
```
[root@rsync_server_156 ~]# echo 'line 1' >156.log
[root@rsync_server_156 ~]# cat 156.log
line 1
```
上述代码表示创建测试文件。

（2）同步156.log文件到远程服务器，代码如下。
```
[root@rsync_server_156 ~]# rsync -avz 156.log 192.168.2.157:/root/
sending incremental file list
156.log

sent 79 bytes  received 31 bytes  220.00 bytes/sec
total size is 7  speedup is 0.06
```
上述代码表示同步本地文件到远程服务器。

（3）查看远程服务器156.log内容，代码如下。
```
[root@rsync_server_156 ~]# ssh 192.168.2.157 cat /root/156.log
line 1
```
上述代码表示查看远程服务器文件是否同步成功。

（4）创建156.log测试文件，写入"line 2"数据进行测试，代码如下。
```
[root@rsync_server_156 ~]# echo 'line 2' >156.log
[root@rsync_server_156 ~]# cat 156.log
line 2
[root@rsync_server_156 ~]# rsync -avz 156.log 192.168.2.157:/root/
```

```
sending incremental file list
156.log

sent 79 bytes  received 37 bytes   232.00 bytes/sec
total size is 7  speedup is 0.06
```

上述代码表示创建测试文件。

（5）查看远程服务器 156.log 内容，代码如下。

```
[root@rsync_server_156 ~]# ssh 192.168.2.157 cat /root/156.log
line 2
```

上述代码表示查看文件内容。

（6）追加"line 3"数据，测试 156.log 文件，代码如下。

```
[root@rsync_server_156 ~]# echo 'line 3' >>156.log
[root@rsync_server_156 ~]# cat 156.log
line 2
line 3
```

上述代码表示创建测试文件。

（7）同步 156.log 文件到远程服务器，代码如下。

```
[root@rsync_server_156 ~]# rsync -avz 156.log 192.168.2.157:/root/
sending incremental file list
156.log

sent 84 bytes  received 37 bytes   242.00 bytes/sec
total size is 14  speedup is 0.12
```

上述代码表示同步本地文件到远程服务器。

（8）查看远程服务器 156.log 内容，代码如下。

```
[root@rsync_server_156 ~]# ssh 192.168.2.157 cat /root/156.log
line 2
line 3
[root@rsync_server_156 ~]#
[root@rsync_server_156 ~]#
[root@rsync_server_156 ~]#
[root@rsync_server_156 ~]# echo 'line 4' >>156.log
[root@rsync_server_156 ~]# cat 156.log
line 2
line 3
line 4
```

上述代码表示查看远程服务器文件内容。

（9）同步 156.log 到远程服务器，代码如下。

```
[root@rsync_server_156 ~]# rsync -v 156.log 192.168.2.157:/root/
156.log

sent 92 bytes  received 37 bytes   258.00 bytes/sec
total size is 21  speedup is 0.16
```

上述代码表示同步本地文件到远程服务器。

（10）查看远程服务器文件内容。

```
[root@rsync_server_156 ~]# ssh 192.168.2.157 cat /root/156.log
line 2
line 3
line 4
[root@rsync_server_156 ~]# ll 156.log
-rw-r--r-- 1 root root 21 6月   8 11:38 156.log
[root@rsync_server_156 ~]# ssh 192.168.2.157 ls -l /root/156.log
-rw-r--r-- 1 root root 21 6月   8 11:38 /root/156.log
```

上述代码表示查看远程服务器文件内容和文件详细信息。

6.2 rsync 配置文件和选项规则

6.2.1 rsync 配置文件

rsyncd.conf 由全局配置和若干模块配置组成，配置文件的语法如下。

（1）模块以[module]开始。

（2）参数配置行的格式是 name = value，其中 value 有以下两种数据类型。

- 字符串（可以不用引号定界字符串）。
- 布尔值（1/0 或 yes/no 或 true/false）。

（3）以"#"或";"开始的行为注释。

（4）"\"为换行符。

1．rsync 全局参数

rsyncd.conf 配置文件中除[module]之外的所有配置行都是全局参数，我们可以在全局参数部分定义模块参数，这时该参数的值是所有模块的默认值，如表 6-3 所示。

表 6-3　　　　　　　　　　　　rsync 全局参数

参数	说明
address	独立运行时指定的服务器 IP 地址
port	指定 rsync 守护进程监听的端口号
motd file	当客户连接服务器时该文件的内容显示给客户
pid file	rsync 的守护进程将其 PID 写入指定的文件
log file	指定 rsync 守护进程的日志文件
syslog facility	rsync 发送日志消息给 syslog 的消息级别
socket options	自定义 TCP 选项

2．rsync 模块参数

模块参数主要用于定义 rsync 服务器中的同步目录，模块参数声明的格式必须为[module]形式，rsync 模块中可以定义以下参数。

（1）rsync 基本模块参数，如表 6-4 所示。

表 6-4　　　　　　　　　　　　rsync 基本模块参数

参数	说明
path	当前模块在 rsync 服务器上的同步路径，该参数必须指定
comment	客户端连接到模块列表时显示客户端的描述信息

（2）rsync 模块控制参数，如表 6-5 所示。

表 6-5　　　　　　　　　　　　rsync 模块控制参数

参数	说明
use chroot	chroot 到 path 参数所指定的目录

续表

参数	说明
uid	指定该模块以指定的 UID 传输文件
gid	指定该模块以指定的 GID 传输文件
max connections	最大并发连接数量，超过限制请求将被限制
lock file	指定支持 max connections 参数的锁文件
list	使用的模块列表是否应该被列出
read only	指定是否允许用户上传文件
write only	指定是否允许用户下载文件
ignore errors	运行 delete 操作时是否忽略 I/O 错误
ignore nonreadable	完全忽略那些用户没有访问权限的文件
timeout	该参数可以覆盖用户指定的 IP 超时时间
dont compress	指定传输之前不进行压缩处理的文件

（3）rsync 模块文件筛选参数，如表 6-6 所示。

表 6-6　　　　　　　　rsync 模块文件筛选参数

参数	说明
exclude	由空格隔开的多个文件或目录（相对路径），添加到 exclude 列表中
exclude from	exclude 规则定义的文件名，服务器读取 exclude 列表定义
include	由空格隔开的多个文件或目录（相对路径），添加到 include 列表中
include from	include 规则定义的文件名，从该文件中读取 include 列表定义

配置 rsync 模块文件筛选参数时注意事项如下。
- 一个模块只能指定一个 exclude 参数、一个 include 参数。
- 结合 include 和 exclude 可以定义复杂的 include/exclude 规则。
- 这几个参数分别与相应的 rsync 客户端指令选项等价，唯一不同的是它们作用在服务器。

（4）rsync 模块用户认证参数，如表 6-7 所示。

表 6-7　　　　　　　　rsync 模块用户认证参数

参数	说明
auth users	指定用户列表允许连接该模块
secrets file	rsync 认证口令文件
strict modes	指定是否监测口令文件的权限

配置 rsync 模块用户认证参数时注意事项如下。
- rsync 认证口令文件的权限一定是 600，否则客户端将不能连接服务器。
- rsync 认证口令文件中每一行指定一个用户名：口令对。
- 用户名：口令对格式为：username:passwd。
- 一般来说口令最好不要超过 8 个字符。若只配置匿名访问的 rsync 服务器，则无须设置上述参数。

（5）rsync 模块访问控制参数，如表 6-8 所示。

表 6-8　rsync 模块访问控制参数

参数	说明
hosts allow	用一个主机列表指定哪些主机客户允许连接该模块
hosts deny	用一个主机列表指定哪些主机客户不允许连接该模块

配置 rsync 模块访问控制参数时注意事项如下。

- 单个 IP 地址。例如 192.168.0.1。
- 整个网段。例如 192.168.0.0/24、192.168.0.0/255.255.255.0。
- 可解析的单个主机名。例如 centos、centos.bsmart.cn。
- 域内的所有主机。例如*.bsmart.cn。
- "*"表示所有。
- 多个列表项要用空格间隔。

（6）rsync 模块日志参数，如表 6-9 所示。

表 6-9　rsync 模块日志参数

参数	说明
transfer logging	rsync 服务器将传输操作记录到传输日志文件
log format	指定传输日志文件的字段

当 rsync 模块访问控制参数中设置了 log file 参数时，在日志每行的开始会添加%t [%p]，可以使用的日志格式定义符如下。

- %a：远程 IP 地址。
- %h：远程主机名。
- %l：文件长度字符数。
- %p：该次 rsync 会话的 PID。
- %o：操作类型为"send"或"recv"。
- %f：文件名。
- %P：模块路径。
- %m：模块名。
- %t：当前时间。
- %u：认证的用户名（匿名时是 null）。
- %b：实际传输的字节数。
- %c：当发送文件时，记录该文件的校验码。

6.2.2　rsync 排除和包含文件规则

【实例 6-8】rsync 同步时排除文件

使用--exclude 选项指定排除规则，排除那些不需要传输的文件，查看测试文件目录结构，代码如下。

```
[root@rsync_server_156 156_test]# tree /root/156_test/
/root/156_test/
├── a
│   └── rsync_server_156.log
├── abc.pdf
```

```
├── a.log
├── a.pdf
├── b
│   └── rsync_server_156.log
├── b.log
├── b.pdf
├── c
│   └── rsync_server_156.log
├── c.log
├── c.pdf
└── def.pdf

6 directories, 8 files
```

上述文件中，所有以.pdf 结尾的文件都不进行同步。首先可以看到，以.pdf 结尾的有文件和目录，可以使用正则表达式进行匹配，代码如下：

```
[root@rsync_server_156 156_test]# rsync -avz *  --exclude="*.pdf"
192.168.2.157:/test_157
sending incremental file list
created directory /test_157
a.log
b.log
c.log
a/
a/rsync_server_156.log
b/
b/rsync_server_156.log
c/
c/rsync_server_156.log

sent 528 bytes  received 138 bytes  444.00 bytes/sec
total size is 101  speedup is 0.15
```

从上述代码可以看到，当前服务器上所有以.pdf 结尾的文件和目录都没有同步到远程服务器上，查看远程服务器同步结果，代码如下：

```
[root@rsync_client_157 ~]# tree /test_157/
/test_157/
├── a
│   └── rsync_server_156.log
├── a.log
├── b
│   └── rsync_server_156.log
├── b.log
├── c
│   └── rsync_server_156.log
└── c.log

3 directories, 6 files
```

从上述代码可以看到，以.pdf 结尾的文件和目录都没有同步过来。

【实例 6-9】rsync 同步时排除指定的文件和目录

（1）进入 c.pdf 目录下，创建以 a~g 开头并且以.pdf 结尾的文件，代码如下：

```
[root@rsync_server_156 c.pdf]# touch {a..g}.pdf
[root@rsync_server_156 c.pdf]# ll
总用量 0
-rw-r--r-- 1 root root 0 6月  8 14:03 a.pdf
-rw-r--r-- 1 root root 0 6月  8 14:03 b.pdf
-rw-r--r-- 1 root root 0 6月  8 14:03 c.pdf
-rw-r--r-- 1 root root 0 6月  8 14:03 d.pdf
-rw-r--r-- 1 root root 0 6月  8 14:03 e.pdf
-rw-r--r-- 1 root root 0 6月  8 14:03 f.pdf
-rw-r--r-- 1 root root 0 6月  8 14:03 g.pdf
```

从上述代码可以看到，.pdf 文件已经成功创建。

```
[root@rsync_server_156 156_test]# rsync -avz *  --exclude="c.pdf" --exclude="c.
pdf/*.pdf"  192.168.2.157:/test_157
sending incremental file list
a.log
abc.pdf
b.log
c.log
def.pdf
a.pdf/
a/
a/rsync_server_156.log
b.pdf/
b/
b/rsync_server_156.log
c/
c/rsync_server_156.log

sent 672 bytes  received 184 bytes   1712.00 bytes/sec
total size is 101  speedup is 0.12
```

上述代码的说明如下。

- rsync -avz * 表示同步当前目录下的所有文件和目录。
- --exclude="c.pdf" 表示不同步此目录。
- --exclude="c.pdf/*.pdf" 表示不同步此目录下所有以.pdf 结尾的文件。

(2) 远程服务器确认同步文件信息，代码如下。

```
[root@rsync_client_157 ~]# ll /test_157/
总用量 24
drwxr-xr-x 2 root root 4096 6月   8 11:22 a
-rw-r--r-- 1 root root    0 6月   8 13:53 abc.pdf
-rw-r--r-- 1 root root   21 6月   8 11:38 a.log
drwxr-xr-x 2 root root 4096 6月   8 13:53 a.pdf
drwxr-xr-x 2 root root 4096 6月   8 11:22 b
-rw-r--r-- 1 root root    0 6月   8 13:53 b.log
drwxr-xr-x 2 root root 4096 6月   8 13:53 b.pdf
drwxr-xr-x 2 root root 4096 6月   8 11:22 c
-rw-r--r-- 1 root root    0 6月   8 13:53 c.log
-rw-r--r-- 1 root root    0 6月   8 13:53 def.pdf
```

从上述代码可以看到，c.pdf 目录和目录下的以.pdf 结尾的文件未被同步过来。

(3) --exclude 使用注意事项。

注意，一个--exclude 只能指定一条排除规则，要指定多条排除规则，需要使用多个--exclude 选项，或者将排除规则写入文件中，然后使用--exclude-from 选项读取该规则文件。另外，除了 --exclude 排除规则，还有--include 包含规则，即筛选出要进行传输的文件，所以--include 规则也称为传输规则。它的使用方法和--exclude 一样。如果一个文件既能匹配排除规则，又能匹配包含规则，则先匹配到的立即生效，生效后就不再进行任何匹配。

6.2.3 rsync 镜像同步

使用--delete 选项后，接收端的 rsync 会先删除目标目录下已经存在的文件。

【实例 6-10】镜像同步文件和目录测试

rsync 的镜像同步文件和目录测试，代码如下。

```
[root@rsync_server_156 ~]# rsync -n -avz  --delete  156_test/ 192.168.2.157:test_157
sending incremental file list
./
deleting 156_test/b/rsync_server_156.log
deleting 156_test/b/
```

```
deleting 156_test/a/rsync_server_156.log
deleting 156_test/a/
deleting 156_test/
a.log
abc.pdf
b.log
c.log
def.pdf
a.pdf/
b.pdf/
b/
b/rsync_server_156.log
c.pdf/
c.pdf/a.pdf
c.pdf/b.pdf
c.pdf/c.pdf
c.pdf/d.pdf
c.pdf/e.pdf
c.pdf/f.pdf
c.pdf/g.pdf
c/
c/rsync_server_156.log

sent 467 bytes   received 78 bytes   1090.00 bytes/sec
total size is 101  speedup is 0.19 (DRY RUN)
```

镜像本机目录 156_test 下的所有文件到远程服务器目标目录。

```
[root@rsync_client_157 ~]# tree test_157/
test_157/
├── 156_test
│   ├── a
│   │   └── rsync_server_156.log
│   └── b
│       └── rsync_server_156.log
└── a
    └── rsync_server_156.log

4 directories, 3 files
```

从上述代码可以看到，远程服务器并没有相关文件。

【实例 6-11】镜像同步文件和目录到远程服务器

1. 镜像同步文件和目录到远程服务器

镜像同步 156_test/目录内所有文件和目录（不包含目录自身）到远程服务器，代码如下。

```
[root@rsync_server_156 ~]# rsync -avz --delete 156_test/ 192.168.2.157:test_157
sending incremental file list
./
deleting 156_test/b/rsync_server_156.log
deleting 156_test/b/
deleting 156_test/a/rsync_server_156.log
deleting 156_test/a/
deleting 156_test/
a.log
abc.pdf
b.log
c.log
def.pdf
a.pdf/
b.pdf/
b/
b/rsync_server_156.log
c.pdf/
c.pdf/a.pdf
c.pdf/b.pdf
c.pdf/c.pdf
```

```
c.pdf/d.pdf
c.pdf/e.pdf
c.pdf/f.pdf
c.pdf/g.pdf
c/
c/rsync_server_156.log

sent 993 bytes   received 302 bytes   2590.00 bytes/sec
total size is 101   speedup is 0.08
```

2. 查看远程服务器上的文件

查看远程服务器上的文件，代码如下。

```
[root@rsync_client_157 ~]# tree test_157/
test_157/
├── a
│   └── rsync_server_156.log
├── abc.pdf
├── a.log
├── a.pdf
├── b
│   └── rsync_server_156.log
├── b.log
├── b.pdf
├── c
│   └── rsync_server_156.log
├── c.log
├── c.pdf
│   ├── a.pdf
│   ├── b.pdf
│   ├── c.pdf
│   ├── d.pdf
│   ├── e.pdf
│   ├── f.pdf
│   └── g.pdf
└── def.pdf

6 directories, 15 files
```

3. 创建测试文件

创建相关测试文件，代码如下。

```
[root@rsync_client_157 ~]# touch test_157/{a..z}.pdf
[root@rsync_client_157 ~]# tree test_157/
test_157/
├── a
│   └── rsync_server_156.log
├── abc.pdf
├── a.log
├── a.pdf
├── b
│   └── rsync_server_156.log
├── b.log
├── b.pdf
├── c
│   └── rsync_server_156.log
├── c.log
├── c.pdf
│   ├── a.pdf
│   ├── b.pdf
│   ├── c.pdf
│   ├── d.pdf
│   ├── e.pdf
│   ├── f.pdf
│   └── g.pdf
├── def.pdf
├── d.pdf
```

```
        ├── e.pdf
        ├── f.pdf
        ├── g.pdf
        ├── h.pdf
        ├── i.pdf
        ├── j.pdf
        ├── k.pdf
        ├── l.pdf
        ├── m.pdf
        ├── n.pdf
        ├── o.pdf
        ├── p.pdf
        ├── q.pdf
        ├── r.pdf
        ├── s.pdf
        ├── t.pdf
        ├── u.pdf
        ├── v.pdf
        ├── w.pdf
        ├── x.pdf
        ├── y.pdf
        └── z.pdf

6 directories, 38 files
```

上述代码创建测试文件和目录。

4. 同步测试文件

本地服务器再次执行同步操作,代码如下。

```
[root@rsync_server_156 ~]# rsync -n -avz --delete 156_test/ 192.168.2.157:test_157
sending incremental file list
./
deleting z.pdf
deleting y.pdf
deleting x.pdf
deleting w.pdf
deleting v.pdf
deleting u.pdf
deleting t.pdf
deleting s.pdf
deleting r.pdf
deleting q.pdf
deleting p.pdf
deleting o.pdf
deleting n.pdf
deleting m.pdf
deleting l.pdf
deleting k.pdf
deleting j.pdf
deleting i.pdf
deleting h.pdf
deleting g.pdf
deleting f.pdf
deleting e.pdf
deleting d.pdf
a.pdf/
b.pdf/
c.pdf/

sent 419 bytes  received 30 bytes  898.00 bytes/sec
total size is 101  speedup is 0.22 (DRY RUN)
[root@rsync_server_156 ~]# rsync -avz --delete 156_test/ 192.168.2.157:test_157
sending incremental file list
./
deleting z.pdf
```

```
        deleting y.pdf
        deleting x.pdf
        deleting w.pdf
        deleting v.pdf
        deleting u.pdf
        deleting t.pdf
        deleting s.pdf
        deleting r.pdf
        deleting q.pdf
        deleting p.pdf
        deleting o.pdf
        deleting n.pdf
        deleting m.pdf
        deleting l.pdf
        deleting k.pdf
        deleting j.pdf
        deleting i.pdf
        deleting h.pdf
        deleting g.pdf
        deleting f.pdf
        deleting e.pdf
        deleting d.pdf
        a.pdf/
        b.pdf/
        c.pdf/

        sent 419 bytes  received 30 bytes   898.00 bytes/sec
        total size is 101  speedup is 0.22
```

上述代码使用- -delete 选项，执行镜像同步。

5. 本地服务器目录和文件与远程服务器目标和文件对比

本地服务器目录和文件与远程服务器目标和文件对比，代码如下。

```
[root@rsync_server_156 ~]# tree 156_test/
156_test/
├── a
│   └── rsync_server_156.log
├── abc.pdf
├── a.log
├── a.pdf
├── b
│   └── rsync_server_156.log
├── b.log
├── b.pdf
├── c
│   └── rsync_server_156.log
├── c.log
├── c.pdf
│   ├── a.pdf
│   ├── b.pdf
│   ├── c.pdf
│   ├── d.pdf
│   ├── e.pdf
│   ├── f.pdf
│   └── g.pdf
└── def.pdf

6 directories, 15 files


[root@rsync_client_157 ~]# tree test_157/
test_157/
├── a
│   └── rsync_server_156.log
├── abc.pdf
```

```
├── a.log
├── a.pdf
├── b
│   └── rsync_server_156.log
├── b.log
├── b.pdf
├── c
│   └── rsync_server_156.log
├── c.log
├── c.pdf
│   ├── a.pdf
│   ├── b.pdf
│   ├── c.pdf
│   ├── d.pdf
│   ├── e.pdf
│   ├── f.pdf
│   └── g.pdf
└── def.pdf

6 directories, 15 files
```

6. 查看本机文件和目录属性信息

查看本机文件和目录属性信息，代码如下。

```
[root@rsync_server_156 ~]# find 156_test/* -type f -o -type d |xargs ls -lhrt --full-time
-rw-r--r-- 1 root root   20 2019-06-08 10:45:35.601034564 +0800 156_test/c/rsync_server_156.log
-rw-r--r-- 1 root root   20 2019-06-08 10:45:35.601034564 +0800 156_test/b/rsync_server_156.log
-rw-r--r-- 1 root root   40 2019-06-08 11:30:09.643030468 +0800 156_test/a/rsync_server_156.log
-rw-r--r-- 1 root root   21 2019-06-08 11:38:35.962034681 +0800 156_test/a.log
-rw-r--r-- 1 root root    0 2019-06-08 13:53:15.008642458 +0800 156_test/def.pdf
-rw-r--r-- 1 root root    0 2019-06-08 13:53:15.008642458 +0800 156_test/abc.pdf
-rw-r--r-- 1 root root    0 2019-06-08 13:53:41.903645912 +0800 156_test/b.log
-rw-r--r-- 1 root root    0 2019-06-08 13:53:44.993648110 +0800 156_test/c.log
-rw-r--r-- 1 root root    0 2019-06-08 14:03:22.463643259 +0800 156_test/c.pdf/g.pdf
-rw-r--r-- 1 root root    0 2019-06-08 14:03:22.463643259 +0800 156_test/c.pdf/f.pdf
-rw-r--r-- 1 root root    0 2019-06-08 14:03:22.463643259 +0800 156_test/c.pdf/e.pdf
-rw-r--r-- 1 root root    0 2019-06-08 14:03:22.463643259 +0800 156_test/c.pdf/d.pdf
-rw-r--r-- 1 root root    0 2019-06-08 14:03:22.463643259 +0800 156_test/c.pdf/c.pdf
-rw-r--r-- 1 root root    0 2019-06-08 14:03:22.463643259 +0800 156_test/c.pdf/b.pdf
-rw-r--r-- 1 root root    0 2019-06-08 14:03:22.463643259 +0800 156_test/c.pdf/a.pdf

156_test/a:
总用量 4.0K
-rw-r--r-- 1 root root 40 2019-06-08 11:30:09.643030468 +0800 rsync_server_156.log

156_test/b:
总用量 4.0K
-rw-r--r-- 1 root root 20 2019-06-08 10:45:35.601034564 +0800 rsync_server_156.log

156_test/c:
总用量 4.0K
-rw-r--r-- 1 root root 20 2019-06-08 10:45:35.601034564 +0800 rsync_server_156.log

156_test/a.pdf:
总用量 0
```

```
156_test/b.pdf:
总用量 0

156_test/c.pdf:
总用量 0
-rw-r--r-- 1 root root 0 2019-06-08 14:03:22.463643259 +0800 g.pdf
-rw-r--r-- 1 root root 0 2019-06-08 14:03:22.463643259 +0800 f.pdf
-rw-r--r-- 1 root root 0 2019-06-08 14:03:22.463643259 +0800 e.pdf
-rw-r--r-- 1 root root 0 2019-06-08 14:03:22.463643259 +0800 d.pdf
-rw-r--r-- 1 root root 0 2019-06-08 14:03:22.463643259 +0800 c.pdf
-rw-r--r-- 1 root root 0 2019-06-08 14:03:22.463643259 +0800 b.pdf
-rw-r--r-- 1 root root 0 2019-06-08 14:03:22.463643259 +0800 a.pdf
```

上述代码使用 find 指令查看本机文件和目录属性信息。

7．查看远程服务器文件同步信息

查看远程服务器文件同步信息，代码如下。

```
[root@rsync_client_157 ~]# find test_157/ -type f -o -type d |xargs ls -lhrt --full-time
-rw-r--r-- 1 root root   20 2019-06-08 10:45:35.000000000 +0800 test_157/c/rsync_server_156.log
-rw-r--r-- 1 root root   20 2019-06-08 10:45:35.000000000 +0800 test_157/b/rsync_server_156.log
-rw-r--r-- 1 root root   40 2019-06-08 11:30:09.000000000 +0800 test_157/a/rsync_server_156.log
-rw-r--r-- 1 root root   21 2019-06-08 11:38:35.000000000 +0800 test_157/a.log
-rw-r--r-- 1 root root    0 2019-06-08 13:53:15.000000000 +0800 test_157/def.pdf
-rw-r--r-- 1 root root    0 2019-06-08 13:53:15.000000000 +0800 test_157/abc.pdf
-rw-r--r-- 1 root root    0 2019-06-08 13:53:41.000000000 +0800 test_157/b.log
-rw-r--r-- 1 root root    0 2019-06-08 13:53:44.000000000 +0800 test_157/c.log
-rw-r--r-- 1 root root    0 2019-06-08 14:03:22.000000000 +0800 test_157/c.pdf/g.pdf
-rw-r--r-- 1 root root    0 2019-06-08 14:03:22.000000000 +0800 test_157/c.pdf/f.pdf
-rw-r--r-- 1 root root    0 2019-06-08 14:03:22.000000000 +0800 test_157/c.pdf/e.pdf
-rw-r--r-- 1 root root    0 2019-06-08 14:03:22.000000000 +0800 test_157/c.pdf/d.pdf
-rw-r--r-- 1 root root    0 2019-06-08 14:03:22.000000000 +0800 test_157/c.pdf/c.pdf
-rw-r--r-- 1 root root    0 2019-06-08 14:03:22.000000000 +0800 test_157/c.pdf/b.pdf
-rw-r--r-- 1 root root    0 2019-06-08 14:03:22.000000000 +0800 test_157/c.pdf/a.pdf

test_157/c:
总用量 4.0K
-rw-r--r-- 1 root root 20 2019-06-08 10:45:35.000000000 +0800 rsync_server_156.log

test_157/b:
总用量 4.0K
-rw-r--r-- 1 root root 20 2019-06-08 10:45:35.000000000 +0800 rsync_server_156.log

test_157/a:
总用量 4.0K
-rw-r--r-- 1 root root 40 2019-06-08 11:30:09.000000000 +0800 rsync_server_156.log

test_157/b.pdf:
总用量 0

test_157/a.pdf:
总用量 0
```

```
test_157/:
总用量 28K
drwxr-xr-x 2 root root 4.0K 2019-06-08 11:22:51.000000000 +0800 c
drwxr-xr-x 2 root root 4.0K 2019-06-08 11:22:51.000000000 +0800 b
drwxr-xr-x 2 root root 4.0K 2019-06-08 11:22:51.000000000 +0800 a
-rw-r--r-- 1 root root   21 2019-06-08 11:38:35.000000000 +0800 a.log
drwxr-xr-x 2 root root 4.0K 2019-06-08 13:53:05.000000000 +0800 b.pdf
drwxr-xr-x 2 root root 4.0K 2019-06-08 13:53:05.000000000 +0800 a.pdf
-rw-r--r-- 1 root root    0 2019-06-08 13:53:15.000000000 +0800 def.pdf
-rw-r--r-- 1 root root    0 2019-06-08 13:53:15.000000000 +0800 abc.pdf
-rw-r--r-- 1 root root    0 2019-06-08 13:53:41.000000000 +0800 b.log
-rw-r--r-- 1 root root    0 2019-06-08 13:53:44.000000000 +0800 c.log
drwxr-xr-x 2 root root 4.0K 2019-06-08 14:03:22.000000000 +0800 c.pdf

test_157/c.pdf:
总用量 0
-rw-r--r-- 1 root root 0 2019-06-08 14:03:22.000000000 +0800 g.pdf
-rw-r--r-- 1 root root 0 2019-06-08 14:03:22.000000000 +0800 f.pdf
-rw-r--r-- 1 root root 0 2019-06-08 14:03:22.000000000 +0800 e.pdf
-rw-r--r-- 1 root root 0 2019-06-08 14:03:22.000000000 +0800 d.pdf
-rw-r--r-- 1 root root 0 2019-06-08 14:03:22.000000000 +0800 c.pdf
-rw-r--r-- 1 root root 0 2019-06-08 14:03:22.000000000 +0800 b.pdf
-rw-r--r-- 1 root root 0 2019-06-08 14:03:22.000000000 +0800 a.pdf
```

上述代码使用 find 指令查看远程服务器文件同步信息。

8．本地修改单一文件内容，然后同步

将字符串写入测试文件中，代码如下。

```
[root@rsync_server_156 156_test]# echo 'This is a test _1' a.log
This is a test _1 a.log
[root@rsync_server_156 156_test]# echo 'This is a test _1' >>a.log
[root@rsync_server_156 156_test]# echo 'This is a test _2' >>a.log
[root@rsync_server_156 156_test]# cd ..
```

9．同步文件到远程服务器

同步文件到远程服务器，代码如下。

```
[root@rsync_server_156 ~]# rsync  -avz  --delete  156_test/ 192.168.2.157:test_157
sending incremental file list
a.log

sent 481 bytes   received 43 bytes   1048.00 bytes/sec
total size is 137   speedup is 0.26
```

10．远程服务器查看同步文件内容

```
[root@rsync_client_157 ~]# cat test_157/a.log
line 2
line 3
line 4
This is a test _1
This is a test _2
```

11．【实例6-11】总结

（1）同步前预览测试。

镜像同步前可使用预览模式，开启同步测试，只需要在原来同步的基础指令上，加-n 选项即可，代码如下。

```
[root@rsync_server_156 ~]# rsync  -avz -n  --delete  156_test/ 192.168.2.157:test_157
sending incremental file list

sent 407 bytes   received 18 bytes   850.00 bytes/sec
total size is 137   speedup is 0.32 (DRY RUN)
```

（2）同步内容确定。

同步目录内所有文件和目录（不包括目录自身），与同步目录内所有文件和目录（包括目录自身）的区别是在源目录上加"/"，测试代码如下。

```
[root@rsync_server_156 ~]# rsync  -avz -n  --delete  156_test 192.168.2.157:test_157
sending incremental file list
156_test/
156_test/a.log
156_test/abc.pdf
156_test/b.log
156_test/c.log
156_test/def.pdf
156_test/a.pdf/
156_test/a/
156_test/a/rsync_server_156.log
156_test/b.pdf/
156_test/b/
156_test/b/rsync_server_156.log
156_test/c.pdf/
156_test/c.pdf/a.pdf
156_test/c.pdf/b.pdf
156_test/c.pdf/c.pdf
156_test/c.pdf/d.pdf
156_test/c.pdf/e.pdf
156_test/c.pdf/f.pdf
156_test/c.pdf/g.pdf
156_test/c/
156_test/c/rsync_server_156.log

sent 496 bytes  received 85 bytes   1162.00 bytes/sec
total size is 137   speedup is 0.24 (DRY RUN)
```

上述代码同步 156_test 目录内的所有文件和目录（包括目录自身）到远程服务器。

同步 156_test 目录内的所有文件和目录（不包括目录自身）到远程服务器，代码如下。

```
[root@rsync_server_156 ~]#
[root@rsync_server_156 ~]# rsync  -avz -n  --delete  156_test/ 192.168.2.157:test_157
sending incremental file list

sent 407 bytes  received 18 bytes   283.33 bytes/sec
total size is 137   speedup is 0.32 (DRY RUN)
```

【实例 6-12】镜像同步时排除特定文件

本地服务器创建测试文件，代码如下。

```
[root@rsync_server_156 ~]# cd 156_test/
[root@rsync_server_156 156_test]# ll
总用量 28
drwxr-xr-x 2 root root 4096 6月   8 11:22 a
-rw-r--r-- 1 root root    0 6月   8 13:53 abc.pdf
-rw-r--r-- 1 root root   57 6月   8 20:08 a.log
drwxr-xr-x 2 root root 4096 6月   8 13:53 a.pdf
drwxr-xr-x 2 root root 4096 6月   8 11:22 b
-rw-r--r-- 1 root root    0 6月   8 13:53 b.log
drwxr-xr-x 2 root root 4096 6月   8 13:53 b.pdf
drwxr-xr-x 2 root root 4096 6月   8 11:22 c
-rw-r--r-- 1 root root    0 6月   8 13:53 c.log
drwxr-xr-x 2 root root 4096 6月   8 14:03 c.pdf
-rw-r--r-- 1 root root    0 6月   8 13:53 def.pdf
[root@rsync_server_156 156_test]# touch {nginx,apache,tomcat}.tar.gz
[root@rsync_server_156 156_test]# ll
总用量 28
drwxr-xr-x 2 root root 4096 6月   8 11:22 a
-rw-r--r-- 1 root root    0 6月   8 13:53 abc.pdf
-rw-r--r-- 1 root root   57 6月   8 20:08 a.log
-rw-r--r-- 1 root root    0 6月   8 20:42 apache.tar.gz
```

```
drwxr-xr-x 2 root root 4096 6月  8 13:53 a.pdf
drwxr-xr-x 2 root root 4096 6月  8 11:22 b
-rw-r--r-- 1 root root    0 6月  8 13:53 b.log
drwxr-xr-x 2 root root 4096 6月  8 13:53 b.pdf
drwxr-xr-x 2 root root 4096 6月  8 11:22 c
-rw-r--r-- 1 root root    0 6月  8 13:53 c.log
drwxr-xr-x 2 root root 4096 6月  8 14:03 c.pdf
-rw-r--r-- 1 root root    0 6月  8 13:53 def.pdf
-rw-r--r-- 1 root root    0 6月  8 20:42 nginx.tar.gz
-rw-r--r-- 1 root root    0 6月  8 20:42 tomcat.tar.gz
[root@rsync_server_156 ~]# touch 156_test/{a..f}.zip
[root@rsync_server_156 ~]# ll 156_test/*.zip
-rw-r--r-- 1 root root 0 6月  8 20:46 156_test/a.zip
-rw-r--r-- 1 root root 0 6月  8 20:46 156_test/b.zip
-rw-r--r-- 1 root root 0 6月  8 20:46 156_test/c.zip
-rw-r--r-- 1 root root 0 6月  8 20:46 156_test/d.zip
-rw-r--r-- 1 root root 0 6月  8 20:46 156_test/e.zip
-rw-r--r-- 1 root root 0 6月  8 20:46 156_test/f.zip
```

上述代码创建*.tar.gz 和*.zip 的测试文件。

```
[root@rsync_server_156 ~]# rsync -avz -n --exclude="*.tar.gz" --delete 156_test/ 192.168.2.157:test_157
sending incremental file list
./
a.zip
b.zip
c.zip
d.zip
e.zip
f.zip

sent 533 bytes  received 39 bytes  1144.00 bytes/sec
total size is 137  speedup is 0.24 (DRY RUN)
```

从上述代码可以看到，*.tar.gz 的文件不会被同步。

如果将--delete 选项和--exclude 选项一起使用，则被排除的文件不会被删除。

6.3 搭建企业级 rsync 备份服务器

6.3.1 为什么需要搭建备份服务器

1. 搭建备份服务器前需要思考的几个问题

（1）备份服务器中的数据及文件。

200 台服务器中的数据及文件如下。

- Web 站点服务数据。
- 系统配置文件，如 Nginx、Tomcat、Apache。
- 负载均衡配置文件。
- MySQL 数据库文件。
- 其他配置文件。
- 日志备份。

（2）备份服务器存在的意义。

- 服务器意外宕机导致文件系统损坏，可以快速恢复宕机服务器上的数据，为用户提供不间断服务。

- 预防数据被攻击者篡改,永远无法恢复的情况。
- 预防 IDC 机房被破坏,全部数据被损毁的情况。
- 预防自然灾害等不可抗因素造成数据损坏的情况。

2. rsync 备份服务器架构信息

(1)服务器架构。

本案例服务器架构环境信息,如表 6-10 所示。

表 6-10　　　　　　　　　　服务器架构环境信息

角色	操作系统	IP 地址	主机名
rsync 服务器	CentOS 6.9 x84_64	192.168.2.156	rsync_server_156
客户端	CentOS 6.9 x84_64	192.168.2.157	rsync_client_157
客户端	CentOS 6.9 x84_64	192.168.2.158	rsync_client_158
客户端	CentOS 6.9 x84_64	192.168.2.159	rsync_client_159

(2)关闭防火墙和 SELinux。

```
[root@rsync_server_156 ~]# ansible -i  /root/host.info all  -m command   -a
"sestatus ;  /etc/init.d/iptables status"
192.168.2.158 | SUCCESS | rc=0 >>
SELINUX status:             disabled

192.168.2.159 | SUCCESS | rc=0 >>
SELINUX status:             disabled

192.168.2.157 | SUCCESS | rc=0 >>
SELINUX status:             disabled

192.168.2.156 | SUCCESS | rc=0 >>
SELINUX status:             disabled
```

上述代码中可以看到防火墙和 SELinux 都已经处于关闭状态。

(3)跨机房备份服务器架构说明。

跨机房备份服务器架构,如图 6-1 所示。

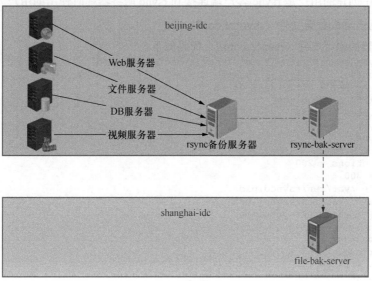

图 6-1　跨机房备份服务器架构

备份服务器架构说明。
- beijing-idc 机房内的所有服务器业务文件全部备份到 rsync 备份服务器。
- beijing-idc rsync 备份服务器和另外一台 rsync-bak-server 服务器保持数据同步。
- beijing-idc rsync-bak-server 和 shanghai-idc file-bak-server 服务器保持同步。

6.3.2 rsync 服务端初始化

1. 安装 rsync 备份服务器

安装 rsync 备份服务器，代码如下。

```
[root@rsync_server_156 ~]# yum install rsync -y >/dev/null  2>&1
[root@rsync_server_156 ~]# echo $?
0
```

确认 rsync 软件包是否安装成功。

```
[root@rsync_server_156 ~]# rsync --version
rsync  version 3.0.6  protocol version 30
Copyright (C) 1996-2009 by Andrew Tridgell, Wayne Davison, and others.
Web site: http://rsync.samba.org/
Capabilities:
    64-bit files, 64-bit inums, 64-bit timestamps, 64-bit long ints,
    socketpairs, hardlinks, symlinks, IPv6, batchfiles, inplace,
    append, ACLs, xattrs, iconv, symtimes

rsync comes with ABSOLUTELY NO WARRANTY.  This is free software, and you
are welcome to redistribute it under certain conditions.  See the GNU
General Public Licence for details.
```

从上述代码可以看到，rsync 软件包已经被成功安装。

2. 创建备份用户

创建备份用户，代码如下。

```
[root@rsync_server_156 ~]# useradd rsync -s /sbin/nologin -M
[root@rsync_server_156 ~]# id rsync
uid=500(rsync) gid=500(rsync) 组=500(rsync)
```

上述代码中创建的用户是 rsync 客户端连接到 rsync 服务器端时使用的用户。

3. 创建服务器的配置文件（rsyncd.conf）

创建服务器的配置文件（rsyncd.conf），代码如下。

```
[root@rsync_server_156 ~]# cat /etc/rsyncd/rsyncd.conf
#rsync_config                 start
#created by 韩艳威 01:23 2019-06-09
#Email:3128743705@qq.com
##rsyncd.conf start##
uid = rsync
gid = rsync
use chroot = no
max connections = 200
timeout = 300
pid file = /var/run/rsyncd.pid
lock file = /var/run/rsync.lock
log file = /var/log/rsyncd.log
[backup]
path = /backup/
ignore errors
read only = false
list = false
hosts allow = 192.168.2.0/24
hosts deny = 0.0.0.0/32
auth users = rsync_backup
```

```
secrets file = /etc/rsyncd/rsync.password
#rsync_config_____end
```

基础的 rsyncd.conf 配置文件如上所示,rsync 服务器配置选项如表 6-11 所示。

表 6-11　　　　　　　　　　rsync 服务器配置选项

选项	说明
uid=rsync	rsync 用户 ID,默认 uid 为 2,通常为 nobody
gid=rsync	rsync 用户组,默认 gid 为 2,通常为 nobody
use chroot=no	rsync 安全配置,内网使用 rsync,不配置也可以
max connections=200	设置最大连接数,默认值为 0,无限制,负值为关闭
timeout=300	默认为 0,意为 no timeout
pid file=/path/*.pid	rsync 守护进程启动后将其进程 PID 写入此文件
lock file=/path/*.lock	默认为/var/run/rsyncd.lock
log file = /path/*.log	rsync 的日志文件
ignore errors	忽略 I/O 错误
read only = false	客户端是否可以上传文件,默认所有模块为 true
list	是否允许客户端查看可用模块列表
hosts allow	允许的客户端主机名或者 IP 地址或者地址段
hosts deny = 0.0.0.0/32	禁用客户端主机名或 IP 地址或地址段
auth users = rsync_backup	指定以空格或逗号分隔的用户可以使用哪些模块
secrets file =file_path	指定用户名和密码存放的文件
[backup]	模块名称,需要用方括号括起来
path = /backup/	在这个模块中,守护进程使用的文件系统或目录

4. 创建备份目录

创建备份目录,代码如下。

```
[root@rsync_server_156 ~]# mkdir /backup
[root@rsync_server_156 ~]# chown rsync.rsync /backup
[root@rsync_server_156 ~]# ls -ld /backup
drwxr-xr-x 2 rsync rsync 4096 6月   8 23:32 /backup
```

一定记得修改文件的用户和组,代码如下。

```
[root@rsync_server_156 ~]# mkdir -pv /etc/rsyncd/
mkdir: 已创建目录 "/etc/rsyncd/"
```

创建 rsyncd 目录。

5. 创建密码文件(根据配置文件生成)

创建密码文件(根据配置文件生成),代码如下。

```
[root@rsync_server_156 ~]# tree /etc/rsyncd/
/etc/rsyncd/

0 directories, 0 files
[root@rsync_server_156 ~]# echo "rsync_backup:hanyanwei" > /etc/rsyncd/rsync.password
[root@rsync_server_156 ~]# chmod 600 /etc/rsyncd/rsync.password
[root@rsync_server_156 ~]# tree /etc/rsyncd/
/etc/rsyncd/
└── rsync.password

0 directories, 1 file
[root@rsync_server_156 ~]# tree /etc/rsyncd/
/etc/rsyncd/
├── rsyncd.conf
└── rsync.password

0 directories, 2 files
```

6. 启动 rsync 服务器

启动 rsync 服务器相当简单，--daemon 是让 rsync 以服务器模式运行，代码如下。

```
/usr/bin/rsync --daemon  --config=/etc/rsyncd/rsyncd.conf
```

7. 查看 rsync 进程和端口

查看 rsync 进程和端口，代码如下。

```
[root@rsync_server_156 ~]# /usr/bin/rsync -daemon
--config=/etc/rsyncd/rsyncd.conf
[root@rsync_server_156 ~]# netstat -ntpl
Active Internet connections (only servers)
Proto Recv-Q Send-Q Local Address          Foreign Address        State       PID/Program name
tcp        0      0 0.0.0.0:873            0.0.0.0:*              LISTEN      2694/rsync
tcp        0      0 0.0.0.0:22             0.0.0.0:*              LISTEN      1214/sshd
tcp        0      0 127.0.0.1:25           0.0.0.0:*              LISTEN      1293/master
tcp        0      0 :::873                 :::*                   LISTEN      2694/rsync
tcp        0      0 :::22                  :::*                   LISTEN      1214/sshd
tcp        0      0 ::1:25                 :::*                   LISTEN      1293/master
[root@rsync_server_156 ~]# ps -ef |grep rsync
root      2694     1  0 23:44 ?        00:00:00 /usr/bin/rsync -daemon
--config=/etc/rsyncd/rsyncd.conf
root      2697  1755  0 23:45 pts/0    00:00:00 grep rsync
```

6.3.3 rsync 客户端配置

1. 配置密码文件

配置密码文件，代码如下。

```
[root@rsync_client_157 ~]# echo "hanyanwei" > /etc/rsync.password
[root@rsync_client_157 ~]# cat /etc/rsync.password
hanyanwei
```

2. 修改密码文件权限

修改密码文件权限，代码如下。

```
[root@rsync_client_157 ~]# chmod 600 /etc/rsync.password
[root@rsync_client_157 ~]# ls -l /etc/rsync.password
-rw------- 1 root root 10 6月   8 23:48 /etc/rsync.password
```

3. 测试

这里的示例都是在客户端（nsf01）上进行操作的，一般的使用场景都是从客户端备份数据到服务器。

rsync 推送和拉取文件代码格式如下。

（1）rsync 推送模式。

```
Pull: rsync [OPTION...] [USER@]HOST::SRC... [DEST]
```

推送数据到服务器，这种模式比较常用，代码如下。

```
rsync [OPTION...] rsync://[USER@]HOST[:PORT]/SRC... [DEST]
```

（2）rsync 拉取模式。

```
Push: rsync [OPTION...] SRC... [USER@]HOST::DEST
```

从服务器拉取数据，这种模式比较常用，代码如下。

```
rsync [OPTION...] SRC... rsync://[USER@]HOST[:PORT]/DEST
```

【实例 6-13】客户端同步文件到服务器端

（1）输入密码验证，代码如下。

```
[root@rsync_client_157 ~]# rsync  delete  avz test_157
rsync_backup@192.168.2.156::backup
Password:
sending incremental file list
```

```
sent 685 bytes  received 30 bytes   204.29 bytes/sec
total size is 137  speedup is 0.19
[root@rsync_client_157 ~]# rsync --delete -avz test_157
rsync_backup@192.168.2.156::backup
Password:
sending incremental file list
test_157/
test_157/a.log
test_157/abc.pdf
test_157/b.log
test_157/c.log
test_157/def.pdf
test_157/a.pdf/
test_157/a/
test_157/a/rsync_server_156.log
test_157/a/1,b/
test_157/a/1,b/1,c/
test_157/a/1,b/2,c/
test_157/a/1,b/3,c/
test_157/a/2,b/
test_157/a/2,b/1,c/
test_157/a/2,b/2,c/
test_157/a/2,b/3,c/
test_157/a/3,b/
test_157/a/3,b/1,c/
test_157/a/3,b/2,c/
test_157/a/3,b/3,c/
test_157/a/a/
test_157/ab/
test_157/ac/
test_157/b.pdf/
test_157/b/
test_157/b/rsync_server_156.log
test_157/c.pdf/
test_157/c.pdf/a.pdf
test_157/c.pdf/b.pdf
test_157/c.pdf/c.pdf
test_157/c.pdf/d.pdf
test_157/c.pdf/e.pdf
test_157/c.pdf/f.pdf
test_157/c.pdf/g.pdf
test_157/c/
test_157/c/rsync_server_156.log

sent 1401 bytes  received 381 bytes   712.80 bytes/sec
total size is 137  speedup is 0.08
```

（2）免密码输入，代码如下。

```
[root@rsync_client_157 ~]# rsync --delete -avz test_157
rsync_backup@192.168.2.156::backup --password-file=/etc/rsync.password
sending incremental file list

sent 685 bytes  received 30 bytes   476.67 bytes/sec
total size is 137  speedup is 0.19
```

上述代码 rsync 指令说明如下。

- rsync -avz 是指令和对应的选项。
- test_157 是指把 test_157 目录下的内容推送到服务器。
- rsync_backup@192.168.2.156::backup 是指服务器的信息，rsync_backup 是指配置文件 rsyncd.conf 中配置的用户名。
- 192.168.2.156 是服务器的 IP 地址，backup 是指/etc/rsyncd.conf 中配置的模块名，这个

地方一定不要弄错。
- --password-file 指定密码文件，不加这个选项的情况下，要手动输入密码。

4．在备份服务器（backup 服务器）查看备份的结果

```
[root@rsync_server_156 ~]# ll /backup/
总用量 4
drwxr-xr-x 10 rsync rsync 4096 6月   8 21:04 test_157
```

【实例 6-14】多目录共享

多目录共享是指客户端可以向服务器端多个目录下进行推送或拉取。这实现起来很简单，就是在配置文件中配置多个模块，每个模块可以指定不同的用户名、密码等信息。如果所有推送的模块基本信息都相同，就可以把配置信息放在多个模块上，模块只配置一个对应的路径即可，代码如下。

```
[appdata]
path = /data/appdatabak

[www]
path = /data/webbak/

[mysql]
path = /data/dbbak

[logs]
path = /data/logsbak

[static]
path = /data/staticbak
```

【实例 6-15】优化传输

（1）修改配置文件

修改 rsyncd.conf 配置文件，代码如下。

```
[root@rsync_server_156 ~]# cat /etc/rsyncd/rsyncd.conf
#rsync_config_____start
#created by 韩艳威 01:23 2019-06-09
#Email:3128743705@qq.com
##rsyncd.conf start##
uid = rsync
gid = rsync
use chroot = no
max connections = 200
timeout = 300
pid file = /var/run/rsyncd.pid
lock file = /var/run/rsync.lock
log file = /var/log/rsyncd.log    #这是 rsync 的日志文件，比较有用
ignore errors
read only = false
list = false
hosts allow = 192.168.2.0/24
hosts deny = 0.0.0.0/32
auth users = rsync_backup
secrets file = /etc/rsyncd/rsync.password



[backup_2_157]
path = /backup/192.168.2.157

#rsync_config_____end
```

rsync 客户端备份文件到 rsync 服务器，需要明确标识是哪台服务器，便于以后 rsync 客户端文件恢复操作。

（2）创建目录并修改目录权限。

创建目录并修改目录权限，代码如下。

```
[root@rsync_server_156 ~]# mkdir -pv /backup/192.168.2.157
mkdir: 已创建目录 "/backup/192.168.2.157"
[root@rsync_server_156 ~]#
[root@rsync_server_156 ~]# ls -ld /backup/192.168.2.157
drwxr-xr-x 2 root root 4096 6月   9 00:12 /backup/192.168.2.157
[root@rsync_server_156 ~]#
[root@rsync_server_156 ~]# chown -R rsync. /backup/192.168.2.157
[root@rsync_server_156 ~]# ls -ld /backup/192.168.2.157/
drwxr-xr-x 3 rsync rsync 4096 6月   9 00:12 /backup/192.168.2.157/
```

修改/backup/192.168.2.157/目录权限为 rsync 用户和组。

（3）rsync 客户端测试。

rsync 客户端测试，代码如下。

```
[root@rsync_client_157 ~]# rsync --delete  -avz test_157 rsync_backup@192.168.2.156::backup_2_157 --password-file=/etc/rsync.password
sending incremental file list
test_157/
test_157/a.log
test_157/abc.pdf
test_157/b.log
test_157/c.log
test_157/def.pdf
test_157/a.pdf/
test_157/a/
test_157/a/rsync_server_156.log
test_157/a/1,b/
test_157/a/1,b/1,c/
test_157/a/1,b/2,c/
test_157/a/1,b/3,c/
test_157/a/2,b/
test_157/a/2,b/1,c/
test_157/a/2,b/2,c/
test_157/a/2,b/3,c/
test_157/a/3,b/
test_157/a/3,b/1,c/
test_157/a/3,b/2,c/
test_157/a/3,b/3,c/
test_157/a/a/
test_157/ab/
test_157/ac/
test_157/b.pdf/
test_157/b/
test_157/b/rsync_server_156.log
test_157/c.pdf/
test_157/c.pdf/a.pdf
test_157/c.pdf/b.pdf
test_157/c.pdf/c.pdf
test_157/c.pdf/d.pdf
test_157/c.pdf/e.pdf
test_157/c.pdf/f.pdf
test_157/c.pdf/g.pdf
test_157/c/
test_157/c/rsync_server_156.log

sent 1401 bytes  received 381 bytes  3564.00 bytes/sec
total size is 137  speedup is 0.08
```

上述代码同步本地 test_157 目录到 rsync 服务器 backup_2_157 模块。

（4）在服务器上查看文件是否存在。

在服务器上查看同步后的文件是否存在。

```
[root@rsync_server_156 ~]# ll /backup/192.168.2.157/
总用量 4
drwxr-xr-x 10 rsync rsync 4096 6月   8 21:04 test_157
```
上述代码输出结果中可以看到文件已经存在。

5．rsync 服务加入开机启动

将 rsync 服务加入开机启动，代码如下。

```
[root@rsync_server_156 ~]# tail -3  /etc/rc.local


/usr/bin/rsync --daemon  --config=/etc/rsyncd/rsyncd.conf
[root@rsync_server_156 ~]# reboot
[root@rsync_server_156 ~]#
Broadcast message from root@rsync_server_156
     (/dev/pts/0) at 8:13 ...

The system is going down for reboot NOW!
Connection closing...Socket close.

Connection closed by foreign host.

Disconnected from remote host(rsync_server_156) at 08:20:52.

Type 'help' to learn how to use Xshell prompt.
```
把 rsync 服务启动指令添加在/etc/rc.local 文件中，重启操作系统，确认 rsync 服务是否可以正常启动。

> **注意**：若 rsync 服务加入开机启动后，重启操作系统发现 rsync 服务并未随着服务器的开机启动而启动对应的服务，此时可以尝试检查/etc/rc.local 文件是否有可执行权限。如果没有可执行权限，可能会导致/etc/rc.local 文件中设置的相关指令或服务不会被执行。

确认 rsync 服务开机启动是否正常，代码如下。

```
[root@rsync_server_156 ~]# netstat -ntpl
Active Internet connections (only servers)
Proto Recv-Q Send-Q Local Address           Foreign Address         State       PID/Program name
tcp        0      0 0.0.0.0:873             0.0.0.0:*               LISTEN      1312/rsync
tcp        0      0 0.0.0.0:22              0.0.0.0:*               LISTEN      1211/sshd
tcp        0      0 127.0.0.1:25            0.0.0.0:*               LISTEN      1290/master
tcp        0      0 :::873                  :::*                    LISTEN      1312/rsync
tcp        0      0 :::22                   :::*                    LISTEN      1211/sshd
tcp        0      0 ::1:25                  :::*                    LISTEN      1290/master
[root@rsync_server_156 ~]# ps -ef |grep rsync
root       1312     1  0 08:14 ?        00:00:00 /usr/bin/rsync –daemon
--config=/etc/rsyncd/rsyncd.conf
root       1356  1339  0 08:14 pts/0    00:00:00 grep rsync
```
上述代码中可以看到，rsync 服务已经被 873 端口监听，且 rsync 进程也存在，表示 rsync 服务开机启动已经被成功执行。

6．初始化 rsync_client_158 客户端

（1）配置密码文件，代码如下。
```
[root@rsync_client_158 ~]# echo "hanyanwei" > /etc/rsync.password
[root@rsync_client_158 ~]# cat /etc/rsync.password
hanyanwei
```
（2）修改密码文件权限，代码如下。
```
[root@rsync_client_158 ~]# chmod 600 /etc/rsync.password
[root@rsync_client_158 ~]# ls -lh /etc/rsync.password
-rw------- 1 root root 10 6月   9 08:29 /etc/rsync.password
```

7. 初始化 rsync_client_159 客户端

（1）配置密码文件，代码如下。

```
[root@rsync_client_159 ~]# echo "hanyanwei" > /etc/rsync.password
[root@rsync_client_159 ~]# cat /etc/rsync.password
hanyanwei
```

（2）修改密码文件权限，代码如下。

```
[root@rsync_client_159 ~]# chmod 600 /etc/rsync.password
[root@rsync_client_159 ~]# ls -lh /etc/rsync.password
-rw------- 1 root root 10 6月   9 08:30 /etc/rsync.password
```

8. rsync 备份服务器设置客户端模块

（1）rsync 备份服务器重新生成配置文件，代码如下。

```
[root@rsync_server_156 ~]# cat /etc/rsyncd/rsyncd.conf
#rsync_config_____start
#created by 韩艳威 01:23 2019-06-09
#Email:3128743705@qq.com
##rsyncd.conf start##
uid = rsync
gid = rsync
use chroot = no
max connections = 200
timeout = 300
pid file = /var/run/rsyncd.pid
lock file = /var/run/rsync.lock
log file = /var/log/rsyncd.log    #这是rsync的日志文件，比较有用
ignore errors
read only = false
list = false
hosts allow = 192.168.2.0/24
hosts deny = 0.0.0.0/32
auth users = rsync_backup
secrets file = /etc/rsyncd/rsync.password


[backup_2_157]
path = /backup/192.168.2.157

[backup_2_158]
path = /backup/192.168.2.158

[backup_2_159]
path = /backup/192.168.2.159

#rsync_config_____end
```

上述代码创建 rsync 服务器配置文件。

（2）创建备份目录，代码如下。

```
[root@rsync_server_156 ~]# mkdir -pv /backup/192.168.2.15{7..9}
mkdir: 已创建目录 "/backup/192.168.2.158"
mkdir: 已创建目录 "/backup/192.168.2.159"
```

（3）修改备份目录权限，代码如下。

```
[root@rsync_server_156 ~]# chown -R rsync. /backup/192.168.2.15{7..9}
[root@rsync_server_156 ~]# ls -ld /backup/192.168.2.15*
drwxr-xr-x 3 rsync rsync 4096 6月   9 00:12 /backup/192.168.2.157
drwxr-xr-x 2 rsync rsync 4096 6月   9 13:07 /backup/192.168.2.158
drwxr-xr-x 2 rsync rsync 4096 6月   9 13:07 /backup/192.168.2.159
```

（4）重启 rsync 服务，代码如下。

```
[root@rsync_server_156 ~]# kill 'at /var/run/rsyncd.pid'
[root@rsync_server_156 ~]# ps -ef |grep rsync
root       1475  1449  0 13:09 pts/1    00:00:00 grep rsync
```

```
[root@rsync_server_156 ~]# /usr/bin/rsync --daemon --config=/etc/rsyncd/rsyncd.conf
[root@rsync_server_156 ~]# ps -ef |grep rsync
root       1478     1  0 13:10 ?        00:00:00 /usr/bin/rsync --daemon
--config=/etc/rsyncd/rsyncd.conf
root       1480  1449  0 13:10 pts/1    00:00:00 grep rsync
```

（5）rsync_client_158 客户端备份文件。

查看当前服务器 root 主目录文件列表，代码如下。

```
[root@rsync_client_158 ~]# ll
总用量 16
-rw-------. 1 root root 1090 5月  18 17:44 anaconda-ks.cfg
-rw-r--r--. 1 root root 8025 5月  18 17:44 install.log
-rw-r--r--. 1 root root 3384 5月  18 17:43 install.log.syslog
```

打包压缩 root 主目录下文件，代码如下。

```
[root@rsync_client_158 ~]# tar czvf 192.168.2.158.tgz  install.log install.log
install.log
install.log
[root@rsync_client_158 ~]# ls -lhrt
总用量 20K
-rw-r--r--. 1 root root 3.4K 5月  18 17:43 install.log.syslog
-rw-r--r--. 1 root root 7.9K 5月  18 17:44 install.log
-rw-------. 1 root root 1.1K 5月  18 17:44 anaconda-ks.cfg
-rw-r--r--  1 root root 2.6K 6月   9 13:19 192.168.2.158.tgz
```

预传输文件到 rsync 服务器测试，代码如下。

```
[root@rsync_client_158 ~]# rsync -avz -n 192.168.2.158.tgz
rsync_backup@192.168.2.156::backup_2_158 --password-file=/etc/rsync.password
sending incremental file list
192.168.2.158.tgz

sent 41 bytes  received 11 bytes  104.00 bytes/sec
total size is 2592  speedup is 49.85 (DRY RUN)
```

备份压缩文件到 rsync 服务器，代码如下。

```
[root@rsync_client_158 ~]# rsync -avz  192.168.2.158.tgz
rsync_backup@192.168.2.156::backup_2_158 --password-file=/etc/rsync.password
sending incremental file list
192.168.2.158.tgz

sent 2674 bytes  received 27 bytes  5402.00 bytes/sec
total size is 2592  speedup is 0.96
```

服务器确认 192.168.2.158.tgz 打包压缩文件是否存在，代码如下。

```
[root@rsync_server_156 ~]# ls -lhrt /backup/192.168.2.158/192.168.2.158.tgz
-rw-r--r-- 1 rsync rsync 2.6K 6月   9 13:19 /backup/192.168.2.158/192.168.2.158.tgz
```

查看 rsync_client_159 root 用户主目录下文件列表，代码如下。

```
[root@rsync_client_159 ~]# ll
总用量 16
-rw-------. 1 root root 1090 5月  18 17:44 anaconda-ks.cfg
-rw-r--r--. 1 root root 8025 5月  18 17:44 install.log
-rw-r--r--. 1 root root 3384 5月  18 17:43 install.log.syslog
```

打包压缩当前目录下相关文件，代码如下。

```
[root@rsync_client_159 ~]# tar czvf 192.168.2.159.tgz anaconda-ks.cfg  install.log
install.log.syslog
anaconda-ks.cfg
install.log
install.log.syslog
```

查看打包压缩文件是否存在，代码如下。

```
[root@rsync_client_159 ~]# ls -lhrt
总用量 20K
-rw-r--r--. 1 root root 3.4K 5月  18 17:43 install.log.syslog
-rw-r--r--. 1 root root 7.9K 5月  18 17:44 install.log
-rw-------. 1 root root 1.1K 5月  18 17:44 anaconda-ks.cfg
-rw-r--r--  1 root root 3.8K 6月   9 13:19 192.168.2.159.tgz
```

预传输文件到 rsync 服务器，代码如下。

```
[root@rsync_client_159 ~]# rsync -avz -n 192.168.2.159.tgz
rsync_backup@192.168.2.156::backup_2_159 --password-file=/etc/rsync.password
sending incremental file list
192.168.2.159.tgz

sent 41 bytes  received 11 bytes  104.00 bytes/sec
total size is 3802  speedup is 73.12 (DRY RUN)
```

备份 192.168.2.159.tgz 文件到 rsync 服务器，代码如下。

```
[root@rsync_client_159 ~]# rsync -avz 192.168.2.159.tgz
rsync_backup@192.168.2.156::backup_2_159 --password-file=/etc/rsync.password
sending incremental file list
192.168.2.159.tgz

sent 3884 bytes  received 27 bytes  7822.00 bytes/sec
total size is 3802  speedup is 0.97
```

在 rsync 服务器查看 192.168.2.159.tgz 文件是否存在，代码如下。

```
[root@rsync_server_156 ~]# ls -lhrt /backup/192.168.2.159/192.168.2.159.tgz
-rw-r--r-- 1 rsync rsync 3.8K 6月   9 13:19 /backup/192.168.2.159/192.168.2.159.tgz
```

6.4 搭建 rsync+inotify 实时备份服务器

随着应用系统规模的不断扩大，对数据的安全性和可靠性也提出了更高的要求。
rsync+inotify 的实时同步解决方案，可以实现数据的实时同步。

6.4.1 企业级主流实时同步工具比较

1. rsync+inotify

生产服务服务器上的同步方案，原先使用的是 sync + inotify，但随着文件数量增大到 100 万，目录下的文件列表就达 20MB。在网络状况不佳或者限速的情况下，变更的文件可能才几 MB，却要发送 20MB 的文件列表，这会严重降低带宽的使用效率和同步效率。更为要紧的是，假如 inotifywait 在 5s 内监控到 10 个小文件发生变化，便会触发 10 个 rsync 同步操作，结果就是真正需要传输的文件仅 2~3MB，比对的文件列表就达 200MB。使用 rsync+inotify 的好处在于，它们都是最基本的软件，可以通过不同选项做到很精确的控制，比如排除同步的目录、同步多个模块或同步到多个主机。

2. sersync

sersync 工具可以提高同步的性能，解决同步大文件时出现异常的问题。sersync 是国内的一个开发者开源出的，使用 C++编写，采用多线程的方式进行同步，失败后还有重传机制，可对临时文件进行过滤，自带 crontab 定时同步功能。

3. Lsyncd

Lsyncd 使用 Lua 封装了 inotify 和 rsync 工具，采用了 Linux 内核（2.6.13 及以上）里的 inotify 触发机制，然后通过 rsync 去差异同步，达到实时的效果。其通过时间延迟或累计触发事件次数完美解决 rsync + inotify 海量文件同步带来的文件频繁发送文件列表的问题。另外，它的配置方式也很简单，Lua 本身就是一种配置语言，可读性非常强。Lsyncd 也有多种工作模式可以选择，本地目录 cp、本地目录 rsync、远程目录 rsyncssh 等。

Lsyncd 一个指令就可实现本地目录同步备份（网络存储挂载也被当作本地目录）。

6.4.2 rsync+inotify 组合基础知识

（1）rsync+inotify 概述。

使用 rsync+inotify 的组合，可以实现数据的实时同步。inotify 是一种强大的、细粒度的、异步的文件系统事件控制机制。

Linux 内核从 2.6.13 起，加入了 inotify，通过 inotify 可以监控文件系统中添加、删除、修改、移动等事件。利用这个内核接口，第三方软件就可以监控文件系统下文件的各种变化情况，而 inotify-tools 正是实施监控的软件，在使用 rsync 首次全量同步后，其结合 inotify 对源目录进行实时监控。只要文件有变动或新文件产生，就会立刻将其同步到目标目录下，非常高效和实用。

CentOS 6 系列操作系统默认已经支持 inotify 事件，使用 ll /proc/sys/fs/inotify 指令有如下第 3～5 行信息输出，如果没有则表示不支持 inotify 事件。

```
1 [root@rsync_client_157 ~]# ll /proc/sys/fs/inotify/
2 总用量 0
3 -rw-r--r-- 1 root root 0 6月   9 21:04 max_queued_events
4 -rw-r--r-- 1 root root 0 6月   9 21:04 max_user_instances
5 -rw-r--r-- 1 root root 0 6月   9 21:04 max_user_watches
```

ll /proc/sys/fs/inotify 指令输出说明如下。

- /proc/sys/fs/inotify/max_queued_evnets 表示调用 inotify_init 时分配给 inotify instance 中可排队的事件数目的最大值，超出这个值的事件被丢弃，但会触发 IN_Q_OVERFLOW 事件。
- /proc/sys/fs/inotify/max_user_instances 表示每一个用户 ID 可创建的 inotify instatnce 的数量上限。
- /proc/sys/fs/inotify/max_user_watches 表示每个 inotify instatnce 可监控的最大目录数量。如果监控的数量巨大，需要根据情况，适当增加此值的大小。

（2）安装 inotify-tools。

inotify-tools 是为 Linux 下 inotify 文件监控工具提供的一套 C 语言的开发接口库函数，同时还提供了一系列的指令行工具，这些工具可以用来监控文件系统的事件。

inotify-tools 是用 C 语言编写的，除了要求内核支持 inotify 外，不依赖于其他环境。inotify-tools 提供如下两种工具。

- inotifywait：它用来监控文件或目录的变化。
- inotifywatch：它用来统计文件系统访问的次数。

使用 yum 指令安装 inotify-tools-3.14-1.el6.x86_64.rpm 软件包。首先安装 epel 源，然后再安装 inotify-tools 软件，代码如下。

```
[root@rsync_client_157 ~]# ll /etc/yum.repos.d/ |grep epel
[root@rsync_client_157 ~]# yum install epel-release -y >/dev/null 2>&1 ; echo $?
0
[root@rsync_client_157 ~]# ll /etc/yum.repos.d/ |grep epel
-rw-r--r-- 1 root root  957 11月  5 2012 epel.repo
-rw-r--r-- 1 root root 1056 11月  5 2012 epel-testing.repo
```

查询 inotify 相关软件包，代码如下。

```
[root@rsync_client_157 ~]# yum list inotify*
已加载插件：fastestmirror
Loading mirror speeds from cached hostfile
epel/metalink
| 7.0 kB     00:00
 * base: mirror.bit.edu.cn
 * epel: mirrors.yun-idc.com
```

```
  * extras: mirror.bit.edu.cn
  * updates: mirror.bit.edu.cn
epel
| 5.3 kB     00:00
epel/primary_db
| 6.0 MB     00:00
可安装的软件包
inotify-tools.i686
3.14-1.el6                                               epel
inotify-tools.x86_64
3.14-1.el6                                               epel
inotify-tools-devel.i686
3.14-1.el6                                               epel
inotify-tools-devel.x86_64
3.14-1.el6                                               epel
```

安装 inotify 软件包，代码如下。

```
[root@rsync_client_157 ~]# yum install inotify-tools-devel inotify -y
已加载插件：fastestmirror
设置安装进程
Loading mirror speeds from cached hostfile
 * base: mirror.bit.edu.cn
 * epel: mirrors.yun-idc.com
 * extras: mirror.bit.edu.cn
 * updates: mirror.bit.edu.cn
（中间代码略）

已安装：
  inotify-tools-devel.x86_64 0:3.14-1.el6

作为依赖被安装：
  inotify-tools.x86_64 0:3.14-1.el6

完毕！
```

（3）查看 inotify-tools 安装的软件组件，代码如下。

```
[root@rsync_client_157 ~]# rpm -ql inotify-tools
/usr/bin/inotifywait
/usr/bin/inotifywatch
/usr/lib64/libinotifytools.so.0
/usr/lib64/libinotifytools.so.0.4.1
/usr/share/doc/inotify-tools-3.14
/usr/share/doc/inotify-tools-3.14/AUTHORS
/usr/share/doc/inotify-tools-3.14/COPYING
/usr/share/doc/inotify-tools-3.14/ChangeLog
/usr/share/doc/inotify-tools-3.14/NEWS
/usr/share/doc/inotify-tools-3.14/README
/usr/share/man/man1/inotifywait.1.gz
/usr/share/man/man1/inotifywatch.1.gz
```

从上述代码可以看到，已安装/usr/bin/inotifywait 和/usr/bin/inotifywatch 这两个实用工具。

6.4.3　inotifywait 实时同步企业级案例

本案例基本环境信息如表 6-12 所示。

表 6-12　　　　　　　　　　　基本环境信息

角色	操作系统	IP 地址	主机名
客户端	CentOS 6.9 x84_64	192.168.2.156	rsync_server_156
服务器	CentOS 6.9 x84_64	192.168.2.157	rsync_client_157
服务器	CentOS 6.9 x84_64	192.168.2.157	rsync_client_158
服务器	CentOS 6.9 x84_64	192.168.2.157	rsync_client_159

1. 查看 inotifywait 帮助信息

查看 inotifywait 帮助信息，代码如下。

```
[root@rsync_client_157 ~]# /usr/bin/inotifywait --help
inotifywait 3.14
Wait for a particular event on a file or set of files.
Usage: inotifywait [ options ] file1 [ file2 ] [ file3 ] [ ... ]
Options:
	-h|--help     	Show this help text.
	@<file>       	Exclude the specified file from being watched.
	--exclude <pattern>
	              	Exclude all events on files matching the
	              	extended regular expression <pattern>.
	--excludei <pattern>
	              	Like --exclude but case insensitive.
	-m|--monitor  	Keep listening for events forever.  Without
	              	this option, inotifywait will exit after one
	              	event is received.
	-d|--daemon   	Same as --monitor, except run in the background
	              	logging events to a file specified by --outfile.
	              	Implies --syslog.
	-r|--recursive	Watch directories recursively.
	--fromfile <file>
	              	Read files to watch from <file> or '-' for stdin.
	-o|--outfile <file>
	              	Print events to <file> rather than stdout.
	-s|--syslog   	Send errors to syslog rather than stderr.
	-q|--quiet    	Print less (only print events).
	-qq           	Print nothing (not even events).
	--format <fmt>	Print using a specified printf-like format
	              	string; read the man page for more details.
	--timefmt <fmt>	strftime-compatible format string for use with
	              	%T in --format string.
	-c|--csv      	Print events in CSV format.
	-t|--timeout <seconds>
	              	When listening for a single event, time out after
	              	waiting for an event for <seconds> seconds.
	              	If <seconds> is 0, inotifywait will never time out.
	-e|--event <event1> [ -e|--event <event2> ... ]
	        Listen for specific event(s).  If omitted, all events are
	        listened for.

Exit status:
	0  - An event you asked to watch for was received.
	1  - An event you did not ask to watch for was received
	     (usually delete_self or unmount), or some error occurred.
	2  - The --timeout option was given and no events occurred
	     in the specified interval of time.

Events:
	access		file or directory contents were read
	modify		file or directory contents were written
	attrib		file or directory attributes changed
	close_write	file or directory closed, after being opened in
	           	writeable mode
	close_nowrite	file or directory closed, after being opened in
	             	read-only mode
	close		file or directory closed, regardless of read/write mode
	open		file or directory opened
	moved_to	file or directory moved to watched directory
	moved_from	file or directory moved from watched directory
	move		file or directory moved to or from watched directory
	create		file or directory created within watched directory
	delete		file or directory deleted within watched directory
	delete_self	file or directory was deleted
	unmount		file system containing file or directory unmounted
```

2. inotifywait 使用示例

监控/root/test_157/目录的变化，代码如下。

```
[root@rsync_client_157 ~]# /usr/bin/inotifywait -mrq --timefmt '%Y/%m/%d-%H:%M:%S'
--format '%T %w %f' \
> -e modify,delete,create,move,attrib /root/test_157/
```

上面的指令表示，持续监听/root/test_157/目录及其子目录的变化，监听事件包括文件修改、删除、创建、移动、属性更改、显示到屏幕等。

开始执行监听程序，代码如下。

```
[root@rsync_client_157 ~]# /usr/bin/inotifywait -mrq --timefmt '%Y/%m/%d-%H:%M:%S'
--format '%T %w %f' \
> -e modify,delete,create,move,attrib /root/test_157/
```

复制文件到/root/test_157/目录，执行完上面的指令后，在/root/test_157/下创建或修改文件都会有信息输出，代码如下。

```
[root@rsync_client_157 ~]# /usr/bin/inotifywait -mrq --timefmt '%Y/%m/%d-%H:%M:%S'
--format '%T %w %f' \
> -e modify,delete,create,move,attrib /root/test_157/
2019/06/09-21:21:00 /root/test_157/ passwd
2019/06/09-21:21:00 /root/test_157/ passwd
2019/06/09-21:21:00 /root/test_157/ passwd
2019/06/09-21:21:00 /root/test_157/ passwd
2019/06/09-21:21:00 /root/test_157/ passwd
2019/06/09-21:21:00 /root/test_157/ shadow
2019/06/09-21:21:00 /root/test_157/ shadow
2019/06/09-21:21:00 /root/test_157/ shadow
2019/06/09-21:21:00 /root/test_157/ shadow
2019/06/09-21:21:00 /root/test_157/ shadow
```

上述代码输出表明，可使用 rsync 获取 inotifywait 监控到的文件列表来做指定的文件同步，而不是每次都由 rsync 做全目录扫描来判断文件是否存在差异。

3. rsync 组合 inotify-tools 完成实时同步

在客户端创建一个脚本 rsync.sh，代码如下。

```
[root@rsync_client_157 test_157]# cat /etc/rsync.sh
#!/bin/bash
/usr/bin/inotifywait -mrq --format '%w%f' -e create,close_write,delete
/root/test_157/ |while read file
do
    cd /root/test_157/ && rsync -az --delete /root/test_157
rsync_backup@192.168.2.156::backup_2_157 --password-file=/etc/rsync.password
done
```

上述代码使用 inotifywait 指令监控本地目录的变化，触发 rsync 指令将变化的文件传输到远程备份服务器上。

4. 创建测试文件和目录

创建测试文件和目录，代码如下。

```
[root@rsync_client_157 test_157]# mkdir abc
[root@rsync_client_157 test_157]# touch handuoduo.log
```

上述代码创建测试文件和目录。

5. 查看服务器同步文件列表信息

查看服务器同步文件列表信息，代码如下。

```
[root@rsync_server_156 ~]# ll /backup/192.168.2.157/test_157/
总用量 60
drwxr-xr-x 6 rsync rsync 4096 6月   8 21:05 a
drwxr-xr-x 2 rsync rsync 4096 6月   8 21:04 ab
drwxr-xr-x 2 rsync rsync 4096 6月   9 22:26 abc
-rw-r--r-- 1 rsync rsync    0 6月   8 13:53 abc.pdf
```

```
drwxr-xr-x 2 rsync rsync 4096 6月  8 21:04 ac
-rw-r--r-- 1 rsync rsync   57 6月  8 20:08 a.log
drwxr-xr-x 2 rsync rsync 4096 6月  8 13:53 a.pdf
drwxr-xr-x 2 rsync rsync 4096 6月  8 21:04 b
-rw-r--r-- 1 rsync rsync    0 6月  8 13:53 b.log
drwxr-xr-x 2 rsync rsync 4096 6月  8 13:53 b.pdf
drwxr-xr-x 2 rsync rsync 4096 6月  8 21:04 c
-rw-r--r-- 1 rsync rsync    0 6月  8 13:53 c.log
drwxr-xr-x 2 rsync rsync 4096 6月  8 14:03 c.pdf
drwxr-xr-x 3 rsync rsync 4096 6月  9 22:05 db
-rw-r--r-- 1 rsync rsync    0 6月  8 13:53 def.pdf
-rw-r--r-- 1 rsync rsync    0 6月  9 22:26 handuoduo.log
drwxr-xr-x 3 rsync rsync 4096 6月  9 22:05 mail
-rw-r--r-- 1 rsync rsync  854 5月 18 17:44 passwd
---------- 1 rsync rsync  592 5月 18 17:44 shadow
drwxr-xr-x 3 rsync rsync 4096 6月  9 22:05 web
```

上述代码中创建的文件和目录都已经存在。

6. 问题和改进

（1）同步文件问题描述。

上述 rsync 程序每次进行全量同步，而且文件列表是以循环形式触发 rsync，即有 10 个文件发生更改，就触发 10 次 rsync 全量同步。rsync 只会输出有差异，且需要同步的文件，在需要同步的源目录文件量很大的情况下，很难做到实时同步。

（2）优化同步文件改进方案。

要做到近乎实时，就必须要减少 rsync 对目录的递归扫描判断，尽可能做到只同步 inotify 监控到已发生更改的文件。结合 rsync 的特性，分开判断实现一个目录的增、删、改、查对应的操作，代码如下。

```bash
[root@rsync_client_157 test_157]# cat /etc/rsync_v2.sh
#!/bin/bash
src=/root/test_157/
des=backup_2_157
#des=/backup/192.168.2.157

rsync_passwd_file=/etc/rsync.password
ip1=192.168.2.156
user=rsync_backup
cd ${src}
/usr/bin/inotifywait -mrq --format '%Xe %w%f' -e modify,create,delete,attrib,close_write,move ./ | while read file
do
        INO_EVENT=$(echo $file | awk '{print $1}')

        INO_FILE=$(echo $file | awk '{print $2}')
        echo "-------------------------------$(date)-----------------------------------"
        echo $file
        if [[ $INO_EVENT =~ 'CREATE' ]] || [[ $INO_EVENT =~ 'MODIFY' ]] || [[ $INO_EVENT =~ 'CLOSE_WRITE' ]] || [[ $INO_EVENT =~ 'MOVED_TO' ]]
        then
                echo 'CREATE or MODIFY or CLOSE_WRITE or MOVED_TO'
                rsync -avzcR --password-file=${rsync_passwd_file} $(dirname ${INO_FILE}) ${user}@${ip1}::${des}
${INO_FILE}) ${user}@${ip2}::${des}
        fi
        if [[ $INO_EVENT =~ 'DELETE' ]] || [[ $INO_EVENT =~ 'MOVED_FROM' ]]
        then
                echo 'DELETE or MOVED_FROM'
                rsync -avzR --delete --password-file=${rsync_passwd_file} $(dirname ${INO_FILE}) ${user}@${ip1}::${des} &&
```

```
                rsync -avzR --delete --password-file=${rsync_passwd_file} $(dirname
${INO_FILE}) ${user}@${ip2}::${des}
        fi
        if [[ $INO_EVENT =~ 'ATTRIB' ]]
        then
                echo 'ATTRIB'
                if [ ! -d "$INO_FILE" ]
                then
                        rsync -avzcR --password-file=${rsync_passwd_file} $(dirname
${INO_FILE}) ${user}@${ip1}::${des} &&
                        rsync -avzcR --password-file=${rsync_passwd_file} $(dirname
${INO_FILE}) ${user}@${ip2}::${des}
                fi
        fi
done
```

上述代码中,每 2h 做 1 次全量同步。因为 inotify 只在启动时会监控目录,没有启动期间的文件发生更改,inotify 是不知道的,所以这里每 2h 做 1 次全量同步,防止各种意外遗漏,保证目录一致。

```
crontab -e
* */2 * * * rsync --delete -avz test_157 rsync_backup@192.168.2.156::backup_2_157
--password-file=/etc/rsync.password
```

inotifywait 选项和事件说明,如表 6-13 和表 6-14 所示。

表 6-13　　　　　　　　　　inotifywait 选项说明

选项名称	说明
-m、-monitor	始终保持事件监听状态
-r、-recursive	递归查询目录
-q、-quiet	只输出监控事件的信息
-exclude	排除文件或目录时,不区分大小写
-t、-timeout	超时时间
-timefmt	指定时间输出格式
-format	指定输出格式
-e、-event	指定删、增、改等事件

表 6-14　　　　　　　　　　inotifywait 事件说明

事件名称	说明
access	访问、读取文件
modify	修改内容
attrib	属性,文件元数据被修改
move	移动,对文件进行移动操作
create	创建,生成新文件
unmount	卸载文件系统
open	打开,对文件进行打开操作
close	关闭,对文件进行关闭操作
delete	删除,文件被删除

7. 优化 inotify

在 /proc/sys/fs/inotify 目录下有 3 个文件,对 inotify 机制有一定的限制。

- max_queued_events 设置 inotify 实例事件队列可容纳的事件数量。
- max_user_instances 设置每个用户可以运行的 inotifywait 或 inotifywatch 指令的进程数。

- max_user_watches 设置 inotifywait 或 inotifywatch 指令可以监视的文件数量（单进程）。

下列代码查看/proc/sys/fs/inotify/目录下的内容。

```
[root@rsync_client_157 ~]# ll /proc/sys/fs/inotify/
总用量 0
-rw-r--r-- 1 root root 0 6月   9 21:04 max_queued_events
-rw-r--r-- 1 root root 0 6月   9 21:04 max_user_instances
-rw-r--r-- 1 root root 0 6月   9 21:04 max_user_watches
```

优化代码如下。

```
[root@rsync_client_157 ~]# echo 60000000 > /proc/sys/fs/inotify/max_user_watches
[root@rsync_client_157 ~]# cat /proc/sys/fs/inotify/max_user_watches
8192
[root@rsync_client_157 ~]# echo 60000000 > /proc/sys/fs/inotify/max_user_watches
[root@rsync_client_157 ~]# cat /proc/sys/fs/inotify/max_queued_events
16384
[root@rsync_client_157 ~]# echo 60000000 > /proc/sys/fs/inotify/max_queued_events
[root@rsync_client_157 ~]# cat /proc/sys/fs/inotify/max_queued_events
60000000
[root@rsync_client_157 ~]# cat /proc/sys/fs/inotify/max_user_watches
60000000
```

加入/etc/rc.local 就可以实现每次重启都生效，代码如下。

```
cat >>/etc/rc.local<<inotify
echo 60000000 > /proc/sys/fs/inotify/max_user_watches
echo 60000000 > /proc/sys/fs/inotify/max_queued_events
inotify
tail -3 /etc/rc.local

[root@rsync_client_157 ~]# cat >>/etc/rc.local<<inotify
> echo 60000000 > /proc/sys/fs/inotify/max_user_watches
> echo 60000000 > /proc/sys/fs/inotify/max_queued_events
> inotify
[root@rsync_client_157 ~]# tail -3 /etc/rc.local
touch /var/lock/subsys/local
echo 60000000 > /proc/sys/fs/inotify/max_user_watches
echo 60000000 > /proc/sys/fs/inotify/max_queued_events
```

6.5 Lsyncd 实时同步详解

Lsyncd 结合 rsync + inotify，Lsyncd 监视本地目录树事件监视器接口（inotify 或 fsevents）。它聚合和组合事件几秒，然后生成一个或多个进程来同步更改。

Lsyncd 不仅可以实现同步，还可以监控某个目录下的文件，根据触发的事件自定义要执行的指令。本实验架构如表 6-15 所示。

表 6-15　　　　　　　　　　　　　实验架构

角色	操作系统	IP 地址	主机名
客户端	CentOS 6.9 x84_64	192.168.2.158	rsync_client_158
客户端	CentOS 6.9 x84_64	192.168.2.159	rsync_client_159

6.5.1 安装 Lsyncd

CentOS 内置的源并没有包括 Lsyncd，我们可以自行编译安装 Lsyncd，更简单的办法是先安装 epel 源，然后直接使用 yum 指令安装 Lsyncd，代码如下。

```
# yum -y install epel-release
# yum -y install lsyncd
```

详细代码输出如下。

```
[root@rsync_client_158 ~]# ll /etc/yum.repos.d/ |grep epel
[root@rsync_client_158 ~]# yum install epel-release -y >/dev/null 2>&1
[root@rsync_client_158 ~]# ll /etc/yum.repos.d/ |grep epel
-rw-r--r-- 1 root root  957 11月  5 2012 epel.repo
-rw-r--r-- 1 root root 1056 11月  5 2012 epel-testing.repo
[root@rsync_client_158 ~]# yum install lsyncd -y >/dev/null 2>&1
[root@rsync_client_158 ~]# rpm -qa lsyncd
lsyncd-2.1.5-0.el6.x86_64
[root@rsync_client_159 ~]# yum -y install epel-release >/dev/null 2>&1
[root@rsync_client_159 ~]# ll /etc/yum.repos.d/ |grep epel
-rw-r--r-- 1 root root  957 11月  5 2012 epel.repo
-rw-r--r-- 1 root root 1056 11月  5 2012 epel-testing.repo
[root@rsync_client_159 ~]#
[root@rsync_client_159 ~]#
[root@rsync_client_159 ~]# yum -y install lsyncd >/dev/null 2>&1
[root@rsync_client_159 ~]# rpm -qa lsyncd
lsyncd-2.1.5-0.el6.x86_64
[root@rsync_client_158 ~]# rpm -ql lsyncd
/etc/logrotate.d/lsyncd
/etc/lsyncd.conf
/etc/rc.d/init.d/lsyncd
/etc/sysconfig/lsyncd
/usr/bin/lsyncd
/usr/share/doc/lsyncd-2.1.5
/usr/share/doc/lsyncd-2.1.5/COPYING
/usr/share/doc/lsyncd-2.1.5/ChangeLog
/usr/share/doc/lsyncd-2.1.5/examples
/usr/share/doc/lsyncd-2.1.5/examples/lbash.lua
/usr/share/doc/lsyncd-2.1.5/examples/lecho.lua
/usr/share/doc/lsyncd-2.1.5/examples/lgforce.lua
/usr/share/doc/lsyncd-2.1.5/examples/limagemagic.lua
/usr/share/doc/lsyncd-2.1.5/examples/lpostcmd.lua
/usr/share/doc/lsyncd-2.1.5/examples/lrsync.lua
/usr/share/doc/lsyncd-2.1.5/examples/lrsyncssh.lua
/usr/share/man/man1/lsyncd.1.gz
/var/log/lsyncd
/var/run/lsyncd
[root@rsync_client_159 ~]# rpm -ql lsyncd
/etc/logrotate.d/lsyncd
/etc/lsyncd.conf
/etc/rc.d/init.d/lsyncd
/etc/sysconfig/lsyncd
/usr/bin/lsyncd
/usr/share/doc/lsyncd-2.1.5
/usr/share/doc/lsyncd-2.1.5/COPYING
/usr/share/doc/lsyncd-2.1.5/ChangeLog
/usr/share/doc/lsyncd-2.1.5/examples
/usr/share/doc/lsyncd-2.1.5/examples/lbash.lua
/usr/share/doc/lsyncd-2.1.5/examples/lecho.lua
/usr/share/doc/lsyncd-2.1.5/examples/lgforce.lua
/usr/share/doc/lsyncd-2.1.5/examples/limagemagic.lua
/usr/share/doc/lsyncd-2.1.5/examples/lpostcmd.lua
/usr/share/doc/lsyncd-2.1.5/examples/lrsync.lua
/usr/share/doc/lsyncd-2.1.5/examples/lrsyncssh.lua
/usr/share/man/man1/lsyncd.1.gz
/var/log/lsyncd
/var/run/lsyncd
```

上述代码使用 yum 指令安装 Lsyncd，并查看 Lsyncd 软件包组件信息。

6.5.2 配置 Lsyncd

以下将演示企业级常用配置案例：本机同步和远程同步。

1. 生成密钥认证文件

生成密钥认证文件，代码如下。

```
[root@rsync_client_158 ~]# ssh-keygen
Generating public/private rsa key pair.
Enter file in which to save the key (/root/.ssh/id_rsa):
Enter passphrase (empty for no passphrase):
Enter same passphrase again:
Your identification has been saved in /root/.ssh/id_rsa.
Your public key has been saved in /root/.ssh/id_rsa.pub.
The key fingerprint is:
40:f5:8c:56:84:e5:68:dd:87:bd:65:8d:6b:a8:0e:e0 root@rsync_client_158
The key's randomart image is:
+--[ RSA 2048]----+
|       ...++     |
|      . .O. o .. |
|       . = = o + +|
|        +    o = |
|        .S   . + |
|         . .   . |
|        E . .    |
|           o     |
|           .     |
+-----------------+
```

上述代码在 192.168.2.158 生成密钥认证文件。

2. 传输公钥文件并进行登录认证

传输公钥文件并进行登录认证，代码如下。

```
[root@rsync_client_158 ~]# ssh-copy-id -i /root/.ssh/id_rsa.pub 192.168.2.159
The authenticity of host '192.168.2.159 (192.168.2.159)' can't be established.
RSA key fingerprint is 82:e8:a0:65:a9:90:8e:f4:60:4d:6e:0e:7d:34:e0:75.
Are you sure you want to continue connecting (yes/no)? yes
Warning: Permanently added '192.168.2.159' (RSA) to the list of known hosts.
root@192.168.2.159's password:
Now try logging into the machine, with "ssh '192.168.2.159'", and check in:

  .ssh/authorized_keys

to make sure we haven't added extra keys that you weren't expecting.

[root@rsync_client_158 ~]# ssh 192.168.2.159
Last login: Mon Jun 10 21:04:17 2019 from 192.168.2.106
[root@rsync_client_159 ~]# logout
Connection to 192.168.2.159 closed.
```

上述代码传输公钥文件并进行登录认证。

3. 查看防火墙和 SELinux 是否关闭

查看防火墙和 SELinux 是否关闭，代码如下。

```
[root@rsync_client_158 ~]# sestatus
SELINUX status:                 disabled
[root@rsync_client_158 ~]# /etc/init.d/iptables status
iptables: 未运行防火墙。
[root@rsync_client_159 ~]# sestatus
SELINUX status:                 disabled
[root@rsync_client_159 ~]# /etc/init.d/iptables status
iptables: 未运行防火墙。
```

上述代码查看防火墙和 SELinux 的状态。

6.5.3 本机同步设置

1. 创建 lsyncd.conf.lua 配置文件

创建 lsyncd.conf.lua 配置文件，代码如下。

```
[root@rsync_client_158 ~]# cat /etc/lsyncd.conf.lua
-- User configuration file for lsyncd.
-- Simple example for default rsync, but executing moves through on the target.
-- For more examples, see /usr/share/doc/lsyncd*/examples/
settings {
    logfile       ="/tmp/lsyncd.log",
    statusFile    ="/tmp/lsyncd.status",
    inotifyMode   = "CloseWrite",
    maxProcesses = 7,
    -- nodaemon =true,
    }
sync {
    default.rsync,
    source    = "/root/source",
    target    = "/root/target",
    rsync     = {
        binary    = "/usr/bin/rsync",
        archive   = true,
        compress  = true,
        verbose   = true,
        perms     = true
        },
    }
```

上述代码创建配置文件。

2. 创建同步测试目录

创建同步测试目录，代码如下。

```
[root@rsync_client_158 ~]# mkdir -pv /root/{source,target}
mkdir: 已创建目录 "/root/source"
mkdir: 已创建目录 "/root/target"
```

上述代码创建同步测试目录。

3. 启动 Lsyncd 服务

启动 Lsyncd 服务，代码如下。

```
[root@rsync_client_158 ~]#  lsyncd -nodaemon /etc/lsyncd.conf.lua
21:08:24 Normal: recursive startup rsync: /root/source/ -> /root/target/
sending incremental file list
./

sent 29 bytes  received 15 bytes  88.00 bytes/sec
total size is 0  speedup is 0.00
21:08:24 Normal: Startup of "/root/source/" finished.
```

上述代码后台启动 Lsyncd 服务。

4. 查看 Lsyncd 进程

查看 Lsyncd 进程，代码如下。

```
[root@rsync_client_158 ~]# ps -ef |grep lsyncd
root      1437  1403  0 21:08 pts/1    00:00:00 lsyncd -nodaemon /etc/lsyncd.conf.lua
root      1457  1441  0 21:08 pts/2    00:00:00 grep lsyncd
```

上述代码查看 Lsyncd 进程信息。

5. 创建同步测试文件

创建同步测试文件，代码如下。

```
[root@rsync_client_158 ~]# cd /root/source/
```

```
[root@rsync_client_158 source]# ll
总用量 0
[root@rsync_client_158 source]# cp -av /etc/passwd .
"/etc/passwd" -> "./passwd"
[root@rsync_client_158 source]# touch abc
[root@rsync_client_158 source]# ll
总用量 4
-rw-r--r-- 1 root root   0 6月  10 21:08 abc
-rw-r--r--. 1 root root 854 5月  18 17:44 passwd
[root@rsync_client_158 source]# ll /root/target/
总用量 0
[root@rsync_client_158 source]# ll /root/target/
总用量 0
[root@rsync_client_158 source]# ll /root/target/
总用量 0
[root@rsync_client_158 source]# ll /root/target/
总用量 0
[root@rsync_client_158 source]# ll /root/target/
总用量 0
[root@rsync_client_158 source]# ll /root/target/
总用量 0
[root@rsync_client_158 source]# ll /root/target/
总用量 0
[root@rsync_client_158 source]# ll /root/target/
总用量 0
[root@rsync_client_158 source]# ll /root/target/
总用量 4
-rw-r--r-- 1 root root   0 6月  10 21:08 abc
-rw-r--r-- 1 root root 854 5月  18 17:44 passwd
```

上述代码创建同步测试文件。

6. 查看 Lsyncd 进程输出内容

查看 Lsyncd 进程输出内容，代码如下。

```
[root@rsync_client_158 ~]# lsyncd -nodaemon /etc/lsyncd.conf.lua
21:08:24 Normal: recursive startup rsync: /root/source/ -> /root/target/
sending incremental file list
./

sent 29 bytes  received 15 bytes  88.00 bytes/sec
total size is 0  speedup is 0.00
21:08:24 Normal: Startup of "/root/source/" finished.
21:09:04 Normal: Calling rsync with filter-list of new/modified files/dirs
/passwd
/
/abc
sending incremental file list
./
abc
passwd

sent 487 bytes  received 53 bytes  1080.00 bytes/sec
total size is 854  speedup is 1.58
21:09:05 Normal: Finished a list after exitcode: 0
```

上述代码查看 Lsyncd 进程输出信息。

6.5.4 远程同步设置

1. 配置 lsyncd.conf.lua 文件

配置 lsyncd.conf.lua 文件，代码如下。

```
[root@rsync_client_158 ~]# cat /etc/lsyncd.conf.lua
-- User configuration file for lsyncd.
-- Simple example for default rsync, but executing moves through on the target.
-- For more examples, see /usr/share/doc/lsyncd*/examples/
settings {
    logfile         ="/tmp/lsyncd.log",
    statusFile      ="/tmp/lsyncd.status",
    inotifyMode     = "CloseWrite",
    maxProcesses    = 7,
    -- nodaemon =true,
    }
sync {
    default.rsync,
    source      = "/root/source",
    target      = "/root/target",
    rsync       = {
        binary      = "/usr/bin/rsync",
        archive     = true,
        compress    = true,
        verbose     = true,
        perms       = true
        },
    }
```

上述代码配置 lsyncd.conf.lua 文件。

2. 创建目录

（1）创建本地测试目录，代码如下。

```
[root@rsync_client_158 ~]# mkdir -pv /root/test_158
mkdir: 已创建目录 "/root/test_158"
```

上述代码创建本地测试目录。

（2）创建远端测试目录，代码如下。

```
[root@rsync_client_159 ~]# mkdir -pv /root/test_159
mkdir: 已创建目录 "/root/test_159"
```

上述代码创建远端测试目录。

3. 启动 Lsyncd 服务

启动 Lsyncd 服务，代码如下。

```
[root@rsync_client_158 ~]#  lsyncd -nodaemon /etc/lsyncd.conf.lua
21:19:13 Normal: recursive startup rsync: /root/test_158/ ->
192.168.2.159:/root/test_159/
```

上述代码启动 Lsyncd 服务。

4. 创建测试文件

```
[root@rsync_client_158 ~]# cd /root/test_158/
[root@rsync_client_158 test_158]# ll
总用量 0
[root@rsync_client_158 test_158]# touch {a..c}.pdf
[root@rsync_client_158 test_158]# ll
总用量 0
-rw-r--r-- 1 root root 0 6月  10 21:19 a.pdf
-rw-r--r-- 1 root root 0 6月  10 21:19 b.pdf
-rw-r--r-- 1 root root 0 6月  10 21:19 c.pdf
[root@rsync_client_158 test_158]# ls -lhrt --full-time
总用量 0
-rw-r--r-- 1 root root 0 2019-06-10 21:19:37.535259516 +0800 c.pdf
```

```
-rw-r--r-- 1 root root 0 2019-06-10 21:19:37.535259516 +0800 b.pdf
-rw-r--r-- 1 root root 0 2019-06-10 21:19:37.535259516 +0800 a.pdf
```
上述代码创建测试文件。

5．查看文件同步信息

查看文件同步信息，代码如下。
```
[root@rsync_client_159 ~]# mkdir -pv /root/test_159
mkdir: 已创建目录 "/root/test_159"
[root@rsync_client_159 ~]# ls -lhrt --full-time /root/test_159/
总用量 0
-rw-r--r-- 1 root root 0 2019-06-10 21:19:37.000000000 +0800 c.pdf
-rw-r--r-- 1 root root 0 2019-06-10 21:19:37.000000000 +0800 b.pdf
-rw-r--r-- 1 root root 0 2019-06-10 21:19:37.000000000 +0800 a.pdf
```
上述代码查看文件同步信息。

6．查看 Lsyncd 服务控制台输出

查看 Lsyncd 服务控制台输出，代码如下。
```
[root@rsync_client_158 ~]# lsyncd -nodaemon /etc/lsyncd.conf.lua
21:19:13 Normal: recursive startup rsync: /root/test_158/ -> 
192.168.2.159:/root/test_159/
21:19:13 Normal: Startup of "/root/test_158/" finished: 0
21:19:52 Normal: rsyncing list
/a.pdf
/b.pdf
/c.pdf
21:19:52 Normal: Finished (list): 0
```
上述代码查看 Lsyncd 服务控制台输出。

7．CentOS 不同版本启动 Lsyncd 的方法

（1）CentOS 7 操作系统使用 systemctl 管理 Lsyncd。

继续输入指令 lsyncd -nodaemon /etc/lsyncd.conf，运行查看是否有报错。如果有报错根据报错情况检查修改。如果没有报错直接退出后输入指令 systemctl start lsyncd，启动 Lsyncd 即可，这里给出 systemctl 管理 Lsyncd 的相关指令。

启动 Lsyncd，代码如下。
```
systemctl start lsyncd
```
停止 Lsyncd，代码如下。
```
systemctl stop lsyncd
```
重启 Lsyncd，代码如下。
```
systemctl restart lsyncd
```
设置 Lsyncd 开机自动启动，代码如下。
```
systemctl enable lsyncd
```
（2）CentOS 6 操作系统使用 system-v 风格脚本管理 Lsyncd。

下面的启动方式，采用自定义 Lsyncd 配置文件的方式进行启动，需要对脚本进行轻微的修改，代码如下。
```
[root@rsync_client_158 ~]# cat    /etc/init.d/lsyncd
#!/bin/bash
#
# chkconfig: - 85 15
# description: Lightweight inotify based sync daemon
#
# processname:    lsyncd
# config:         /etc/lsyncd.conf
# config:         /etc/sysconfig/lsyncd
# pidfile:        /var/run/lsyncd.pid
```

```bash
# Source function library
. /etc/init.d/functions

# Source networking configuration.
. /etc/sysconfig/network

# Check that networking is up.
[ "$NETWORKING" = "no" ] && exit 0

LSYNCD_OPTIONS="-pidfile /var/run/lsyncd.pid /etc/lsyncd.conf.lua"

if [ -e /etc/sysconfig/lsyncd ]; then
  . /etc/sysconfig/lsyncd
fi

RETVAL=0

prog="lsyncd"
thelock=/var/lock/subsys/lsyncd

start() {
    [ -f /etc/lsyncd.conf ] || exit 6
        echo -n $"Starting $prog: "
        if [ $UID -ne 0 ]; then
                RETVAL=1
                failure
        else
                daemon ${LSYNCD_USER:+--user ${LSYNCD_USER}} /usr/bin/lsyncd $LSYNCD_OPTIONS
                RETVAL=$?
                [ $RETVAL -eq 0 ] && touch $thelock
        fi;
        echo
        return $RETVAL
}

stop() {
        echo -n $"Stopping $prog: "
        if [ $UID -ne 0 ]; then
                RETVAL=1
                failure
        else
                killproc lsyncd
                RETVAL=$?
                [ $RETVAL -eq 0 ] && rm -f $thelock
        fi;
        echo
        return $RETVAL
}

reload(){
        echo -n $"Reloading $prog: "
        killproc lsyncd -HUP
        RETVAL=$?
        echo
        return $RETVAL
}

restart(){
        stop
        start
}

condrestart(){
    [ -e $thelock ] && restart
```

```
        return 0
}

case "$1" in
  start)
        start
        ;;
  stop)
        stop
        ;;
  restart)
        restart
        ;;
  reload)
        reload
        ;;
  condrestart)
        condrestart
        ;;
  status)
        status lsyncd
        RETVAL=$?
        ;;
  *)
        echo $"Usage: $0 {start|stop|status|restart|condrestart|reload}"
        RETVAL=1
esac

exit $RETVAL
```

重启、停止和启动 Lsyncd 服务，测试代码如下。

```
[root@rsync_client_158 ~]# /etc/init.d/lsyncd restart
停止 lsyncd:                                              [确定]
正在启动 lsyncd:                                          [确定]
[root@rsync_client_158 ~]# /etc/init.d/lsyncd stop
停止 lsyncd:                                              [确定]
[root@rsync_client_158 ~]# /etc/init.d/lsyncd start
正在启动 lsyncd:                                          [确定]
```

上述代码测试重启、停止和启动 Lsyncd 服务。

第 7 章

SFTP 服务

前文讲解了 vsftpd 服务，其配置非常灵活，也很强大，足够满足大家在局域网内的使用需求。但 vsftpd 本身传输的数据是明文的，如果需要传输一些很重要的数据，比如金融数据、银行卡交易记录、合同信息或其他支付记录时会存在很大的安全隐患，比如被非法嗅探、数据被劫持等，这会给企业和用户带来难以挽回的损失。因此，企业在端到端或在公网传输文件时，必须采用支持加密的传输协议或者服务，SFTP 服务的出现可以解决文件跨网传输安全性的问题。

本章重点讲解 SFTP 服务器的单机版本和金融领域常用的基于跨机房的 SFTP 双机房密钥认证服务搭建。

7.1 构建 SFTP 服务运行环境

SFTP 是 Secure File Transfer Protocol 的英文缩写，即安全文件传送协议。SFTP 可以为传输文件提供一种安全的加密方法。SFTP 与 FTP 有着几乎一样的语法和功能。

SFTP 为 SSH 的一部分，是一种安全传输文件的方式。在 SSH 软件包中，已经包含了 SFTP 的安全文件传输子系统。

SFTP 本身没有单独的守护进程，它必须使用 sshd 守护进程（端口号默认是 22）来完成相应的连接操作。从某种意义上来说，SFTP 并不像一个服务器程序，而更像一个客户端程序，同样是使用加密传输认证信息和数据，使用 SFTP 是非常安全的。但由于这种传输方式使用了加密和解密技术，因此传输效率比普通的 FTP 要低得多。如果对网络安全性要求更高时，可以使用 SFTP 代替 FTP。本次实验操作环境如表 7-1 所示。

表 7-1　　　　　　　　　　　　　　实验操作环境

操作系统	IP 地址	主机名	SSH 版本
CentOS 6.9 x86_64	192.168.2.171	sftp_171	OpenSSH_5.3p1

7.1.1 初始化 SFTP 服务器

1．关闭防火墙和 SELinux

关闭防火墙和 SELinux，代码如下。

```
[root@sftp_171 ~]# /etc/init.d/iptables stop
[root@sftp_171 ~]# chkconfig iptables off
[root@sftp_171 ~]# sed -i 's#SELINUX=enforcing#SELINUX=disabled#g' /etc/sysconfig/SELINUX
```

上述代码表示，关闭防火墙和 SELinux。

```
[root@sftp_171 ~]# grep ^SELINUX /etc/sysconfig/SELINUX
SELINUX=disabled
SELINUXTYPE=targeted
[root@sftp_171 ~]# setenforce 0
[root@sftp_171 ~]# sestatus
SELINUX status:                 enabled
SELINUXfs mount:                /SELINUX
Current mode:                   permissive
Mode from config file:          enforcing
Policy version:                 24
Policy from config file:        targeted
[root@sftp_171 ~]# /etc/init.d/iptables status
iptables：未运行防火墙。
```

7.1 构建 SFTP 服务运行环境

上述代码表示，防火墙和 SELinux 都已经处于关闭状态。

2. 查看操作系统版本和 IP 地址

查看操作系统版本和 IP 地址，代码如下。

```
[root@sftp_171 ~]# ip ro |grep src |awk '{print $NF}'
192.168.2.171
[root@sftp_171 ~]# cat /etc/redhat-release
CentOS release 6.9 (Final)
[root@sftp_171 ~]# uname -rm
2.6.32-696.el6.x86_64 x86_64
[root@sftp_171 ~]# ping jd*.com -c 3
PING jd*.com (120.52.148.118) 56(84) bytes of data.
64 bytes from 120.52.148.118: icmp_seq=1 ttl=50 time=6.11 ms
64 bytes from 120.52.148.118: icmp_seq=2 ttl=50 time=6.73 ms
64 bytes from 120.52.148.118: icmp_seq=3 ttl=50 time=7.66 ms

--- jd*.com ping statistics ---
3 packets transmitted, 3 received, 0% packet loss, time 2011ms
rtt min/avg/max/mdev = 6.110/6.837/7.669/0.640 ms
```

上述代码表示查看当前主机的 IP 地址、操作系统版本、网络连通性测试等信息。

3. 安装 epel 源

使用 yum 指令安装 epel 源，代码如下。

```
[root@sftp_171 ~]# yum install epel-release
已加载插件：fastestmirror
设置安装进程
Determining fastest mirrors
 * base: mirror.bit.edu.cn
 * extras: mirror.bit.edu.cn
 * updates: mirror.bit.edu.cn
（中间代码略）

已安装：
  epel-release.noarch 0:6-8

完毕！
```

上述代码表示 epel 源文件已经存在 /etc/yum.repos.d/ 目录下。

4. 查看 OpenSSH 软件版本信息

查看 OpenSSH 软件版本信息，代码如下。

```
[root@sftp_171 ~]# ssh -V
OpenSSH_5.3p1, OpenSSL 1.0.1e-fips 11 Feb 2013
OpenSSH_5.3p1, OpenSSL 1.0.1e-fips 11 Feb 2013
[root@sftp_171 ~]# yum list OpenSSH*
已加载插件：fastestmirror
Loading mirror speeds from cached hostfile
epel/metalink
| 6.5 kB     00:00
 * base: mirror.bit.edu.cn
 * epel: mirrors.yun-idc.com
 * extras: mirror.bit.edu.cn
 * updates: mirror.bit.edu.cn
epel
| 4.7 kB     00:00
epel/primary_db
| 6.1 MB     00:00
已安装的软件包
OpenSSH.x86_64    5.3p1-122.el6
@anaconda-CentOS-201703281317.x86_64/6.9
OpenSSH-clients.x86_64 5.3p1-122.el6
@anaconda-CentOS-201703281317.x86_64/6.9
OpenSSH-server.x86_64      5.3p1-122.el6
```

```
@anaconda-CentOS-201703281317.x86_64/6.9
可安装的软件包
OpenSSH.x86_64                  5.3p1-124.el6_10         updates
OpenSSH-askpass.x86_64          5.3p1-124.el6_10         updates
OpenSSH-clients.x86_64          5.3p1-124.el6_10         updates
OpenSSH-ldap.x86_64             5.3p1-124.el6_10         updates
OpenSSH-server.x86_64           5.3p1-124.el6_10         updates
```

CentOS 下可以使用如下指令安装 OpenSSH 的客户端。
```
yum install OpenSSH-clients
```

Ubuntu 下可以使用如下指令安装 OpenSSH 的服务器和客户端。
```
sudo apt-get update
sudo apt-get install openssh-server
```

5. 升级 OpenSSH 相关软件

升级 OpenSSH 相关软件，代码如下。
```
[root@sftp_171 ~]# yum update openssh-clients openssh-server
已加载插件：fastestmirror
设置更新进程
Loading mirror speeds from cached hostfile
 * base: mirror.bit.edu.cn
 * epel: mirrors.yun-idc.com
 * extras: mirror.bit.edu.cn
 * updates: mirror.bit.edu.cn
（中间代码略）
更新完毕：
  openssh-clients.x86_64 0:5.3p1-124.el6_10
openssh-server.x86_64 0:5.3p1-124.el6_10

作为依赖被升级：
  openssh.x86_64 0:5.3p1-124.el6_10

完毕！
```
上述代码表示 OpenSSH-clients、OpenSSH-server 软件包已经升级完成。

6. 备份 SFTP 服务配置文件

（1）查看 OpenSSH-clients 配置文件信息，代码如下。
```
[root@sftp_171 ~]# rpm -ql openssh-clients |grep config
/etc/ssh/ssh_config
/usr/share/man/man5/ssh_config.5.gz
```
上述代码使用 grep 指令过滤 OpenSSH-clients 配置文件。

（2）查看 OpenSSH-server 配置文件信息，代码如下。
```
[root@sftp_171 ~]# rpm -ql openssh-server
/etc/pam.d/ssh-keycat
/etc/pam.d/sshd
/etc/rc.d/init.d/sshd
/etc/ssh/sshd_config
/etc/sysconfig/sshd
/usr/libexec/OpenSSH/sftp-server
/usr/libexec/OpenSSH/ssh-keycat
/usr/sbin/.sshd.hmac
/usr/sbin/sshd
/usr/share/doc/OpenSSH-server-5.3p1
/usr/share/doc/OpenSSH-server-5.3p1/HOWTO.ssh-keycat
/usr/share/man/man5/moduli.5.gz
/usr/share/man/man5/sshd_config.5.gz
/usr/share/man/man8/sftp-server.8.gz
/usr/share/man/man8/sshd.8.gz
```
上述代码查看 OpenSSH-server 软件包安装后生成的所有文件信息。

（3）备份 OpenSSH-server 配置文件，代码如下。
```
[root@sftp_171 ~]# cp -av /etc/ssh/sshd_config /etc/ssh/sshd_config_bak_ori_1
```

```
"/etc/ssh/sshd_config" -> "/etc/ssh/sshd_config_bak_ori_1"
```
读者也可以使用如下指令对 OpenSSH-server 配置文件进行备份。
```
[root@sftp_171 ~]# cp -av /etc/ssh/sshd_config{,_bak_ori_2}
"/etc/ssh/sshd_config" -> "/etc/ssh/sshd_config_bak_ori_2"
```
上述代码表示在/etc/ssh/目录下备份同名配置文件，并以_bak 为结尾，使用此方法备份配置文件较为方便，推荐读者使用。

7.1.2 初始化 SFTP 用户运行环境

1. 创建 duoduo_sftp 用户、sftp 用户组

（1）创建用户和用户组，代码如下。
```
[root@sftp_171 ~]# groupadd sftp && useradd -g sftp duoduo_sftp
[root@sftp_171 ~]# echo 'duoduo_sftp' |passwd --stdin duoduo_sftp
更改用户 duoduo_sftp 的密码。
passwd: 所有的身份验证令牌已经成功更新。
```
上述代码表示用户和用户组已经创建成功。

（2）确认用户和用户组信息是否正确，代码如下。
```
[root@sftp_171 ~]# grep duoduo_sftp /etc/passwd
duoduo_sftp:x:500:500::/home/duoduo_sftp:/bin/bash
[root@sftp_171 ~]# grep sftp /etc/group
sftp:x:500:
```
上述代码表示 duoduo_sftp 用户和 sftp 用户组已经创建成功。

2. 创建用户密钥对

切换到 duoduo_sftp 用户，并创建用户密钥对，代码如下。
```
[root@sftp_171 ~]# su - duoduo_sftp
[duoduo_sftp@sftp_171 ~]$ ssh-keygen
Generating public/private rsa key pair.
Enter file in which to save the key (/home/duoduo_sftp/.ssh/id_rsa):
Created directory '/home/duoduo_sftp/.ssh'.
Enter passphrase (empty for no passphrase):
Enter same passphrase again:
Your identification has been saved in /home/duoduo_sftp/.ssh/id_rsa.
Your public key has been saved in /home/duoduo_sftp/.ssh/id_rsa.pub.
The key fingerprint is:
5e:45:4c:57:f1:21:6f:47:2d:a9:aa:b1:af:8d:be:0b duoduo_sftp@sftp_171
The key's randomart image is:
+--[ RSA 2048]----+
|         oo..==|
|         ...=.+|
|          .. +o|
|          .. ..|
|        S ..   |
|         ....  |
|      E .+     |
|       .oo     |
|       .*=o    |
+-----------------+
[duoduo_sftp@sftp_171 ~]$ ll /home/duoduo_sftp/.ssh/id*
-rw-------. 1 duoduo_sftp sftp 1671 5月  20 22:48 /home/duoduo_sftp/.ssh/id_rsa
-rw-r--r--. 1 duoduo_sftp sftp 402 5月  20 22:48 /home/duoduo_sftp/.ssh/id_rsa.pub
```
上述代码表示，duoduo_sftp 用户密钥对已经创建成功。

3. 指定用户数据目录

sftp 用户组的用户的 home 目录统一指定到/data/sftp 下，按用户名区分。这里先新建一个 duoduo_sftp 目录，然后指定 duoduo_sftp 的 home 目录为/data/sftp/duoduo_sftp，代码如下。

```
[root@sftp_171 ~]# mkdir -pv /data/sftp/duoduo_sftp
mkdir: 已创建目录 "/data"
mkdir: 已创建目录 "/data/sftp"
mkdir: 已创建目录 "/data/sftp/duoduo_sftp"
```

上述代码表示 duoduo_sftp 用户的数据目录为/data/sftp/duoduo_sftp。

4．查看用户权限

查看 duoduo_sftp 用户目录.ssh 权限。

```
[root@sftp_171 ~]# ls -ld /home/duoduo_sftp/.ssh
drwx------. 2 duoduo_sftp sftp 4096 5月  20 22:48 /home/duoduo_sftp/.ssh
```

上述代码表示/home/duoduo_sftp/.ssh 的权限为 700。

7.2 搭建 SFTP 服务

7.2.1 基本配置

1．备份/etc/ssh/sshd_config 配置文件

```
[root@sftp_171 ~]# cp -av /etc/ssh/sshd_config{,_bak}
"/etc/ssh/sshd_config" -> "/etc/ssh/sshd_config_bak"
```

上述代码表示备份/etc/ssh/sshd_config 文件为/etc/ssh/sshd_config_bak。

2．编辑/etc/ssh/sshd_config

```
# vim +132 /etc/ssh/sshd_config
```

找到下面这行，并注释掉。
```
Subsystem sftp /usr/libexec/OpenSSH/sftp-server
```

添加如下几行代码。
```
Subsystem sftp internal-sftp
Match Group sftp
ChrootDirectory /data/sftp/%u
ForceCommand internal-sftp
AllowTcpForwarding no
X11Forwarding no
```

SSH 开启密钥登录，代码如下。
```
 43 PubkeyAuthentication yes
 55 RSAAuthentication yes
 57 AuthorizedKeysFile    #.ssh/authorized_keys
```

完整代码配置，如下。
```
#Subsystem # sftp #/usr/libexec/OpenSSH/sftp-server

# Example of overriding settings on a per-user basis
#Match User anoncvs
#       X11Forwarding no
#       AllowTcpForwarding no
#       ForceCommand cvs server

# SFTP 服务基本配置   begin #
AuthorizedKeysFile      .ssh/authorized_keys

Subsystem sftp internal-sftp
Match Group sftp
ChrootDirectory /data/sftp/%u
ForceCommand internal-sftp
AllowTcpForwarding no
X11Forwarding no
PubkeyAuthentication yes
RSAAuthentication yes
```

```
# SFTP服务基本配置    end #
```
SFTP 服务基本配置如图 7-1 所示。

图 7-1　SFTP 服务基本配置

7.2.2　安全设置

1．设置 chroot 权限

设置 chroot 权限，代码如下。

```
[root@sftp_171 ~]# chown root:sftp /data/sftp/duoduo_sftp
[root@sftp_171 ~]# chmod 755 /data/sftp/duoduo_sftp
[root@sftp_171 ~]# ls -ld /data/sftp/duoduo_sftp
drwxr-xr-x. 2 root sftp 4096 5月  20 22:52 /data/sftp/duoduo_sftp
```

错误的目录权限设置会导致在日志文件中出现"fatal: bad ownership or modes for chroot directory"的内容，目录的权限设置有两个要点。

- 从 ChrootDirectory 指定的目录开始一直往上，到系统根目录为止的目录拥有者都只能是 root。
- 从 ChrootDirectory 指定的目录开始一直往上，到系统根目录为止都不可以具有群组写入权限。

遵循以上两个要点设置 SFTP 服务。

（1）将/data/sftp/duoduo_sftp 的所有者设置为 root，所有组设置为 sftp。

（2）将/data/sftp/duoduo_sftp 的权限设置为 755，所有者 root 有写入权限，而所有组 sftp 无写入权限。

2．设置目录权限

建立 SFTP 用户登录后可写入的目录，照上面设置后，在重启 sshd 服务后，用户 duoduo_sftp 已经可以登录；但使用 chroot 指定根目录后，根应该是无法写入的，所以要新建一个目录供 duoduo_sftp 上传文件。这个目录所有者为 duoduo_sftp，所有组为 sftp，所有者有写入权限，而所有组无写入权限，完整代码如下。

```
[root@sftp_171 ~]# ls -ld /data/sftp/duoduo_sftp
drwxr-xr-x. 2 root sftp 4096 5月  20 22:52 /data/sftp/duoduo_sftp
[root@sftp_171 ~]# mkdir -pv /data/sftp/duoduo_sftp/webank
mkdir: 已创建目录 "/data/sftp/duoduo_sftp/webank"
[root@sftp_171 ~]# chown duoduo_sftp:sftp /data/sftp/duoduo_sftp/webank
[root@sftp_171 ~]# chmod 755 /data/sftp/duoduo_sftp/webank
[root@sftp_171 ~]# ls -ld /data/sftp/duoduo_sftp
drwxr-xr-x. 3 root sftp 4096 5月  20 23:22 /data/sftp/duoduo_sftp
[root@sftp_171 ~]# ls -ld /data/sftp/duoduo_sftp/webank/
drwxr-xr-x. 2 duoduo_sftp sftp 4096 5月  20 23:22 /data/sftp/duoduo_sftp/webank/
```

注意对比/data/sftp/duoduo_sftp 目录和/data/sftp/duoduo_sftp/webank 目录的权限。

3. 重启 sshd 服务

```
# service sshd restart
```

到这里，duoduo_sftp 已经可以通过 SFTP 客户端登录并可以上传文件到 webank 目录。

如果还是不能在此目录下上传文件，提示没有权限，检查 SELinux 是否关闭，可以使用如下方式关闭 SELinux。

修改/etc/SELINUX/config 文件中的 SELINUX 为 disabled，然后重启。

```
# setenforce 0
```

上述代码关闭 SELinux。

4. 核心配置文件注释

```
Subsystem sftp internal-sftp
```

上述代码使用 SFTP 服务使用系统自带的 internal-sftp。

```
Match Group sftp
```

上述代码匹配 sftp 所有组的用户。如果要匹配多个组，多个组之间用逗号分隔。当然，也可以匹配用户，代码如下。

```
Match User duoduo_sftp
```

匹配用户时，多个用户名之间也用逗号分隔，但这里按组匹配更灵活和方便，代码如下。

```
ChrootDirectory /data/sftp/%u
```

用 chroot 将用户的根目录指定到/data/sftp/%u，%u 代表用户名。这样用户就只能在/data/sftp/%u 下活动。

5. 设置密钥认证文件权限

设置密钥认证文件权限基本流程如下。

- 进入密钥对所在目录。
- 安装公钥文件至系统。

对公钥和目录设置权限。

- 设置 600 权限。
- 为目录设置 700 权限。

完整代码如下。

```
[root@sftp_171 ~]# cd /home/duoduo_sftp/.ssh
[root@sftp_171 .ssh]# pwd
/home/duoduo_sftp/.ssh
[root@sftp_171 .ssh]# ls -lhrt
总用量 8.0K
-rw-r--r--. 1 duoduo_sftp sftp  402 5月  20 22:48 id_rsa.pub
-rw-------. 1 duoduo_sftp sftp 1.7K 5月  20 22:48 id_rsa
[root@sftp_171 .ssh]# cat id_rsa.pub >> authorized_keys
[root@sftp_171 .ssh]# ls -lhrt
总用量 12K
-rw-r--r--. 1 duoduo_sftp sftp  402 5月  20 22:48 id_rsa.pub
-rw-------. 1 duoduo_sftp sftp 1.7K 5月  20 22:48 id_rsa
```

```
-rw-r--r--. 1 root         root    402 5月  20 23:30 authorized_keys
[root@sftp_171 .ssh]# chown -R duoduo_sftp:sftp *
[root@sftp_171 .ssh]# ls -lhrt
总用量 12K
-rw-r--r--. 1 duoduo_sftp sftp  402 5月  20 22:48 id_rsa.pub
-rw-------. 1 duoduo_sftp sftp 1.7K 5月  20 22:48 id_rsa
-rw-r--r--. 1 duoduo_sftp sftp  402 5月  20 23:30 authorized_keys


[root@sftp_171 .ssh]# sftp -oIdentityFile=/home/duoduo_sftp/.ssh/id_rsa -oPort=22
duoduo_sftp@192.168.2.171
Connecting to 192.168.2.171...
The authenticity of host '192.168.2.171 (192.168.2.171)' can't be established.
RSA key fingerprint is 82:e8:a0:65:a9:90:8e:f4:60:4d:6e:0e:7d:34:e0:75.
Are you sure you want to continue connecting (yes/no)? yes
Warning: Permanently added '192.168.2.171' (RSA) to the list of known hosts.
sftp> ls
sftp> pwd
Remote working directory: /home/duoduo_sftp
sftp> ls
sftp> exit
```

使用上述代码测试是否可以使用 duoduo_sftp 用户登录系统。

6. 设置用户登录权限

设置 duoduo_sftp 用户不可直接登录操作系统，只能通过 SFTP 的方式登录操作系统。

```
[root@sftp_171 .ssh]# grep duoduo_sftp /etc/passwd
duoduo_sftp:x:500:500::/home/duoduo_sftp:/bin/bash
[root@sftp_171 .ssh]# usermod -s  /sbin/nologin  duoduo_sftp
[root@sftp_171 .ssh]# grep duoduo_sftp /etc/passwd
duoduo_sftp:x:500:500::/home/duoduo_sftp:/sbin/nologin
```

上述代码表示 duoduo_sftp 用户不可直接登录操作系统。

7. 关闭防火墙

重启 sshd 服务，代码如下。

```
[root@sftp_171 .ssh]# /etc/init.d/sshd restart
停止 sshd:                                              [确定]
正在启动 sshd:                                          [确定]
```

duoduo_sftp 已经可以通过 SFTP 客户端登录操作系统并可以上传文件到 webank 目录。

如果还是不能在此目录下上传文件，提示没有权限，检查 SELinux 是否关闭。可以修改/etc/SELINUX/config 文件中的 SELINUX 为 disabled，然后重启，或者使用如下指令进行关闭操作。

```
# setenforce 0
```

7.2.3 验证 SFTP 环境

1. 验证是否可以登录

用 duoduo_sftp 用户登录，按 "Enter" 键输入密码。

```
[root@sftp_171 .ssh]# sftp   -oIdentityFile=/home/duoduo_sftp/.ssh/id_rsa -oPort=22
duoduo_sftp@192.168.2.171
Connecting to 192.168.2.171...
sftp> ls
webank
sftp> exit
```

上述代码显示 sftp> exit，则表示 SFTP 服务器搭建成功。

2. 验证目录和文件是否创建成功

```
[root@sftp_171 .ssh]# sftp   -oIdentityFile=/home/duoduo_sftp/.ssh/id_rsa -oPort=22
duoduo_sftp@192.168.2.171
Connecting to 192.168.2.171...
```

```
sftp> ls
webank
sftp> cd webank
sftp> ls
sftp> pwd
Remote working directory: /webank
sftp> mkdir day_1
sftp> ls
day_1
sftp> put /etc/passwd .
Uploading /etc/passwd to /webank/./passwd
/etc/passwd
100%   909     0.9KB/s   00:00
sftp> ls
day_1    passwd
sftp> ls -l
drwxr-xr-x   2 500       500        4096 May 20 16:03 day_1
-rw-r--r--   1 500       500         909 May 20 16:04 passwd
sftp> exit
```

从上述代码中可以看到,创建 day_1 目录,并上传本地的 passwd 文件到 webank 目录中。

3. 使用客户端登录工具

使用 FileZilla SFTP 客户端连接 SFTP 服务器,到如下地址下载客户端软件。

https://www.filezilla.cn/download/client

输入主机 IP 地址、用户名、密码、端口连接 SFTP 服务器(见图 7-2),默认为 22 端口。

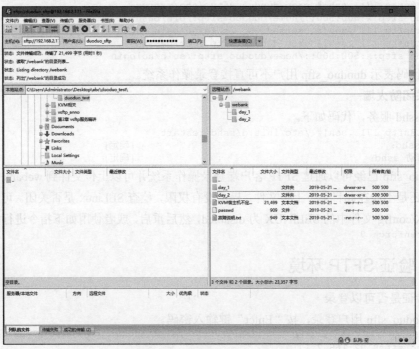

图 7-2 客户端连接 SFTP 服务器

7.2.4 开启 SFTP 服务日志记录

通过 SFTP 上传、删除文件,怎样记录日志呢?可以采取如下方式。

1. 修改 SSH 配置

修改 SSH 配置,代码如下。

```
vim /etc/ssh/sshd_config
```

```
Subsystem sftp internal-sftp -l INFO -f local5
LogLevel INFO
```

2. 修改 rsyslog

修改 rsyslog，增加如下内容。

```
vim /etc/rsyslog.conf
```

增加一行如下内容。

```
auth,authpriv.*    # /var/log/sftp.log
```

3. 查看/var/log/sftp.log 日志文件

查看/var/log/sftp.log 日志文件，代码如下。

```
[root@sftp_171 ~]# /etc/init.d/sshd restart
停止 sshd:                                                 [确定]
正在启动 sshd:                                             [确定]
[root@sftp_171 ~]# /etc/init.d/rsyslog restart
关闭系统日志记录器：                                       [确定]
启动系统日志记录器：                                       [确定]
[root@sftp_171 ~]# tailf /var/log/sftp.log
May 21 00:17:08 sftp_171 sshd[2051]: subsystem request for sftp
May 21 00:17:08 sftp_171 sshd[2051]: error: subsystem: cannot stat
/usr/libexec/OpenSSH/sftp-server: No such file or directory
May 21 00:17:08 sftp_171 sshd[2051]: subsystem request for sftp failed,
subsystem not found
May 21 00:17:08 sftp_171 sshd[2048]: pam_unix(sshd:session): session closed
for user duoduo_sftp
May 21 00:17:53 sftp_171 sshd[2016]: Received signal 15; terminating.
May 21 00:18:49 sftp_171 sshd[2152]: Server listening on 0.0.0.0 port 22.
May 21 00:18:49 sftp_171 sshd[2152]: Server listening on :: port 22.
May 21 00:22:04 sftp_171 sshd[2152]: Received signal 15; terminating.
May 21 00:22:04 sftp_171 sshd[2175]: Server listening on 0.0.0.0 port 22.
May 21 00:22:04 sftp_171 sshd[2175]: Server listening on :: port 22.
```

从上述代码中可以看到，日志相关的配置已经生效。

4. 查看登录日志/var/log/sftp.log

查看登录日志/var/log/sftp.log，代码如下。

```
[root@sftp_171 ~]# tailf  /var/log/sftp.log
May 21 00:17:08 sftp_171 sshd[2051]: subsystem request for sftp
May 21 00:17:08 sftp_171 sshd[2051]: error: subsystem: cannot stat
/usr/libexec/OpenSSH/sftp-server: No such file or directory
May 21 00:17:08 sftp_171 sshd[2051]: subsystem request for sftp failed,
subsystem not found
May 21 00:17:08 sftp_171 sshd[2048]: pam_unix(sshd:session): session closed
for user duoduo_sftp
May 21 00:17:53 sftp_171 sshd[2016]: Received signal 15; terminating.
May 21 00:18:49 sftp_171 sshd[2152]: Server listening on 0.0.0.0 port 22.
May 21 00:18:49 sftp_171 sshd[2152]: Server listening on :: port 22.
May 21 00:22:04 sftp_171 sshd[2152]: Received signal 15; terminating.
May 21 00:22:04 sftp_171 sshd[2175]: Server listening on 0.0.0.0 port 22.
May 21 00:22:04 sftp_171 sshd[2175]: Server listening on :: port 22.
May 21 00:23:37 sftp_171 sshd[2200]: Accepted password for duoduo_sftp from
192.168.2.105 port 52010 ssh2
May 21 00:23:37 sftp_171 sshd[2200]: pam_unix(sshd:session): session opened
for user duoduo_sftp by (uid=0)
May 21 00:23:37 sftp_171 sshd[2204]: subsystem request for sftp
```

通过客户端登录 SFTP 服务器，上传一些文件，可以看到日志已经记录了。

7.2.5 SFTP 服务基础环境初始化

1. 准备实验服务器

使用两台服务器进行演示，服务器相关环境如表 7-2 所示。

表 7-2　　　　　　　　　　　　SFTP 服务器相关环境

操作系统	IP 地址	主机名	SSH 版本
CentOS 6.9 x86_64	192.168.2.172	sftp_172	OpenSSH_5.3p1
CentOS 6.9 x86_64	192.168.2.173	sftp_173	OpenSSH_5.3p1

2. 关闭防火墙

关闭防火墙和 SELinux，代码如下。

```
[root@sftp_172 ~]# sestatus
SELINUX status:                 disabled
[root@sftp_172 ~]# /etc/init.d/iptables status
iptables: 未运行防火墙。
[root@sftp_173 ~]# sestatus
SELINUX status:                 disabled
[root@sftp_173 ~]# /etc/init.d/iptables status
iptables: 未运行防火墙。
```

3. 查看 SFTP 服务器配置信息

查看 SFTP 服务器的主机名称、IP 地址等信息，代码如下。

```
[root@sftp_172 ~]# ifconfig eth0
eth0      Link encap:Ethernet  HWaddr 00:0C:29:28:43:4A
          inet addr:192.168.2.172  Bcast:192.168.2.255  Mask:255.255.255.0
          inet6 addr: fe80::20c:29ff:fe28:434a/64 Scope:Link
          UP BROADCAST RUNNING MULTICAST  MTU:1500  Metric:1
          RX packets:302 errors:0 dropped:0 overruns:0 frame:0
          TX packets:142 errors:0 dropped:0 overruns:0 carrier:0
          collisions:0 txqueuelen:1000
          RX bytes:24544 (23.9 KiB)  TX bytes:15351 (14.9 KiB)

[root@sftp_172 ~]# hostname
sftp_172
[root@sftp_172 ~]# uname -r
2.6.32-696.el6.x86_64
[root@sftp_173 ~]# ifconfig eth0
eth0      Link encap:Ethernet  HWaddr 00:0C:29:0B:80:CC
          inet addr:192.168.2.173  Bcast:192.168.2.255  Mask:255.255.255.0
          inet6 addr: fe80::20c:29ff:fe0b:80cc/64 Scope:Link
          UP BROADCAST RUNNING MULTICAST  MTU:1500  Metric:1
          RX packets:161 errors:0 dropped:0 overruns:0 frame:0
          TX packets:100 errors:0 dropped:0 overruns:0 carrier:0
          collisions:0 txqueuelen:1000
          RX bytes:15300 (14.9 KiB)  TX bytes:12209 (11.9 KiB)

[root@sftp_173 ~]# hostname
sftp_173
[root@sftp_173 ~]# uname -r
2.6.32-696.el6.x86_64
```

4. 安装基础组件

安装 OpenSSH-clients、OpenSSH-server、epel-release 等基础组件，代码如下。

```
[root@sftp_172 ~]# yum -y install epel-release >/dev/null 2>&1
[root@sftp_172 ~]# yum -y update OpenSSH-clients OpenSSH-server >/dev/null 2>&1
[root@sftp_172 ~]# echo $?
0
[root@sftp_173 ~]# yum -y install epel-release >/dev/null 2>&1
[root@sftp_173 ~]# yum -y update OpenSSH-clients OpenSSH-server >/dev/null 2>&1
[root@sftp_173 ~]# echo $?
0
```

上述代码表示 OpenSSH、epel-release 等相关组件已经安装完毕。

7.2.6　192.168.2.172 搭建 SFTP 服务

1. 创建用户和组

创建 sftp 组，并将 id_user 用户加入 sftp 组，代码如下。

```
[root@sftp_172 ~]# groupadd group_sftp &&  useradd jd_user -g  group_sftp
[root@sftp_172 ~]# echo 'jd_user' | passwd --stdin jd_user
更改用户 jd_user 的密码。
passwd：所有的身份验证令牌已经成功更新。
```

2. 确认 id_user 用户和 sftp 组的信息是否正确

确认用户和组的信息是否正确，代码如下。

```
[root@sftp_172 ~]# grep jd_user /etc/passwd
jd_user:x:500:500::/home/jd_user:/bin/bash
[root@sftp_172 ~]# grep group_sftp /etc/group
group_sftp:x:500:
```

3. 创建用户密钥对

切换到 jd_user 用户，并创建用户密钥对，代码如下。

```
[root@sftp_172 ~]# su - jd_user
[jd_user@sftp_172 ~]$  ssh-keygen
Generating public/private rsa key pair.
Enter file in which to save the key (/home/jd_user/.ssh/id_rsa):
Created directory '/home/jd_user/.ssh'.
Enter passphrase (empty for no passphrase):
Enter same passphrase again:
Your identification has been saved in /home/jd_user/.ssh/id_rsa.
Your public key has been saved in /home/jd_user/.ssh/id_rsa.pub.
The key fingerprint is:
34:83:02:e0:da:0a:4a:fa:00:c3:1e:e4:d1:df:4b:d8 jd_user@sftp_172
The key's randomart image is:
+--[ RSA 2048]----+
|...              |
|. ..   .         |
| + ... +         |
|=.. ..+. o       |
|==.  o ES        |
|Bo.  . .  |
|=.     .         |
| o              |
| .              |
+-----------------+
```

4. 指定用户数据目录

切换到 root 用户，把 group_sftp 组的用户的 home 目录统一指定到 /data/sftp 下，按用户名区分。这里先新建一个 jd_user 目录，然后指定 jd_user 的 home 为 /data/sftp/jd_user。

```
[root@sftp_172 ~]# mkdir -pv /data/sftp/jd_user
mkdir: 已创建目录 "/data"
mkdir: 已创建目录 "/data/sftp"
mkdir: 已创建目录 "/data/sftp/jd_user"
```

5. 查看用户权限

查看用户权限，代码如下。

```
[root@sftp_172 ~]# ls -ld /home/jd_user/.ssh/
drwx------ 2 jd_user group_sftp 4096 5月  23 23:16 /home/jd_user/.ssh/
```

上述代码表示 /home/jd_user/.ssh/ 目录的权限为 700。

6. 备份 OpenSSH 服务配置文件

备份配置文件，代码如下。

```
[root@sftp_172 ~]# cp -av /etc/ssh/sshd_config{,_ori_bak}
```

"/etc/ssh/sshd_config" -> "/etc/ssh/sshd_config_ori_bak"

7. SFTP 服务核心配置文件

创建的 SFTP 服务用户登录后，按照下面的设置，重启 sshd 服务后，用户 jd_user 可以登录。

```
###SFTP 配置文件 2019/05/23 hanyw --- beging ###

AuthorizedKeysFile      .ssh/authorized_keys
#Subsystem      sftp/usr/libexec/OpenSSH/sftp-server
Subsystem sftp internal-sftp
Match Group group_sftp
ChrootDirectory /data/sftp/%u
ForceCommand internal-sftp
AllowTcpForwarding no
X11Forwarding no
PubkeyAuthentication yes
RSAAuthentication yes
###SFTP 配置文件 2019/05/23 hanyw --- beging ###
```

8. 设定 chroot 权限

设定 chroot 权限的代码如下。

```
[root@sftp_172 ~]# ls -ld /data/sftp/user_data/
drwxr-xr-x 2 root root 4096 5月  23 23:20 /data/sftp/user_data/
[root@sftp_172 ~]#
[root@sftp_172 ~]# chown  root:group_sftp /data/sftp/user_data/
[root@sftp_172 ~]# chmod 755 /data/sftp/user_data/
[root@sftp_172 ~]# ls -ld /data/sftp/user_data/
drwxr-xr-x 2 root group_sftp 4096 5月  23 23:20 /data/sftp/user_data/
```

但使用 chroot 指定根目录后，根应该是无法写入的，所以要新建一个目录供 jd_user 目录上传文件。这个目录的所有者为 jd_user，所有组为 group_sftp，所有者有写入权限，而所有组无写入权限，完整代码如下。

```
[root@sftp_172 sftp]# ll /data/sftp/jd_user/
总用量 0
[root@sftp_172 sftp]# ls -ld  /data/sftp/jd_user/
drwxr-xr-x 2 root root 4096 5月  24 00:20 /data/sftp/jd_user/
[root@sftp_172 sftp]# chown root:group_sftp /data/sftp/jd_user/
[root@sftp_172 sftp]#
[root@sftp_172 sftp]#
[root@sftp_172 sftp]# mkdir -pv /data/sftp/jd_user/jd_user_data
mkdir: 已创建目录 "/data/sftp/jd_user/jd_user_data"
[root@sftp_172 sftp]# chown  jd_user:group_sftp /data/sftp/jd_user/jd_user_data
[root@sftp_172 sftp]#
[root@sftp_172 sftp]#
[root@sftp_172 sftp]# ls -ld /data/sftp/jd_user/jd_user_data
drwxr-xr-x 2 jd_user group_sftp 4096 5月  24 00:22 /data/sftp/jd_user/jd_user_data
```

9. 设置密钥认证文件权限

设置密钥认证文件权限，代码如下。

```
[root@sftp_172 ~]# cd /home/jd_user/.ssh/
[root@sftp_172 .ssh]# ls -la
总用量 16
drwx------ 2 jd_user group_sftp 4096 5月  23 23:16 .
drwx------ 3 jd_user group_sftp 4096 5月  23 23:19 ..
-rw------- 1 jd_user group_sftp 1675 5月  23 23:16 id_rsa
-rw-r--r-- 1 jd_user group_sftp  398 5月  23 23:16 id_rsa.pub
[root@sftp_172 .ssh]# cat id_rsa.pub >> authorized_keys
[root@sftp_172 .ssh]# ls -lhrt
总用量 12K
-rw-r--r-- 1 jd_user group_sftp  398 5月  23 23:16 id_rsa.pub
-rw------- 1 jd_user group_sftp 1.7K 5月  23 23:16 id_rsa
-rw-r--r-- 1 root    root        398 5月  23 23:35 authorized_keys
[root@sftp_172 .ssh]# chown -R jd_user:group_sftp *
```

```
[root@sftp_172 .ssh]# ls -lhrt
总用量 12K
-rw-r--r-- 1 jd_user group_sftp  398 5月  23 23:16 id_rsa.pub
-rw------- 1 jd_user group_sftp 1.7K 5月  23 23:16 id_rsa
-rw-r--r-- 1 jd_user group_sftp  398 5月  23 23:35 authorized_keys
```

上述代码中 id_rsa.pub 文件权限为 644，id_rsa 文件权限为 600，authorized_keys 文件权限为 644。

10. 重启 sshd 服务

重启 sshd 服务，代码如下。

```
[root@sftp_172 .ssh]# /etc/init.d/sshd restart
停止 sshd:                                                [确定]
正在启动 sshd:                                            [确定]
```

11. 本地登录测试

使用 sftp 登录，代码如下。

```
[root@sftp_172 sftp]# sftp   -oIdentityFile=/home/jd_user/.ssh/id_rsa   -oPort=22
jd_user@192.168.2.172
Connecting to 192.168.2.172...
sftp>
sftp>
sftp> ls
jd_user_data
sftp> cd jd_user_data
sftp> ls
sftp> mkdir a
sftp> ls
a
sftp>
sftp>
sftp>
sftp> exit
```

上述代码使用 sftp 登录，成功创建目录和文件。

7.2.7 创建 SFTP 服务的用户和组

1. 创建用户和组

创建用户和组，代码如下。

```
[root@sftp_173 ~]# groupadd group_sftp &&  useradd huawei_user -g  group_sftp
[root@sftp_173 ~]# echo 'huawei_user' | passwd --stdin huawei_user
更改用户 huawei_user 的密码。
```

passwd 表示所有的身份验证令牌已经成功更新。

2. 确认用户和组的信息是否正确

确认用户和组的信息是否正确，代码如下。

```
[root@sftp_173 ~]# grep huawei_user /etc/passwd && grep group_sftp /etc/group
huawei_user:x:500:500::/home/huawei_user:/bin/bash
group_sftp:x:500:
```

3. 创建用户密钥对

切换到 huawei_user 用户，并创建用户密钥对，代码如下。

```
[root@sftp_173 ~]# su - huawei_user
[huawei_user@sftp_173 ~]$ ssh-keygen
Generating public/private rsa key pair.
Enter file in which to save the key (/home/huawei_user/.ssh/id_rsa):
Created directory '/home/huawei_user/.ssh'.
Enter passphrase (empty for no passphrase):
Enter same passphrase again:
Your identification has been saved in /home/huawei_user/.ssh/id_rsa.
```

```
Your public key has been saved in /home/huawei_user/.ssh/id_rsa.pub.
The key fingerprint is:
73:09:74:a3:1f:2f:77:2c:3a:d1:fb:0b:14:86:4d:34 huawei_user@sftp_173
The key's randomart image is:
+--[ RSA 2048]----+
|        . o.E    |
|       . o = .   |
|        o o +    |
|         o * o   |
|        S * * o  |
|         o * +   |
|          o o    |
|         . .     |
|          o .    |
+-----------------+
```

4．指定用户数据目录

切换到 root 用户，把 group_sftp 组的用户的 home 目录统一指定到/data/sftp 下，按用户名区分。这里先新建一个 huawei_user 目录，然后指定 huawei_user 的 home 为/data/sftp/ huawei_user，代码如下。

```
[huawei_user@sftp_173 ~]$ exit
logout
[root@sftp_173 ~]# mkdir -pv /data/sftp/huawei_user
mkdir: 已创建目录 "/data"
mkdir: 已创建目录 "/data/sftp"
mkdir: 已创建目录 "/data/sftp/huawei_user"
```

上述代码表示 huawei_user 用户的数据目录为/data/sftp/huawei_user。

5．查看用户权限

用户数据目录/home/huawei_user/.ssh/下的权限设置为 700，代码如下。

```
[root@sftp_173 ~]# ls -ld /home/huawei_user/.ssh/
drwx------ 2 huawei_user group_sftp 4096 5月  24 00:31 /home/huawei_user/.ssh/
```

上述代码表示/home/huawei_user/.ssh/目录的权限为 700。

6．配置 SFTP 服务

备份配置文件，代码如下。

```
[root@sftp_173 ~]#  cp -av /etc/ssh/sshd_config{,_ori_bak}
"/etc/ssh/sshd_config" -> "/etc/ssh/sshd_config_ori_bak"
```

核心配置代码如下。

```
###SFTP 配置文件 2019/05/23 hanyw --- beging ###

AuthorizedKeysFile    .ssh/authorized_keys
#Subsystem   sftp/usr/libexec/OpenSSH/sftp-server
Subsystem sftp internal-sftp
Match Group group_sftp
ChrootDirectory /data/sftp/%u
ForceCommand internal-sftp
AllowTcpForwarding no
X11Forwarding no
PubkeyAuthentication yes
RSAAuthentication yes
###SFTP 配置文件 2019/05/23 hanyw --- beging ###
```

7．设定 chroot 权限

设定 SFTP 服务的 chroot 权限，代码如下。

```
[root@sftp_173 ~]# ls -ld /data/sftp/huawei_user/
drwxr-xr-x 2 root root 4096 5月  24 00:32 /data/sftp/huawei_user/
[root@sftp_173 ~]# chown  root:group_sftp /data/sftp/huawei_user
[root@sftp_173 ~]# ls -ld /data/sftp/huawei_user/
drwxr-xr-x 2 root group_sftp 4096 5月  24 00:32 /data/sftp/huawei_user/
```

创建 SFTP 服务用户登录后可写入的目录。按照上面的设置，重启 sshd 服务后，用户 huawei_user 已经可以登录，但使用 chroot 指定根目录后，根应该是无法写入的，所以要新建一个目录供 huawei_user 目录上传文件。这个目录所有者为 huawei_user，所有组为 group_sftp，所有者有写入权限，而所有组无写入权限，完整代码如下。

```
[root@sftp_173 ~]# mkdir -pv /data/sftp/huawei_user/huawei_user_data
mkdir: 已创建目录 "/data/sftp/huawei_user/huawei_user_data"
[root@sftp_173 ~]# ls -ld /data/sftp/huawei_user/huawei_user_data
drwxr-xr-x 2 root root 4096 5月  24 00:45 /data/sftp/huawei_user/huawei_user_data
[root@sftp_173 ~]# chown huawei_user:group_sftp /data/sftp/huawei_user/huawei_user_data
[root@sftp_173 ~]# ls -ld /data/sftp/huawei_user/huawei_user_data
drwxr-xr-x 2 huawei_user group_sftp 4096 5月  24 00:45 /data/sftp/huaweiuser/huawei_user_data
```

8. 设置密钥认证文件权限

设置 SFTP 密钥认证文件权限，代码如下。

```
[root@sftp_173 ~]#  cd /home/huawei_user/.ssh/
[root@sftp_173 .ssh]# ls -lhrt
总用量 8.0K
-rw-r--r-- 1 huawei_user group_sftp  402 5月  24 00:31 id_rsa.pub
-rw------- 1 huawei_user group_sftp 1.7K 5月  24 00:31 id_rsa
[root@sftp_173 .ssh]# cat id_rsa.pub >> authorized_keys
[root@sftp_173 .ssh]# ls -lhrt
总用量 12K
-rw-r--r-- 1 huawei_user group_sftp  402 5月  24 00:31 id_rsa.pub
-rw------- 1 huawei_user group_sftp 1.7K 5月  24 00:31 id_rsa
-rw-r--r-- 1 root        root       402 5月  24 00:46 authorized_keys
[root@sftp_173 .ssh]# chown -R huawei_user:group_sftp *
[root@sftp_173 .ssh]# ls -lhrt
总用量 12K
-rw-r--r-- 1 huawei_user group_sftp  402 5月  24 00:31 id_rsa.pub
-rw------- 1 huawei_user group_sftp 1.7K 5月  24 00:31 id_rsa
-rw-r--r-- 1 huawei_user group_sftp  402 5月  24 00:46 authorized_keys
```

上述代码设置密钥认证文件权限。

9. 重启服务

重启 SSH 服务，使 SFTP 配置生效，代码如下。

```
[root@sftp_173 .ssh]# /etc/init.d/sshd restart
停止 sshd:                                                 [确定]
正在启动 sshd:                                             [确定]
```

上述代码重启 SSH 服务。

10. 本地登录测试

本地测试登录 SFTP 服务，代码如下。

```
[root@sftp_173 .ssh]# sftp   -oIdentityFile=/home/huawei_user/.ssh/id_rsa
-oPort=22  huawei_user@192.168.2.173
Connecting to 192.168.2.173...
The authenticity of host '192.168.2.173 (192.168.2.173)' can't be established.
RSA key fingerprint is 82:e8:a0:65:a9:90:8e:f4:60:4d:6e:0e:7d:34:e0:75.
Are you sure you want to continue connecting (yes/no)? yes
Warning: Permanently added '192.168.2.173' (RSA) to the list of known hosts.
sftp> ls
huawei_user_data
sftp> cd huawei_user_data
sftp> mkdir a
sftp> cd a
sftp> touch aaa
Invalid command.
sftp> mkdir aaa
sftp> ls
```

```
aaa
sftp> exit
```

从上述代码输出可以看到，登录用户已成功创建目录和文件。

7.2.8 配置双机互信

配置 SFTP 服务器之间的密钥，并分别复制密钥文件到对方主机。

1. 传输密钥文件

```
[root@sftp_172 .ssh]# cat /home/jd_user/.ssh/id_rsa.pub
ssh-rsa AAAAB3NzaC1yc2EAAAABIwAAAQEAxWmgveY4eCw48/dCK0C/MydBSKKpiZqNtnNUW8HSYG7
oZzHRnk7z6/UPgS76mpWgkh42rLUw3bzSk5JqdcMxYKpcymRTEn2CKemFBE/1H8wmKXf1Q8tpqDRJpZ
QQrlwV9db50AEkaJuvYrf+FyWhqtD4oiijs2+0UryCdSC7b6483E8hh/hDGl1J1w3WmI1O4VLZ70W1Q
rnvl8IZ2mNjdouFL5psc1/9rMOMqcYfXITeTJLYf1tHVBBK2JYu797u7XsrUauGUCLIqMId+3I3+Msc
oJiLtSREnkL9pdYA7xCLTTziXkRGXLhGjMBsON9IaXXHUUJp+nYCDDtnsIjVuw== jd_user@sftp_172
[root@sftp_172 .ssh]# cat /home/jd_user/.ssh/authorized_keys
ssh-rsa AAAAB3NzaC1yc2EAAAABIwAAAQEAxWmgveY4eCw48/dCK0C/MydBSKKpiZqNtnNUW8HSYG7
oZzHRnk7z6/UPgS76mpWgkh42rLUw3bzSk5JqdcMxYKpcymRTEn2CKemFBE/1H8wmKXf1Q8tpqDRJpZ
QQrlwV9db50AEkaJuvYrf+FyWhqtD4oiijs2+0UryCdSC7b6483E8hh/hDGl1J1w3WmI1O4VLZ70W1Q
rnvl8IZ2mNjdouFL5psc1/9rMOMqcYfXITeTJLYf1tHVBBK2JYu797u7XsrUauGUCLIqMId+3I3+Msc
oJiLtSREnkL9pdYA7xCLTTziXkRGXLhGjMBsON9IaXXHUUJp+nYCDDtnsIjVuw== jd_user@sftp_172
ssh-rsa AAAAB3NzaC1yc2EAAAABIwAAAQEAwjMzZD7uJxnhygqg+LAXeRPCXyCfw130upByHcTpQJk
Dy0iZvxdOQlECKXZ80mu4oYHFX6timapZqXdY2OKCis5uCNZ54ItvM3Pg74k8JN0IMPm2OlMsQ89I69
UFIo1aXTro+/k3nDFFZuE//Lo7hEr7qu2TmNNSQgs0QBajf8TdWVsAnGzIpPeKC9cRUdMEYcooKVGdK
XJYaovzmL3NZrtlomxEiZOFEtiFZloggb92mEZ8dBTsoszwYoTXKbAEQNCJXa45wQDbTJCDzFyEPfRR
4RhYi5buwSxr5r1e0GDDIjOOO2ecxNpjnieNYQ/BmZK5UwmEKwfpEoAWyMiteQ== huawei_user@sftp_173
[root@sftp_173 .ssh]# cat /home/huawei_user/.ssh/id_rsa.pub
ssh-rsa AAAAB3NzaC1yc2EAAAABIwAAAQEAwjMzZD7uJxnhygqg+LAXeRPCXyCfw130upByHcTpQJk
Dy0iZvxdOQlECKXZ80mu4oYHFX6timapZqXdY2OKCis5uCNZ54ItvM3Pg74k8JN0IMPm2OlMsQ89I69
UFIo1aXTro+/k3nDFFZuE//Lo7hEr7qu2TmNNSQgs0QBajf8TdWVsAnGzIpPeKC9cRUdMEYcooKVGdK
XJYaovzmL3NZrtlomxEiZOFEtiFZloggb92mEZ8dBTsoszwYoTXKbAEQNCJXa45wQDbTJCDzFyEPfRR
4RhYi5buwSxr5r1e0GDDIjOOO2ecxNpjnieNYQ/BmZK5UwmEKwfpEoAWyMiteQ== huawei_user@sftp_173
[root@sftp_173 .ssh]# cat /home/huawei_user/.ssh/authorized_keys
ssh-rsa AAAAB3NzaC1yc2EAAAABIwAAAQEAwjMzZD7uJxnhygqg+LAXeRPCXyCfw130upByHcTpQJk
Dy0iZvxdOQlECKXZ80mu4oYHFX6timapZqXdY2OKCis5uCNZ54ItvM3Pg74k8JN0IMPm2OlMsQ89I69
UFIo1aXTro+/k3nDFFZuE//Lo7hEr7qu2TmNNSQgs0QBajf8TdWVsAnGzIpPeKC9cRUdMEYcooKVGdK
XJYaovzmL3NZrtlomxEiZOFEtiFZloggb92mEZ8dBTsoszwYoTXKbAEQNCJXa45wQDbTJCDzFyEPfRR
4RhYi5buwSxr5r1e0GDDIjOOO2ecxNpjnieNYQ/BmZK5UwmEKwfpEoAWyMiteQ== huawei_user@sftp_173
ssh-rsa AAAAB3NzaC1yc2EAAAABIwAAAQEAxWmgveY4eCw48/dCK0C/MydBSKKpiZqNtnNUW8HSYG7
oZzHRnk7z6/UPgS76mpWgkh42rLUw3bzSk5JqdcMxYKpcymRTEn2CKemFBE/1H8wmKXf1Q8tpqDRJpZ
QQrlwV9db50AEkaJuvYrf+FyWhqtD4oiijs2+0UryCdSC7b6483E8hh/hDGl1J1w3WmI1O4VLZ70W1Q
rnvl8IZ2mNjdouFL5psc1/9rMOMqcYfXITeTJLYf1tHVBBK2JYu797u7XsrUauGUCLIqMId+3I3+Msc
oJiLtSREnkL9pdYA7xCLTTziXkRGXLhGjMBsON9IaXXHUUJp+nYCDDtnsIjVuw== jd_user@sftp_172
```

上述代码传输密钥文件。

2. 验证登录

以下代码验证登录是否正确。

```
[root@sftp_172 .ssh]# sftp  -oIdentityFile=/home/jd_user/.ssh/id_rsa  -oPort=22
huawei_user@192.168.2.173
Connecting to 192.168.2.173...
The authenticity of host '192.168.2.173 (192.168.2.173)' can't be established.
RSA key fingerprint is 82:e8:a0:65:a9:90:8e:f4:60:4d:6e:0e:7d:34:e0:75.
Are you sure you want to continue connecting (yes/no)? yes
Warning: Permanently added '192.168.2.173' (RSA) to the list of known hosts.
sftp> ls
huawei_user_data
sftp> cd huawei_user_data
sftp> mkdir abc
sftp> ls
a     abc
sftp> exit
[root@sftp_172 .ssh]#
[root@sftp_173 .ssh]# sftp   -oIdentityFile=/home/huawei_user/.ssh/id_rsa
```

```
    -oPort=22  jd_user@192.168.2.172
Connecting to 192.168.2.172...
sftp> ls
jd_user_data
sftp> cd jd_user_data
sftp> mkdir test
sftp> cd test
sftp> mkdir a
Invalid command.
sftp> mkdir a
sftp> exit
```

7.3 SFTP 服务配置文件对比

7.3.1 192.168.2.171 配置文件

查看 192.168.2.171 配置文件，代码如下。

```
[root@sftp_171 ~]# tree -L 2  /data/sftp/
/data/sftp/
└── duoduo_sftp
    └── webank

2 directories, 0 files
[root@sftp_171 ~]# ls -ld /data/sftp/
drwxr-xr-x. 3 root root 4096 5月  20 22:52 /data/sftp/
[root@sftp_171 ~]# ls -ld /data/sftp/duoduo_sftp/
drwxr-xr-x. 3 root sftp 4096 5月  20 23:22 /data/sftp/duoduo_sftp/
[root@sftp_171 ~]# ls -ld /data/sftp/duoduo_sftp/webank/
drwxr-xr-x. 4 duoduo_sftp sftp 4096 5月  21 00:23 /data/sftp/duoduo_sftp/webank/
[root@sftp_171 ~]#
[root@sftp_171 ~]# cat /etc/ssh/sshd_config
#    $OpenBSD: sshd_config,v 1.80 2008/07/02 02:24:18 djm Exp $

# This is the sshd server system-wide configuration file.  See
# sshd_config(5) for more information.

# This sshd was compiled with PATH=/usr/local/bin:/bin:/usr/bin

# The strategy used for options in the default sshd_config shipped with
# OpenSSH is to specify options with their default value where
# possible, but leave them commented.  Uncommented options change a
# default value.

# Port 22
# AddressFamily any
# ListenAddress 0.0.0.0
# ListenAddress ::

# Disable legacy (protocol version 1) support in the server for new
# installations. In future the default will change to require explicit
# activation of protocol 1
Protocol 2

# HostKey for protocol version 1
# HostKey /etc/ssh/ssh_host_key
# HostKeys for protocol version 2
# HostKey /etc/ssh/ssh_host_rsa_key
# HostKey /etc/ssh/ssh_host_dsa_key

# Lifetime and size of ephemeral version 1 server key
```

```
#KeyRegenerationInterval 1h
#ServerKeyBits 1024

# Logging
# obsoletes QuietMode and FascistLogging
#SyslogFacility AUTH
SyslogFacility AUTHPRIV
#LogLevel INFO

# Authentication:

#LoginGraceTime 2m
#PermitRootLogin yes
#StrictModes yes
#MaxAuthTries 6
#MaxSessions 10

#RSAAuthentication yes
#PubkeyAuthentication yes
#AuthorizedKeysCommand none
#AuthorizedKeysCommandRunAs nobody

# For this to work you will also need host keys in /etc/ssh/ssh_known_hosts
#RhostsRSAAuthentication no
# similar for protocol version 2
#HostbasedAuthentication no
# Change to yes if you don't trust ~/.ssh/known_hosts for
# RhostsRSAAuthentication and HostbasedAuthentication
#IgnoreUserKnownHosts no
# Don't read the user's ~/.rhosts and ~/.shosts files
#IgnoreRhosts yes

# To disable tunneled clear text passwords, change to no here!
#PasswordAuthentication yes
#PermitEmptyPasswords no
PasswordAuthentication yes

# Change to no to disable s/key passwords
#ChallengeResponseAuthentication yes
ChallengeResponseAuthentication no

# Kerberos options
#KerberosAuthentication no
#KerberosOrLocalPasswd yes
#KerberosTicketCleanup yes
#KerberosGetAFSToken no
#KerberosUseKuserok yes

# GSSAPI options
#GSSAPIAuthentication no
GSSAPIAuthentication yes
#GSSAPICleanupCredentials yes
GSSAPICleanupCredentials yes
#GSSAPIStrictAcceptorCheck yes
#GSSAPIKeyExchange no

# Set this to 'yes' to enable PAM authentication, account processing,
# and session processing. If this is enabled, PAM authentication will
# be allowed through the ChallengeResponseAuthentication and
# PasswordAuthentication.  Depending on your PAM configuration,
# PAM authentication via ChallengeResponseAuthentication may bypass
# the setting of "PermitRootLogin without-password".
# If you just want the PAM account and session checks to run without
# PAM authentication, then enable this but set PasswordAuthentication
# and ChallengeResponseAuthentication to 'no'.
```

```
# UsePAM no
UsePAM yes

# Accept locale-related environment variables
AcceptEnv LANG LC_CTYPE LC_NUMERIC LC_TIME LC_COLLATE LC_MONETARY LC_MESSAGES
AcceptEnv LC_PAPER LC_NAME LC_ADDRESS LC_TELEPHONE LC_MEASUREMENT
AcceptEnv LC_IDENTIFICATION LC_ALL LANGUAGE
AcceptEnv XMODIFIERS

# AllowAgentForwarding yes
# AllowTcpForwarding yes
# GatewayPorts no
# X11Forwarding no
X11Forwarding yes
# X11DisplayOffset 10
# X11UseLocalhost yes
# PrintMotd yes
# PrintLastLog yes
# TCPKeepAlive yes
# UseLogin no
# UsePrivilegeSeparation yes
# PermitUserEnvironment no
# Compression delayed
# ClientAliveInterval 0
# ClientAliveCountMax 3
# ShowPatchLevel no
# UseDNS yes
# PidFile /var/run/sshd.pid
# MaxStartups 10:30:100
# PermitTunnel no
# ChrootDirectory none

# no default banner path
# Banner none

# override default of no subsystems
# Subsystem      sftp/usr/libexec/OpenSSH/sftp-server

# Example of overriding settings on a per-user basis
# Match User anoncvs
#       X11Forwarding no
#       AllowTcpForwarding no
#       ForceCommand cvs server


# SFTP服务基本配置    begin #
AuthorizedKeysFile       .ssh/authorized_keys
Subsystem sftp internal-sftp -l INFO -f local5
LogLevel INFO
Match Group sftp
ChrootDirectory /data/sftp/%u
ForceCommand internal-sftp
AllowTcpForwarding no
X11Forwarding no
PubkeyAuthentication yes
RSAAuthentication yes

# SFTP服务基本配置    end #
```

7.3.2　192.168.2.172 配置文件

查看 192.168.2.172 配置文件，代码如下。

```
[root@sftp_172 ~]# tree /data/sftp/ -L 2
/data/sftp/
└── jd_user
    └── jd_user_data

2 directories, 0 files
[root@sftp_172 ~]# ls -ld /data/sftp/
drwxr-xr-x 3 root group_sftp 4096 5月  24 00:20 /data/sftp/
[root@sftp_172 ~]# ls -ld /data/sftp/jd_user/
drwxr-xr-x 3 root group_sftp 4096 5月  24 00:22 /data/sftp/jd_user/
[root@sftp_172 ~]# ls -ld /data/sftp/jd_user/jd_user_data/
drwxr-xr-x 4 jd_user group_sftp 4096 5月  24 00:57 /data/sftp/jd_user/jd_user_data/
[root@sftp_172 ~]# cat /etc/ssh/sshd_config
#    $OpenBSD: sshd_config,v 1.80 2008/07/02 02:24:18 djm Exp $

# This is the sshd server system-wide configuration file.  See
# sshd_config(5) for more information.

# This sshd was compiled with PATH=/usr/local/bin:/bin:/usr/bin

# The strategy used for options in the default sshd_config shipped with
# OpenSSH is to specify options with their default value where
# possible, but leave them commented.  Uncommented options change a
# default value.

# Port 22
# AddressFamily any
# ListenAddress 0.0.0.0
# ListenAddress ::

# Disable legacy (protocol version 1) support in the server for new
# installations. In future the default will change to require explicit
# activation of protocol 1
Protocol 2

# HostKey for protocol version 1
# HostKey /etc/ssh/ssh_host_key
# HostKeys for protocol version 2
# HostKey /etc/ssh/ssh_host_rsa_key
# HostKey /etc/ssh/ssh_host_dsa_key

# Lifetime and size of ephemeral version 1 server key
# KeyRegenerationInterval 1h
# ServerKeyBits 1024

# Logging
# obsoletes QuietMode and FascistLogging
# SyslogFacility AUTH
SyslogFacility AUTHPRIV
# LogLevel INFO

# Authentication:

# LoginGraceTime 2m
# PermitRootLogin yes
# StrictModes yes
# MaxAuthTries 6
# MaxSessions 10

# RSAAuthentication yes
# PubkeyAuthentication yes
# AuthorizedKeysFile     .ssh/authorized_keys
# AuthorizedKeysCommand none
# AuthorizedKeysCommandRunAs nobody
```

```
# For this to work you will also need host keys in /etc/ssh/ssh_known_hosts
# RhostsRSAAuthentication no
# similar for protocol version 2
# HostbasedAuthentication no
# Change to yes if you don't trust ~/.ssh/known_hosts for
# RhostsRSAAuthentication and HostbasedAuthentication
# IgnoreUserKnownHosts no
# Don't read the user's ~/.rhosts and ~/.shosts files
# IgnoreRhosts yes

# To disable tunneled clear text passwords, change to no here!
# PasswordAuthentication yes
# PermitEmptyPasswords no
PasswordAuthentication yes

# Change to no to disable s/key passwords
# ChallengeResponseAuthentication yes
ChallengeResponseAuthentication no

# Kerberos options
# KerberosAuthentication no
# KerberosOrLocalPasswd yes
# KerberosTicketCleanup yes
# KerberosGetAFSToken no
# KerberosUseKuserok yes

# GSSAPI options
# GSSAPIAuthentication no
GSSAPIAuthentication yes
# GSSAPICleanupCredentials yes
GSSAPICleanupCredentials yes
# GSSAPIStrictAcceptorCheck yes
# GSSAPIKeyExchange no

# Set this to 'yes' to enable PAM authentication, account processing,
# and session processing. If this is enabled, PAM authentication will
# be allowed through the ChallengeResponseAuthentication and
# PasswordAuthentication.  Depending on your PAM configuration,
# PAM authentication via ChallengeResponseAuthentication may bypass
# the setting of "PermitRootLogin without-password".
# If you just want the PAM account and session checks to run without
# PAM authentication, then enable this but set PasswordAuthentication
# and ChallengeResponseAuthentication to 'no'.
# UsePAM no
UsePAM yes

# Accept locale-related environment variables
AcceptEnv LANG LC_CTYPE LC_NUMERIC LC_TIME LC_COLLATE LC_MONETARY LC_MESSAGES
AcceptEnv LC_PAPER LC_NAME LC_ADDRESS LC_TELEPHONE LC_MEASUREMENT
AcceptEnv LC_IDENTIFICATION LC_ALL LANGUAGE
AcceptEnv XMODIFIERS

# AllowAgentForwarding yes
# AllowTcpForwarding yes
# GatewayPorts no
# X11Forwarding no
X11Forwarding yes
# X11DisplayOffset 10
# X11UseLocalhost yes
# PrintMotd yes
# PrintLastLog yes
# TCPKeepAlive yes
# UseLogin no
# UsePrivilegeSeparation yes
# PermitUserEnvironment no
```

```
# Compression delayed
# ClientAliveInterval 0
# ClientAliveCountMax 3
# ShowPatchLevel no
# UseDNS yes
# PidFile /var/run/sshd.pid
# MaxStartups 10:30:100
# PermitTunnel no
# ChrootDirectory none

# no default banner path
# Banner none

# override default of no subsystems

# Example of overriding settings on a per-user basis
# Match User anoncvs
#   X11Forwarding no
#   AllowTcpForwarding no
#   ForceCommand cvs server


###SFTP 配置文件 2019/05/23 hanyw   --- beging ###

AuthorizedKeysFile      .ssh/authorized_keys
# Subsystem     sftp/usr/libexec/OpenSSH/sftp-server
Subsystem sftp internal-sftp
Match Group group_sftp
ChrootDirectory /data/sftp/%u
ForceCommand internal-sftp
AllowTcpForwarding no
X11Forwarding no
PubkeyAuthentication yes
RSAAuthentication yes
###SFTP 配置文件 2019/05/23 hanyw --- beging ###
```

7.3.3　192.168.2.173 配置文件

查看 192.168.2.173 配置文件，代码如下。

```
[root@sftp_173 ~]# tree /data/sftp/ -L 2
/data/sftp/
└── huawei_user
    └── huawei_user_data

2 directories, 0 files
[root@sftp_173 ~]# ls -ld /data/sftp/
drwxr-xr-x 3 root root 4096 5月  24 00:32 /data/sftp/
[root@sftp_173 ~]# ls -ld /data/sftp/huawei_user/
drwxr-xr-x 3 root group_sftp 4096 5月  24 00:45 /data/sftp/huawei_user/
[root@sftp_173 ~]# ls -ld /data/sftp/huawei_user/huawei_user_data/
drwxr-xr-x 4 huawei_user group_sftp 4096 5月  24 00:56 /data/sftp/huawei_user/huawei_user_data/
[root@sftp_173 ~]#
[root@sftp_173 ~]#
[root@sftp_173 ~]#
[root@sftp_173 ~]# cat /etc/ssh/sshd_config
#       $OpenBSD: sshd_config,v 1.80 2008/07/02 02:24:18 djm Exp $

# This is the sshd server system-wide configuration file.  See
# sshd_config(5) for more information.

# This sshd was compiled with PATH=/usr/local/bin:/bin:/usr/bin

# The strategy used for options in the default sshd_config shipped with
# OpenSSH is to specify options with their default value where
```

```
# possible, but leave them commented.  Uncommented options change a
# default value.

# Port 22
# AddressFamily any
# ListenAddress 0.0.0.0
# ListenAddress ::

# Disable legacy (protocol version 1) support in the server for new
# installations. In future the default will change to require explicit
# activation of protocol 1
Protocol 2

# HostKey for protocol version 1
# HostKey /etc/ssh/ssh_host_key
# HostKeys for protocol version 2
# HostKey /etc/ssh/ssh_host_rsa_key
# HostKey /etc/ssh/ssh_host_dsa_key

# Lifetime and size of ephemeral version 1 server key
# KeyRegenerationInterval 1h
# ServerKeyBits 1024

# Logging
# obsoletes QuietMode and FascistLogging
# SyslogFacility AUTH
SyslogFacility AUTHPRIV
# LogLevel INFO

# Authentication:

# LoginGraceTime 2m
# PermitRootLogin yes
# StrictModes yes
# MaxAuthTries 6
# MaxSessions 10

# RSAAuthentication yes
# PubkeyAuthentication yes
# AuthorizedKeysFile     .ssh/authorized_keys
# AuthorizedKeysCommand none
# AuthorizedKeysCommandRunAs nobody

# For this to work you will also need host keys in /etc/ssh/ssh_known_hosts
# RhostsRSAAuthentication no
# similar for protocol version 2
# HostbasedAuthentication no
# Change to yes if you don't trust ~/.ssh/known_hosts for
# RhostsRSAAuthentication and HostbasedAuthentication
# IgnoreUserKnownHosts no
# Don't read the user's ~/.rhosts and ~/.shosts files
# IgnoreRhosts yes

# To disable tunneled clear text passwords, change to no here!
# PasswordAuthentication yes
# PermitEmptyPasswords no
PasswordAuthentication yes

# Change to no to disable s/key passwords
# ChallengeResponseAuthentication yes
ChallengeResponseAuthentication no

# Kerberos options
# KerberosAuthentication no
# KerberosOrLocalPasswd yes
# KerberosTicketCleanup yes
```

```
# KerberosGetAFSToken no
# KerberosUseKuserok yes

# GSSAPI options
# GSSAPIAuthentication no
GSSAPIAuthentication yes
# GSSAPICleanupCredentials yes
GSSAPICleanupCredentials yes
# GSSAPIStrictAcceptorCheck yes
# GSSAPIKeyExchange no

# Set this to 'yes' to enable PAM authentication, account processing,
# and session processing. If this is enabled, PAM authentication will
# be allowed through the ChallengeResponseAuthentication and
# PasswordAuthentication.  Depending on your PAM configuration,
# PAM authentication via ChallengeResponseAuthentication may bypass
# the setting of "PermitRootLogin without-password".
# If you just want the PAM account and session checks to run without
# PAM authentication, then enable this but set PasswordAuthentication
# and ChallengeResponseAuthentication to 'no'.
# UsePAM no
UsePAM yes

# Accept locale-related environment variables
AcceptEnv LANG LC_CTYPE LC_NUMERIC LC_TIME LC_COLLATE LC_MONETARY LC_MESSAGES
AcceptEnv LC_PAPER LC_NAME LC_ADDRESS LC_TELEPHONE LC_MEASUREMENT
AcceptEnv LC_IDENTIFICATION LC_ALL LANGUAGE
AcceptEnv XMODIFIERS

# AllowAgentForwarding yes
# AllowTcpForwarding yes
# GatewayPorts no
# X11Forwarding no
X11Forwarding yes
# X11DisplayOffset 10
# X11UseLocalhost yes
# PrintMotd yes
# PrintLastLog yes
# TCPKeepAlive yes
# UseLogin no
# UsePrivilegeSeparation yes
# PermitUserEnvironment no
# Compression delayed
# ClientAliveInterval 0
# ClientAliveCountMax 3
# ShowPatchLevel no
# UseDNS yes
# PidFile /var/run/sshd.pid
# MaxStartups 10:30:100
# PermitTunnel no
# ChrootDirectory none

# no default banner path
# Banner none

# override default of no subsystems

# Example of overriding settings on a per-user basis
# Match User anoncvs
#   X11Forwarding no
#   AllowTcpForwarding no
#   ForceCommand cvs server
```

```
###SFTP 配置文件 2019/05/23 hanyw  --- beging ###

AuthorizedKeysFile     .ssh/authorized_keys
# Subsystem   sftp/usr/libexec/OpenSSH/sftp-server
Subsystem sftp internal-sftp
Match Group group_sftp
ChrootDirectory /data/sftp/%u
ForceCommand internal-sftp
AllowTcpForwarding no
X11Forwarding no
PubkeyAuthentication yes
RSAAuthentication yes
###SFTP 配置文件 2019/05/23 hanyw --- beging ###
```

第 8 章
Samba 服务

8.1 搭建基本的 Samba 服务器

8.1.1 Samba 简介

Samba 是基于信息服务块（Server Message Block，SMB）协议的开源软件。

SMB 协议是一种在 Linux、UNIX 操作系统上可用于共享文件和打印机等资源的协议。这种协议是基于 C/S 架构的协议，客户端可以通过 SMB 访问服务器上的共享资源。当客户端的操作系统是 Windows，服务器的操作系统是 CentOS 时，通过 Samba 就可以实现 Windows 访问 CentOS 的资源，实现两个操作系统间的数据交互。

Samba 最大的功能就是可以用于 Linux 与 Windows 操作系统之间直接的文件共享和打印机共享。Samba 既可以用于 Windows 与 Linux 之间的资源共享，也可以用于 Linux 与 Linux 之间的资源共享。由于 NFS 可以很好地完成 Linux 与 Linux 之间的资源共享，所以 Samba 较多地用于 Linux 与 Windows 之间的资源共享。

组成 Samba 的服务有两个，即 SMB 服务和 NMB 服务。

（1）SMB 服务是 Samba 的核心启动。

SMB 服务主要负责建立 Linux Samba 服务器与 Samba 客户端之间的对话，验证用户身份并提供对文件和输出系统的访问。只有 SMB 服务启动，才能实现文件的共享，监听 139 TCP 端口。

（2）NMB 服务负责解析。

NMB 服务类似于 DNS，NMB 服务可以把 Linux 系统共享的工作组名称与其 IP 地址对应起来。如果 NMB 服务没有启动，就只能通过 IP 地址来访问共享文件，监听 137 和 138 UDP 端口。

8.1.2 构建 Samba 服务器环境

1．基本环境

Samba 服务器基本环境信息如表 8-1 所示。

表 8-1　　　　　　　　　　　Samba 服务器基本环境信息

角色	操作系统	IP 地址	主机名
Samba 服务器	CentOS 6.9 x86_64	192.168.2.150	smb_share_150

关闭防火墙和 SELinux，代码如下。

```
[root@smb_share_150 ~]# sed -i '#SELINUX=enforcing#SELINUX=disabled#' /etc/sysconfig/seLinux
[root@smb_share_150 ~]# setenforce 0
[root@smb_share_150 ~]# /etc/init.d/iptables stop
[root@smb_share_150 ~]# chkconfig iptables off
[root@smb_share_150 ~]# /etc/init.d/iptables status
iptables: 未运行防火墙。
[root@smb_share_150 ~]# sestatus
SELinux status:                 enabled
SELinuxfs mount:                /seLinux
Current mode:                   permissive
Mode from config file:          enforcing
Policy version:                 24
Policy from config file:        targeted
```

查看系统信息和 IP 地址。

```
[root@smb_share_150 ~]# cat /etc/redhat-release
CentOS release 6.9 (Final)
[root@smb_share_150 ~]# uname  -r
2.6.32-696.el6.x86_64
[root@smb_share_150 ~]# ifconfig
eth0      Link encap:Ethernet  HWaddr 00:0C:29:FC:67:2A
          inet addr:192.168.2.150  Bcast:192.168.2.255  Mask:255.255.255.0
          inet6 addr: fe80::20c:29ff:fefc:672a/64 Scope:Link
          UP BROADCAST RUNNING MULTICAST  MTU:1500  Metric:1
          RX packets:12607 errors:0 dropped:0 overruns:0 frame:0
          TX packets:1304 errors:0 dropped:0 overruns:0 carrier:0
          collisions:0 txqueuelen:1000
          RX bytes:800236 (787.4 KiB)  TX bytes:117081 (114.3 KiB)

lo        Link encap:Local Loopback
          inet addr:127.0.0.1  Mask:255.0.0.0
          inet6 addr: ::1/128 Scope:Host
          UP LOOPBACK RUNNING  MTU:65536  Metric:1
          RX packets:0 errors:0 dropped:0 overruns:0 frame:0
          TX packets:0 errors:0 dropped:0 overruns:0 carrier:0
          collisions:0 txqueuelen:0
          RX bytes:0 (0.0 b)  TX bytes:0 (0.0 b)
```

从上述代码可以看到，操作系统版本、IP 地址信息均和表 8-1 中设定的一致。

安装 epel 源头文件，代码如下。

```
[root@smb_share_150 ~]# yum install epel-release -y >/dev/null 2>&1
[root@smb_share_150 ~]# ll /etc/yum.repos.d/
总用量 32
-rw-r--r--. 1 root root 1991 3月  28 2017 CentOS-Base.repo
-rw-r--r--. 1 root root  647 3月  28 2017 CentOS-Debuginfo.repo
-rw-r--r--. 1 root root  289 3月  28 2017 CentOS-fasttrack.repo
-rw-r--r--. 1 root root  630 3月  28 2017 CentOS-Media.repo
-rw-r--r--. 1 root root 7989 3月  28 2017 CentOS-Vault.repo
-rw-r--r--. 1 root root  957 11月  5 2012 epel.repo
-rw-r--r--. 1 root root 1056 11月  5 2012 epel-testing.repo
```

从上述代码可以看到，epel.repo 文件已经被成功安装。

2. 部署 Samba 服务

（1）安装 Samba 软件。

CentOS 使用 yum 指令一次性安装所有依赖的软件包。

```
[root@smb_share_150 ~]# yum list Samba*
已加载插件：fastestmirror
Loading mirror speeds from cached hostfile
epel/metalink
| 7.6 kB     00:00
 * base: mirror.jdcloud.com
 * epel: mirrors.yun-idc.com
 * extras: mirrors.neusoft.edu.cn
 * updates: mirrors.neusoft.edu.cn
epel
| 4.7 kB     00:00
epel/primary_db
| 6.0 MB     00:00
可安装的软件包
Samba.x86_64 # 3.6.23-51.el6 # base
Samba-client.x86_64 base
Samba-common.i686 # 3.6.23-51.el6 # base
Samba-common.x86_64                      3.6.23-51.el6        base
Samba-doc.x86_64                         3.6.23-51.el6        base
Samba-domainjoin-gui.x86_64              3.6.23-51.el6        base
Samba-glusterfs.x86_64                   3.6.23-51.el6        base
Samba-swat.x86_64                        3.6.23-51.el6        base
```

```
Samba-winbind.x86_64                         3.6.23-51.el6        base
Samba-winbind-clients.i686                   3.6.23-51.el6        base
Samba-winbind-clients.x86_64                 3.6.23-51.el6        base
Samba-winbind-devel.i686                     3.6.23-51.el6        base
Samba-winbind-devel.x86_64                   3.6.23-51.el6        base
Samba-winbind-krb5-locator.x86_64            3.6.23-51.el6        base
Samba4.x86_64                                4.2.10-15.el6        base
Samba4-client.x86_64                         4.2.10-15.el6        base
Samba4-common.x86_64                         4.2.10-15.el6        base
Samba4-dc.x86_64                             4.2.10-15.el6        base
Samba4-dc-libs.x86_64                        4.2.10-15.el6        base
Samba4-devel.x86_64                          4.2.10-15.el6        base
Samba4-libs.x86_64                           4.2.10-15.el6        base
Samba4-pidl.x86_64                           4.2.10-15.el6        base
Samba4-python.x86_64                         4.2.10-15.el6        base
Samba4-test.x86_64                           4.2.10-15.el6        base
Samba4-winbind.x86_64                        4.2.10-15.el6        base
Samba4-winbind-clients.x86_64                4.2.10-15.el6        base
Samba4-winbind-krb5-locator.x86_64           4.2.10-15.el6        base
```

上述代码列出所有的 Samba 相关的软件包。

安装 Samba，代码如下。

```
[root@smb_share_150 ~]# yum install Samba -y
已加载插件：fastestmirror
设置安装进程
Loading mirror speeds from cached hostfile
 * base: mirror.jdcloud.com
（中间代码略）
已安装:
  Samba.x86_64 0:3.6.23-51.el6

作为依赖被安装:
  avahi-libs.x86_64 0:0.6.25-17.el6    cups-libs.x86_64 1:7.4.2-81.el6_10
gnutls.x86_64 0:2.12.23-22.el6         libjpeg-turbo.x86_64 0:7.2.1-3.el6_5
  libpng.x86_64 2:7.2.49-2.el6_7       libtalloc.x86_64 0:2.7.5-1.el6_7
libtdb.x86_64 0:7.3.8-3.el6_8.2        libtevent.x86_64 0:0.9.26-2.el6_7
  libtiff.x86_64 0:3.9.4-21.el6_8      Samba-common.x86_64 0:3.6.23-51.el6
Samba-winbind.x86_64 0:3.6.23-51.el6   Samba-winbind-clients.x86_64 0:3.6.23-51.el6

完毕!
```

（2）查看 Samba 安装是否成功，代码如下。

```
[root@smb_share_150 ~]# rpm -qa Samba
Samba-3.6.23-51.el6.x86_64
```

上述代码表示，Samba 软件已经被成功安装。

```
[root@smb_share_150 ~]# yum check-update Samba
已加载插件：fastestmirror
Loading mirror speeds from cached hostfile
 * base: mirror.jdcloud.com
 * epel: mirrors.yun-idc.com
 * extras: mirrors.neusoft.edu.cn
 * updates: mirrors.neusoft.edu.cn
```

上述代码检查是否有可更新的软件包。

```
[root@smb_share_150 ~]# yum update Samba
已加载插件：fastestmirror
设置更新进程
Loading mirror speeds from cached hostfile
 * base: mirror.jdcloud.com
 * epel: mirrors.yun-idc.com
 * extras: mirrors.neusoft.edu.cn
 * updates: mirrors.neusoft.edu.cn
不升级任何软件包
```

使用 yum -y install Samba 指令不能安装 Samba-client 和 Samba-swat 软件，但可以使用如下

指令进行安装。
```
yum install  Samba-client -y
yum install Samba-swat -y
```
（3）安装包说明，代码如下。

使用如下指令查询 Samba 安装包。
```
[root@smb_share_150 ~]# rpm -qa Samba*
Samba-winbind-clients-3.6.23-51.el6.x86_64
Samba-3.6.23-51.el6.x86_64
Samba-winbind-3.6.23-51.el6.x86_64
Samba-common-3.6.23-51.el6.x86_64
Samba-client-3.6.23-51.el6.x86_64
Samba-swat-3.6.23-51.el6.x86_64
```
Samba 安装包说明如下。

Samba-common-3.6.23-51.el6.x86_64

主要提供 Samba 服务器的设置文件与设置文件语法检验程序 testparm。

Samba-client-3.6.23-51.el6.x86_64

客户端软件，主要提供 Linux 主机作为客户端时所需要的工具指令集。

Samba-swat-3.6.23-51.el6.x86_64

基于 HTTPS 的 Samba 服务器 Web 界面配置。

Samba-3.6.23-51.el6.x86_64

提供 Samba 服务器的守护程序、共享文档、开机默认选项，Samba 服务器安装完毕，会生成配置文件目录/etc/Samba/，/etc/Samba/smb.conf 是 Samba 的核心配置文件。

（4）查看 SMB 服务状态，代码如下。
```
[root@smb_share_150 ~]# /etc/init.d/smb status
smbd 已停
```
由于还未启动 SMB 服务，SMB 服务默认处于关闭状态。

（5）启动 SMB 服务，代码如下。
```
[root@smb_share_150 ~]# /etc/init.d/smb start
启动 SMB 服务：                                       [确定]
[root@smb_share_150 ~]# /etc/init.d/smb status
smbd (pid  1547) 正在运行...
[root@smb_share_150 ~]# /etc/init.d/nmb restart
关闭 NMB 服务：                                       [失败]
启动 NMB 服务：                                       [确定]
[root@smb_share_150 ~]# /etc/init.d/nmb status
nmbd (pid  1430) 正在运行...
```
启动 SMB 服务后，查看其运行状态。

（6）设置 SMB 服务开机自动启动，代码如下。
```
[root@smb_share_150 ~]# chkconfig smb on
[root@smb_share_150 ~]# chkconfig --list |grep smb
smb              0:关闭  1:关闭  2:启用  3:启用  4:启用  5:启用  6:关闭
[root@smb_share_150 ~]# chkconfig nmb on
[root@smb_share_150 ~]# chkconfig --list |grep nmb
nmb              0:关闭  1:关闭  2:启用  3:启用  4:启用  5:启用  6:关闭
```
从上述代码可以看到，SMB 服务开机启动设置已经生效。

（7）查看 SMB 服务端口。
```
[root@smb_share_150 ~]# netstat -ntpl
Active Internet connections (only servers)
Proto Recv-Q Send-Q Local Address           Foreign Address         State       PID/Program name
tcp        0      0 0.0.0.0:139             0.0.0.0:*               LISTEN      1547/smbd
tcp        0      0 0.0.0.0:22              0.0.0.0:*               LISTEN      1229/sshd
tcp        0      0 127.0.0.1:25            0.0.0.0:*               LISTEN      1308/master
tcp        0      0 0.0.0.0:445             0.0.0.0:*               LISTEN      1547/smbd
tcp        0      0 :::139                  :::*                    LISTEN      1547/smbd
```

```
tcp        0      0 :::22                :::*                    LISTEN      1229/sshd
tcp        0      0 ::1:25               :::*                    LISTEN      1308/master
tcp        0      0 :::445               :::*                    LISTEN      1547/smbd
```

从上述代码可以看到，SMB 服务端口已经被成功监听。

8.1.3 Samba 服务器组件说明

Samba 有两个主要的进程：smbd 和 nmbd。

smbd 进程提供文件和输出服务，而 nmbd 则提供 NetBIOS 名称和浏览支持服务，帮助 SMB 客户端定位服务器，处理所有基于 UDP 的协议。

1. Samba 服务器相关的配置文件

（1）/etc/Samba/smb.conf。

/etc/Samba/smb.conf 是 Samba 的主要配置文件，主要的设置包括服务器全局设置。

（2）/etc/Samba/lmhosts。

Samba 会使用本机名称（hostname）作为 NetBIOS 名称。

（3）/etc/sysconfig/Samba。

用于提供启动 smbd、nmbd 时的相关服务参数。

（4）/etc/Samba/smbusers。

Windows 与 Linux 的管理员与访客的用户名称不一致，为了对应这两者之间的用户关系，可使用该文件进行设定。

（5）/var/lib/Samba/private/{passdb.tdb,secrets.tdb}。

管理 Samba 的用户账号和密码时会用到的数据库文件。

（6）/usr/share/doc/Samba-<版本信息>。

该文件包含了 Samba 的使用说明信息。

2. Samba 服务器常用脚本

Samba 脚本包含的服务器与客户端功能如下。

（1）/usr/sbin/{smbd,nmbd}。

服务器功能，主要包括权限管理（smbd）以及 NetBIOS 名称查询（nmbd）服务程序。

（2）/usr/bin/{tdbdump,tdbtool}。

服务器功能，在 Samba 3.0 以后的版本中，用户的账号与密码保存在数据库。

（3）/usr/bin/smbstatus。

服务器功能，可以列出目前 Samba 的联机状况，包括每一条 Samba 联机的 PID、分享的资源、用户来源等。

（4）/usr/bin/{smbpasswd,pdbedit}。

服务器功能，在管理 Samba 的用户账号和密码时，早期使用 smbpasswd 指令，后来使用 TDB 数据库，因此建议使用新的 pdbedit 指令来管理用户数据。

（5）/usr/bin/testparm。

服务器功能，检验配置文件 smb.conf 的语法是否正确。

（6）/sbin/mount.cifs。

客户端功能，在 Windows 操作系统上设定网络驱动器连接主机 Linux 操作系统，通过 mount（mount.cifs）将远程主机共享的目录挂载到 Linux 主机。

（7）/usr/bin/smbclient。

客户端功能，类似 Windows 操作系统的网络邻居。

（8）/usr/bin/nmblookup。

客户端功能，类似 nslookup，可以查询 NetBIOS 名称。

（9）/usr/bin/smbtree。

客户端功能，可查询工作组与计算机名称的树状目录分布图。

8.1.4 配置 Samba 服务器

1. Samba 服务主配置文件详解

Samba 的主配置文件为/etc/Samba/smb.conf。主配置文件由以下两部分构成。

Global Settings：该设置都是与 Samba 服务整体运行环境有关的选项，它的设置项目是针对所有共享资源的。

Share Definitions：该设置针对的是共享目录的个别设置，只对当前的共享资源起作用，全局参数配置说明如下。

（1）config file = /usr/local/Samba/lib/smb.conf.%m

config file 配置可以让 Samba 服务管理员使用另一个配置文件来覆盖默认的配置文件。

（2）workgroup = WORKGROUP。

设定 Samba 服务器所要加入的工作组或者域。

（3）server string = Samba Server Version %v。

设定 Samba 服务器的注释，可以是任何字符串，也可以不填。%v 显示 Samba 的版本号。

（4）netbios name = smbserver。

设置 Samba 服务器的 NetBIOS 名称。

（5）interfaces = lo eth0 192.168.12.2/24 192.168.13.2/24。

设置 Samba 服务器监听哪些网卡，可以写网卡名称或该网卡设置的 IP 地址。

（6）hosts allow = 127. 192.168.1 192.168.10.1。

表示允许连接到 Samba 服务器的客户端，多个参数以空格隔开。可以用一个 IP 地址表示，也可以用一个网段表示。

（7）max connections = 0。

max connections 用来指定连接 Samba 服务器的最大连接数目。如果超出连接数目，则新的连接请求将被拒绝，0 表示不限制。

（8）deadtime = 0。

deadtime 用来设置切断一个没有打开任何文件的连接的时间。单位是分钟，0 代表 Samba 服务不自动切断任何连接。

（9）time server = yes/no。

time server 是把 nmdb 设置为 Windows 客户端的时间服务器。

（10）log file = /var/log/Samba/log.%m。

设置 Samba 服务器日志文件的存储位置以及日志文件名称。在文件名后加%m（主机名），表示对每台访问 Samba 服务器的计算机都单独记录一个日志文件。

(11) max log size = 50。

设置 Samba 服务器日志文件的最大容量,单位为 KB,0 代表不限制。

(12) security = user。

设置用户访问 Samba 服务器的验证方式,一共有 4 种验证方式。

- share:用户访问 Samba 服务器不需要提供用户名和口令,安全性能较低。
- user:Samba 服务器的共享目录只能被授权用户访问,由 Samba 服务器负责检查账号和密码的正确性,账号和密码要在本 Samba 服务器中建立。
- server:依靠其他系统或 Samba 服务器来验证用户的账号和密码,是一种代理验证。此安全模式下,系统管理员可以把所有的 Windows 用户和口令集中到另一个 Windows 操作系统上,使用 Windows 操作系统进行 Samba 认证;远程服务器可以自动认证全部用户和口令,如果认证失败,Samba 将使用用户级安全模式作为替代的方式。
- domain:域安全级别,使用主域控制器来完成认证。

(13) passdb backend = tdbsam。

passdb backend 即用户后台,目前有 3 种后台:smbpasswd、tdbsam 和 ldapsam。sam 是 security account manager(安全用户管理)的缩写。

- smbpasswd:该方式使用 SMB 自己的工具 smbpasswd 设置 Samba 密码,客户端使用该密码访问 Samba 的资源。
- ldapsam:该方式则基于轻量目录访问协议(Lightweight Directory Access Protocol,LDAP)的用户管理方式来验证用户,首先要建立 LDAP 服务,然后设置 "passdb backend = ldapsam:ldap://LDAP Server"。
- tdbsam:该方式使用一个数据库文件来建立用户数据库。

pdbedit 指令的核心参数如下。

```
pdbedit -a username
```
新建 Samba 用户。
```
pdbedit -x username
```
删除 Samba 用户。
```
pdbedit -L
```
列出 Samba 用户列表,读取 passdb.tdb 数据库文件。
```
pdbedit -Lv
```
列出 Samba 用户列表的详细信息。
```
pdbedit -c "[D]" -u username
```
暂停该 Samba 用户的账号。
```
pdbedit -c "[]" -u username
```
恢复该 Samba 用户的账号。

(14) encrypt passwords = yes/no。

是否将认证密码加密。

(15) smb passwd file = /etc/Samba/smbpasswd。

定义 Samba 用户的密码文件。

(16) username map = /etc/Samba/smbusers。

定义用户名映射。

(17) guest account = nobody。

设置 guest 用户名。

（18）socket options = TCP_NODELAY SO_RCVBUF=8192 SO_SNDBUF=8192

设置服务器和客户端之间会话的 socket 选项，可以提高传输速度。

（19）domain master = yes/no。

设置 Samba 服务器是否要成为网域主浏览器，网域主浏览器可以管理跨子网域的浏览服务。

（20）local master = yes/no。

local master 用来指定 Samba 服务器是否试图成为本地网域主浏览器。

（21）preferred master = yes/no。

设置 Samba 服务器一开机就强迫进行主浏览器选择。

（22）os level = 200。

设置 Samba 服务器的 os level。

（23）domain logons = yes/no。

设置 Samba 服务器是否要作为本地域控制器。

（24）logon script = %u.bat。

登录相关设置。

（25）wins support = yes/no。

设置 Samba 服务器是否提供 Windows 服务。

（26）wins proxy = yes/no。

设置 Samba 服务器是否开启 Windows 代理服务。

（27）dns proxy = yes/no。

设置 Samba 服务器是否开启 DNS 代理服务。

（28）load printers = yes/no。

设置是否在启动 Samba 服务器时就共享打印机。

（29）printcap name = cups。

设置共享打印机的配置文件。

（30）printing = cups。

设置 Samba 共享打印机的类型。

2．Samba 服务共享参数详解

（1）[comment 服务共享名称]。

comment = 任意字符串，comment 是对该共享的描述。

（2）path = 共享目录路径。

path 用来指定共享目录的路径。

（3）browseable = yes/no。

browseable 用来指定该共享是否可以浏览。

（4）writable = yes/no。

writable 用来指定该共享路径是否可写。

（5）available = yes/no。

available 用来指定该共享资源是否可用。

（6）admin users = 该共享的管理者。

admin users 用来指定该共享的管理者。

（7）valid users = 允许访问该共享资源的用户。

valid users 用来指定允许访问该共享资源的用户。

（8）invalid users = 禁止访问该共享资源的用户。

invalid users 用来指定不允许访问该共享资源的用户。

（9）write list = 允许写入该共享的用户。

write list 用来指定可以在该共享下写入文件的用户。

（10）guest ok = yes/no。

意义同"public"（意为公共的）。

3．几个特殊共享

安装好 Samba 服务器后，使用 testparm 指令可以测试 smb.conf 配置是否正确。使用 testparm –v 指令可以详细列出 smb.conf 支持的配置参数，代码如下。

```
[root@smb_share_150 ~]# testparm
Load smb config files from /etc/Samba/smb.conf
rlimit_max: increasing rlimit_max (1024) to minimum Windows limit (16384)
Processing section "[homes]"
Processing section "[printers]"
Loaded services file OK.
Server role: ROLE_STANDALONE
Press enter to see a dump of your service definitions
```

输入 testparm，按"Enter"键后，显示内容如下。

```
[global]
    workgroup = MYGROUP
    server string = Samba Server Version %v
    log file = /var/log/Samba/log.%m
    max log size = 50
    client signing = required
    idmap config * : backend = tdb
    cups options = raw


[homes]
    comment = Home Directories
    read only = No
    browseable = No

[printers]
    comment = All Printers
    path = /var/spool/Samba
    printable = Yes
    print ok = Yes
    browseable = No
```

使用 testparm 指令可以简单测试 Samba 的配置文件。假如测试结果无误，Samba 常驻服务就能正确载入为其设置的值，但并不保证其后的操作如预期般一切正常，代码如下。

```
[root@smb_share_150 ~]# testparm -s
Load smb config files from /etc/Samba/smb.conf
rlimit_max: increasing rlimit_max (1024) to minimum Windows limit (16384)
Processing section "[homes]"
Processing section "[printers]"
Loaded services file OK.
Server role: ROLE_STANDALONE
[global]
    workgroup = MYGROUP
    server string = Samba Server Version %v
    log file = /var/log/Samba/log.%m
    max log size = 50
```

```
            client signing = required
            idmap config * : backend = tdb
            cups options = raw

    [homes]
            comment = Home Directories
            read only = No
            browseable = No

    [printers]
            comment = All Printers
            path = /var/spool/Samba
            printable = Yes
            print ok = Yes
            browseable = No
```

testparm 指令的-s 选项，表示不显示提示符号，等待用户按 "Enter" 键，就直接列出 Samba 服务定义信息。

8.1.5 用户权限与配置文件

1．创建 Samba 服务授权用户

注意在建立 Samba 账号之前，一定要先建立一个与 Samba 账号同名的系统账号。账号相关的配置如下。

```
    security = user
    username map = /etc/Samba/smbusers
    encrypt passwords = true
    passdb backend = smbpasswd
    smb passwd file =/etc/Samba/smbpasswd
```

配置完毕后，再添加用户，代码如下。

```
[root@smb_share_150 ~]# useradd  handuoduo
[root@smb_share_150 ~]# echo "1" |passwd --stdin handuoduo
更改用户 handuoduo 的密码 。
passwd:  所有的身份验证令牌已经成功更新。
```

Linux 的用户密码和 Samba 的用户密码并不是一回事，只是 Samba 的用户必须是 Linux 的用户，因此需要将 handuoduo 这个用户添加到 Samba 的用户数据库，执行如下指令。

```
[root@smb_share_150 ~]# smbpasswd -a handuoduo
New SMB password:
Retype new SMB password:
Added user handuoduo.
```

查看密码文件，代码如下。

```
[root@smb_share_150 Samba]# cat /etc/Samba/smbpasswd
handuoduo:500:XXXXXXXXXXXXXXXXXXXXXXXXXXXXXXXX:69943C5E63B4D2C104DBBCC15138B72B:
[U       ]:LCT-5CE9031B:
```

上述代码表示用户文件已经建立完毕。

2．创建目录并授权给组

```
[root@smb_share_150 ~]# mkdir -pv /data/web50
mkdir: 已创建目录 "/data"
mkdir: 已创建目录 "/data/web50"
[root@smb_share_150 ~]# ls -ld /data/bj-web70
drwxr-xr-x. 2 root root 4096 5月  25 13:53 /data/web50
[root@smb_share_150 Samba]# chmod u=rwx,g=rwx,o=--- /data/web50
[root@smb_share_150 Samba]# ls -ld /data/web50/
drwxrwx---. 3 root handuoduo 4096 5月  25 16:57 /data/web50/
```

这里设置/data/web50 目录所属组具备 w 权限（写入、删除、查看）。

3. 客户端测试

使用 Windows 图标+R 组合键打开运行对话框，输入\\192.168.2.150\web50，如图 8-1 所示。输入 Samba 服务共享地址后，连接认证如图 8-2 和图 8-3 所示。

图 8-1　输入 Samba 服务共享地址　　图 8-2　连接 Samba 服务认证　　图 8-3　输入用户名和密码

上传文件和目录测试，如图 8-4 所示。

图 8-4　上传文件和目录测试

4．新建文件夹权限

尝试在/data/web50 里面新建文件和文件夹，确认是否有权限。如果新建成功，再在 Samba 服务器端查看新建的文件和文件夹，如图 8-5 所示。

图 8-5　查看文件和文件夹

新版的 Samba 要在[global]后面追加文件和文件夹权限，放在 smb.conf 后是无效的，在[global]后面添加文件和文件夹权限，如图 8-6 所示。

图 8-6　添加相关权限

5．Samba 设置用户名和密码

Window 操作系统连接远程主机 Linux，需要在 Samba 里添加一个新用户，代码如下。

```
Linux-06bq:/usr/local/services/Samba/bin # ./smbpasswd -a sunjing
New SMB password:
Retype new SMB password:
Failed to add entry for user sunjing.
```

添加的 Samba 新用户首先必须是 Linux 用户。

(1) Samba 服务添加用户。

创建的 Samba 用户必须是 Linux 服务器上实际存在的用户,代码如下。

```
$ sudo smbpasswd -a root
New SMB password:
Retype new SMB password:
Added user root.
```

根据提示输入 Samba 用户的密码,当 Samba 服务成功安装、启动后,通过 Windows 操作系统访问服务器共享目录时,就要输入这里配置的用户名、密码。

查看 Samba 服务器中已拥有哪些用户,可以使用如下指令。

```
[root@smb_share_150 web50]# pdbedit -L
handuoduo:500:
```

删除 Samba 服务中的某个用户,可以使用如下指令。

```
smbpasswd -x 用户名
```

(2) Samba 服务密码文件。

Samba 服务器发布共享资源后,客户端访问 Samba 服务器,需要提交用户名和密码进行身份验证,验证合格后才可以登录。

Samba 服务为了实现用户身份验证功能,将用户名和密码信息存放在/etc/Samba/smbpasswd 中。在客户端访问时,将用户提交的资料与 smbpasswd 中存放的信息进行比对。比对结果为相同,并且 Samba 服务器其他安全设置允许时,客户端与 Samba 服务器连接才能成功建立。

在 Samba 服务器中建立 Samba 账号之前必须先添加相对应的系统账号,使用 useradd 指令建立账号 hanmeimei,然后执行 passwd 指令为账号 hanmeimei 设置密码。最后添加 hanmeimei 用户的 Samba 账号,执行 smbpasswd 添加账号 hanmeimei 到 Samba 配置文件中,代码如下。

```
[root@smb_share_150 ~]# useradd hanmeimei
[root@smb_share_150 ~]# passwd hanmeimei
Changing password for user hanmeimei.
New UNIX password:
Retype new UNIX password:
passwd:all authentication tokens updated successfully.
[root@smb_share_150 ~]# cd/etc/Samba/
[root@smb_share_150 ~]# ls
lmhosts secrets.tdb smb.conf smbpasswd smbusers
[root@rusky2 Samba]# smbpasswd --help
smbpasswd: invalid option -- -
When run by root:
smbpasswd [options] [username]
otherwise:
smbpasswd [options]
options:
  -L                   local mode (must be first option)
  -h                   print this usage message
  -s                   use stdin for password prompt
  -c smb.conf file     Use the given path to the smb.conf file
  -D LEVEL             debug level
  -r MACHINE           remote machine
  -U USER              remote username
extra options when run by root or in local mode:
  -a                   add user
  -d                   disable user
  -e                   enable user
  -i                   interdomain trust account
  -m                   machine trust account
  -n                   set no password
  -W                   use stdin ldap admin password
  -w PASSWORD          ldap admin password
  -x                   delete user
```

```
    -R ORDER name resolve order
[root@smb_share_150 ~]# smbpasswd-a hanmeimei
New SMB password:
Retype new SMB password:
Added user hanmeimei.
```

经过上面的设置，再次访问 Samba 共享文件时就可以使用 hanmeimei 账号访问了。smbpasswd 指令常用选项详解和用法如下。

```
[root@smb_share_150 ~]# useradd hanmeimei
[root@smb_share_150 ~]# echo 1 |passwd --stdin hanmeimei
更改用户 hanmeimei 的密码。
passwd: 所有的身份验证令牌已经成功更新。
```

Samba 的配置文件中设置 security 为 user 级别时，客户端访问 Samba 服务器时需要提交用户名和密码进行身份验证，此时就需要用到 smbpasswd 指令了。

smbpasswd 指令添加、删除 Samba 用户，语法格式如下。

```
smbpasswd [options] USERNAME
```

如果要建立一个 Samba 用户，其必须是在/etc/shadow 中（当前 Linux 操作系统中）存在的用户，代码如下。

```
[root@smb_share_150 ~]# smbpasswd -a hanmeimei
New SMB password:
Retype new SMB password:
Added user hanmeimei.
```

创建一个 Samba 用户，用户名为 hanmeimei，输入此用户密码（注意：此密码不必和操作系统中 hanmeimei 用户的密码一样，Samba 密码是独立的），提示"Added user hanmeimei."，表示创建成功。

使用-d 选项将禁用 Samba 用户，代码如下。

```
[root@smb_share_150 ~]# smbpasswd -d hanmeimei
Disabled user hanmeimei.
```

使用-e 选项将启用 Samba 用户，代码如下。

```
[root@smb_share_150 ~]# smbpasswd -e hanmeimei
Enabled user hanmeimei.
```

使用-n 选项将指定用户的密码置空，代码如下。

```
[root@smb_share_150 ~]# smbpasswd -n hanmeimei
User hanmeimei password set to none.
```

使用-x 选项将删除 Samba 用户，代码如下。

```
[root@smb_share_150 ~]# smbpasswd -x hanmeimei
Deleted user hanmeimei.
```

上述代码表示 hanmeimei 用户删除成功。

6. Samba 服务完整配置文件

Samba 服务完整配置文件，代码如下。

```
[root@smb_share_150 web50]# cat /etc/Samba/smb.conf
        [global]
        workgroup = WORKGROUP
        server string = Samba server on ubuntu
        netbios name = CentOS_50_smb_server
        interfaces = 192.168.2.0/24 eth0
        hosts allow = 192.168.2.*
        security = user
        username map = /etc/Samba/smbusers
        encrypt passwords = true
        passdb backend = smbpasswd
        smb passwd file =/etc/Samba/smbpasswd
        log file = /var/log/Samba/log.%m
        max open files = 1000
        socket options = TCP_NODELAY
```

第 8 章 Samba 服务

```
            force create mode = 0775
            directory mask = 0775
            force directory mode = 0775


            [web50]
            comment = This is a test share
            path = /data/web50
            valid users = handuoduo,root
            write list = handuoduo,root
            read only = No
            create mask = 0755
            directory mask = 0755
            guest ok = Yes
            writable = yes
            browseable = yes
            available = yes
```

至此，最基本的 Samba 服务已经配置完成了。

8.1.6　Windows 客户端访问 Samba 服务器

Windows 10 客户端访问 Samba 服务器出现如图 8-7 所示的错误时，解决方案如下。

1．启用安全组策略配置

启用安全组相关配置，如图 8-8～图 8-10 所示。

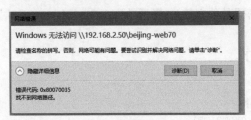
图 8-7　访问错误

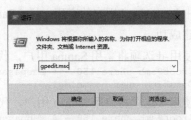
图 8-8　输入组策略快捷指令

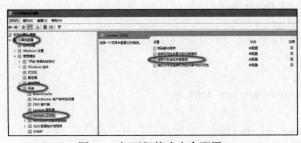
图 8-9　打开组策略安全配置

图 8-10　启用组策略安全配置

2．Windows 客户端先启用网络发现、文件和打印机共享

Windows 客户端先启用网络发现、文件和打印机共享、使用用户账户和密码连接到其他计算机，如图 8-11 所示。

3．检查系统服务

在服务中查看 Server、Workstation、Print Spooler、TCP/IP NetBIOS Helper 的属性，是否为

启动的状态，如图 8-12～图 8-15 所示。

图 8-11　启用网络发现和打印机共享

图 8-12　查看 Server 的属性

图 8-13　查看 Workstation 的属性

图 8-14　查看 Print Spooler 的属性

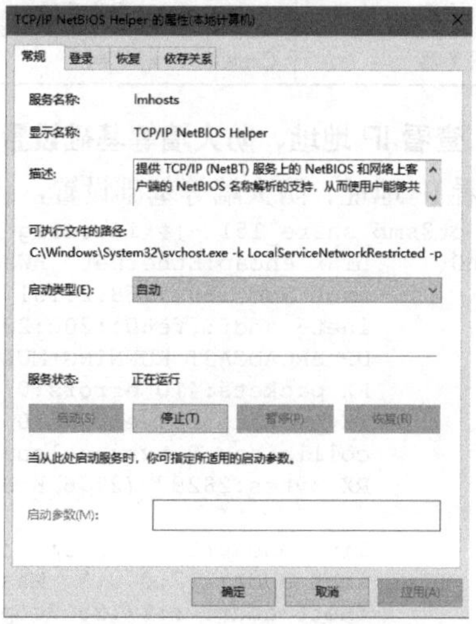

图 8-15　查看 TCP/IP NetBIOS Helper 的属性

8.2 Samba 服务之 user 配置案例

8.2.1 案例需求及其分析

案例需求：公司现有多个部门，因工作需要，将 IT 部的资料存放在 Samba 服务器的 /data/IT 目录中集中管理，以便 IT 部员工浏览，并且该目录只允许 IT 部员工访问。

需求分析：需要在公司内部搭建一台 Samba 服务器，为公司网络内的客户端计算机提供 user 级别 Samba 服务，具体配置描述如表 8-2 所示。

表 8-2　　　　　　　　　　　　　具体配置描述

需求	选项
Samba 服务器所在工作组	workgroup
Samba 服务器描述信息	IT Files Share
Samba 服务器 NetBIOS 名称	CentOS_51_smb_server
Samba 服务器日志文件路径	/var/log/Samba/log.%m
Samba 服务器日志文件大小	10000KB
Samba 服务器安全模式	user
Samba 服务器密码	加密
Samba 服务器密码数据库类型	tdbsam
Samba 服务器共享目录	/data/IT
Samba 服务器共享目录权限	属主和属组均为 smb_it
客户端访问 Samba 服务器权限	读写

8.2.2 初始化 Samba 服务器

Samba 服务器基础环境信息如表 8-3 所示。

表 8-3　　　　　　　　　　　　Samba 服务器基础环境信息

角色	操作系统	IP 地址	主机名
Samba 服务器	CentOS 6.9 x86_64	192.168.2.151	smb_share_151

1. 查看 IP 地址、防火墙等基础设置

查看 IP 地址、防火墙等基础设置，代码如下。

```
[root@smb_share_151 ~]# ifconfig
eth0      Link encap:Ethernet  HWaddr 00:0C:29:FD:CB:3A
          inet addr:192.168.2.151  Bcast:192.168.2.255  Mask:255.255.255.0
          inet6 addr: fe80::20c:29ff:fefd:cb3a/64 Scope:Link
          UP BROADCAST RUNNING MULTICAST  MTU:1500  Metric:1
          RX packets:410 errors:0 dropped:0 overruns:0 frame:0
          TX packets:137 errors:0 dropped:0 overruns:0 carrier:0
          collisions:0 txqueuelen:1000
          RX bytes:28297 (27.6 KiB)  TX bytes:11776 (11.5 KiB)

lo        Link encap:Local Loopback
          inet addr:127.0.0.1  Mask:255.0.0.0
          inet6 addr: ::1/128 Scope:Host
          UP LOOPBACK RUNNING  MTU:65536  Metric:1
```

```
              RX packets:0 errors:0 dropped:0 overruns:0 frame:0
              TX packets:0 errors:0 dropped:0 overruns:0 carrier:0
              collisions:0 txqueuelen:0
              RX bytes:0 (0.0 b)  TX bytes:0 (0.0 b)

[root@smb_share_151 ~]# sestatus
SELinux status:                 disabled
[root@smb_share_151 ~]# /etc/init.d/iptables status
```

未运行 iptables 防火墙，且 SELinux 状态为 disabled。

2. 查看 Samba 软件

查看 Samba 4 相关的软件包，代码如下。

```
[root@smb_share_151 ~]# yum list Samba4*
已加载插件: fastestmirror
Loading mirror speeds from cached hostfile
 * base: mirror.bit.edu.cn
 * extras: mirror.bit.edu.cn
 * updates: mirror.bit.edu.cn
可安装的软件包
Samba4.x86_64                           4.2.10-15.el6              base
Samba4-client.x86_64                    4.2.10-15.el6              base
Samba4-common.x86_64                    4.2.10-15.el6              base
Samba4-dc.x86_64                        4.2.10-15.el6              base
Samba4-dc-libs.x86_64                   4.2.10-15.el6              base
Samba4-devel.x86_64                     4.2.10-15.el6              base
Samba4-libs.x86_64                      4.2.10-15.el6              base
Samba4-pidl.x86_64                      4.2.10-15.el6              base
Samba4-python.x86_64                    4.2.10-15.el6              base
Samba4-test.x86_64                      4.2.10-15.el6              base
Samba4-winbind.x86_64                   4.2.10-15.el6              base
Samba4-winbind-clients.x86_64           4.2.10-15.el6              base
Samba4-winbind-krb5-locator.x86_64      4.2.10-15.el6              base
```

3. 安装 Samba 4 相关软件包

安装 Samba 4 软件包，代码如下。

```
[root@smb_share_151 ~]# yum install Samba4 Samba4-common Samba4-client Samba4-libs
已加载插件: fastestmirror
设置安装进程
(中间代码略)
已安装:
  Samba4.x86_64 0:4.2.10-15.el6      Samba4-client.x86_64 0:4.2.10-15.el6
Samba4-common.x86_64 0:4.2.10-15.el6  Samba4-libs.x86_64 0:4.2.10-15.el6

作为依赖被安装:
  avahi-libs.x86_64 0:0.6.25-17.el6     cups-libs.x86_64 1:1.4.2-81.el6_10
gnutls.x86_64 0:2.12.23-22.el6        libjpeg-turbo.x86_64 0:1.2.1-3.el6_5
  libldb.x86_64 0:1.1.25-2.el6_7      libpng.x86_64 2:1.2.49-2.el6_7
libtalloc.x86_64 0:2.1.5-1.el6_7      libtdb.x86_64 0:1.3.8-3.el6_8.2
  libtevent.x86_64 0:0.9.26-2.el6_7   libtiff.x86_64 0:3.9.4-21.el6_8
pytalloc.x86_64 0:2.1.5-1.el6_7

完毕!
```

4. 确认 Samba 4 软件包是否安装成功

确认 Samba 4 软件包是否安装成功，代码如下。

```
[root@smb_share_151 ~]# rpm -qa Samba4*
Samba4-libs-4.2.10-15.el6.x86_64
Samba4-4.2.10-15.el6.x86_64
Samba4-common-4.2.10-15.el6.x86_64
Samba4-client-4.2.10-15.el6.x86_64
```

上述代码表示，Samba 4 相关软件包已经被成功安装。

8.2.3 配置 Samba 服务器

1. 添加 IT 部用户和组

（1）创建 Linux 系统用户和组。

创建用户的同时加入相应的组中的方式：useradd -g 组名 用户名。具体指令如下。

```
[root@smb_share_151 ~]# groupadd smb_it && useradd -a -g smb_it hanyanwei
[root@smb_share_151 ~]# id hanyanwei
uid=500(hanyanwei) gid=500(smb_it) 组=500(smb_it)
[root@smb_share_151 ~]# echo hanyanwei |passwd --stdin hanyanwei
更改用户 hanyanwei 的密码。
passwd: 所有的身份验证令牌已经成功更新。
```

查看用户信息是否正确，代码如下。

```
[root@smb_share_151 ~]# grep hanyanwei /etc/passwd
hanyanwei:x:500:500:::/home/hanyanwei:/bin/bash
[root@smb_share_151 ~]# id hanyanwei
uid=500(hanyanwei) gid=500(smb_it) 组=500(smb_it)
```

从上述代码可以看到，hanyanwei 用户已经创建成功。

（2）创建 Samba 服务用户。

```
[root@smb_share_151 ~]# smbpasswd -a hanyanwei
New SMB password:
Retype new SMB password:
Added user hanyanwei.
    [root@smb_share_151 ~]# pdbedit -L
hanyanwei:500:
```

从上述代码可以看到，hanyanwei 用户已经成功添加到 Samba 服务的数据成员信息库中。

2. 在根目录下建立/IT 文件夹并授权

在根目录下建立/IT 文件夹并授权，代码如下。

```
[root@smb_share_151 ~]# mkdir  -pv /data/IT
mkdir: 已创建目录 "/data"
mkdir: 已创建目录 "/data/IT"
[root@smb_share_151 ~]# ls -ld /data/IT
drwxr-xr-x 2 root root 4096 5月  25 23:23 /data/IT
[root@smb_share_151 ~]# chown root:smb_it /data/IT
[root@smb_share_151 ~]# ls -ld /data/IT
drwxr-xr-x 2 root smb_it 4096 5月  25 23:23 /data/IT
[root@smb_share_151 ~]# chmod g+w /data/IT
[root@smb_share_151 ~]# ls -ld /data/IT
drwxrwxr-x 2 root smb_it 4096 5月  25 23:23 /data/IT
```

上述代码修改权限所有者为 root，所属用户组为 smb_it。

3. 备份与设置/etc/Samba/smb.conf 文件

备份/etc/Samba/smb.conf 文件，代码如下。

```
[root@smb_share_151 ~]# cp -av /etc/Samba/smb.conf{,_bak}
"/etc/Samba/smb.conf" -> "/etc/Samba/smb.conf_bak"
[root@smb_share_151 ~]# vim /etc/Samba/smb.conf
[root@smb_share_151 ~]#
设置/etc/Samba/smb.conf 文件。
[root@smb_share_151 ~]# cat /etc/Samba/smb.conf
[global]
workgroup = workgroup
server string =    IT Files Share
netbios name = CentOS_151_smb_server
interfaces = eth0 192.168.2.151
hosts allow = 192.168.2.0/24
log file = /var/log/Samba/log.%m
max log size = 10000
security = user
```

```
passdb backend = tdbsam
encrypt passwords = yes


[IT]
comment = IT Files Share
path = /data/IT
public = no
writable = yes
```

4. 重启 Samba 服务

重启 Samba 服务，代码如下。

```
[root@smb_share_151 ~]# /etc/init.d/smb restart
关闭 SMB 服务:                                      [失败]
启动 SMB 服务:                                      [确定]
[root@smb_share_151 ~]# /etc/init.d/nmb restart
关闭 NMB 服务:                                      [失败]
启动 NMB 服务:                                      [确定]
[root@smb_share_151 ~]# netstat -ntpl
Active Internet connections (only servers)
Proto Recv-Q Send-Q Local Address       Foreign Address      State       PID/Program name
tcp        0      0 0.0.0.0:139         0.0.0.0:*            LISTEN      1482/smbd
tcp        0      0 0.0.0.0:22          0.0.0.0:*            LISTEN      1211/sshd
tcp        0      0 127.0.0.1:25        0.0.0.0:*            LISTEN      1290/master
tcp        0      0 0.0.0.0:445         0.0.0.0:*            LISTEN      1482/smbd
tcp        0      0 :::139              :::*                 LISTEN      1482/smbd
tcp        0      0 :::22               :::*                 LISTEN      1211/sshd
tcp        0      0 :::1:25             :::*                 LISTEN      1290/master
tcp        0      0 :::445              :::*                 LISTEN      1482/smbd
```

5. Windows 10 客户端测试

打开运行对话框，输入\\192.168.2.151\IT，测试过程如图 8-16、图 8-17 所示。

图 8-16　打开共享文件

图 8-17　新建目录

6. 新增 Samba 服务用户

新增 Samba 服务用户，代码如下。

```
[root@smb_share_151 ~]# useradd -g smb_it hanbing
[root@smb_share_151 ~]# id hanbing
uid=501(hanbing) gid=500(smb_it) 组=500(smb_it)
[root@smb_share_151 ~]# groups hanbing
hanbing : smb_it
[root@smb_share_151 ~]# groups hanyanwei
hanyanwei : smb_it
[root@smb_share_151 ~]# smbpasswd -a hanbing
New SMB password:
Retype new SMB password:
[root@smb_share_151 ~]#
[root@smb_share_151 ~]# pdbedit -L
hanyanwei:500:
```

```
hanbing:501:
```
上述代码表示新增用户。

7．访问 Samba 服务

下面代码表示访问 IT 目录。
```
\\192.168.2.151\IT
```

8.3 Samba 服务之 share 配置案例

8.3.1 Samba 服务需求及分析

案例需求：公司现有多个部门，因工作需要，将剪辑部的资料存放在 Samba 服务器的 /data/movie 目录中集中管理，以便剪辑部员工浏览，并且该目录只允许剪辑部员工访问。

需求分析：需要在公司内部搭建一台 Samba 服务器，为公司网络内的客户端计算机提供 share 级别的 Samba 服务，具体配置描述如表 8-4 所示。

表 8-4　　　　　　　　　　　　具体配置描述

需求	选项
Samba 服务器所在工作组	workgroup
Samba 服务器 NetBIOS 名称	CentOS_52_smb_server
Samba 服务器日志文件路径	/var/log/Samba/log.%m
Samba 服务器日志文件大小	10000KB
Samba 服务器安全模式	share
Samba 服务器共享目录	/data/movie
Samba 服务器共享目录权限	root
客户端访问 Samba 服务器权限	读写

8.3.2 初始化 Samba 服务器

Samba 服务器基础环境信息如表 8-5 所示。

表 8-5　　　　　　　　　　Samba 服务器基础环境信息

角色	操作系统	IP 地址	主机名
Samba 服务器	CentOS 6.9 x86_64	192.168.2.152	smb_share_152

1．初始化 Samba 服务运行环境

查看 IP 地址、防火墙等基础设置，代码如下。
```
[root@smb_share_152 ~]# ifconfig
eth0      Link encap:Ethernet  HWaddr 00:0C:29:C3:CE:60
          inet addr:192.168.2.152  Bcast:192.168.2.255  Mask:255.255.255.0
          inet6 addr: fe80::20c:29ff:fec3:ce60/64 Scope:Link
          UP BROADCAST RUNNING MULTICAST  MTU:1500  Metric:1
          RX packets:11033 errors:0 dropped:0 overruns:0 frame:0
          TX packets:581 errors:0 dropped:0 overruns:0 carrier:0
          collisions:0 txqueuelen:1000
          RX bytes:709082 (692.4 KiB)  TX bytes:41549 (40.5 KiB)

lo        Link encap:Local Loopback
          inet addr:127.0.0.1  Mask:255.0.0.0
```

```
          inet6 addr: ::1/128 Scope:Host
          UP LOOPBACK RUNNING  MTU:65536  Metric:1
          RX packets:0 errors:0 dropped:0 overruns:0 frame:0
          TX packets:0 errors:0 dropped:0 overruns:0 carrier:0
          collisions:0 txqueuelen:0
          RX bytes:0 (0.0 b)  TX bytes:0 (0.0 b)

[root@smb_share_152 ~]# sestatus
SELinux status:                 disabled
[root@smb_share_152 ~]# /etc/init.d/iptables status
iptables: 未运行防火墙。
```

2. 查看 Samba 软件

查看 Samba 相关的软件包，代码如下。

```
[root@smb_share_152 ~]# yum list Samba4*
已加载插件：fastestmirror
Determining fastest mirrors
 * base: mirror.jdcloud.com
 * extras: mirrors.aliyun.com
 * updates: mirror.jdcloud.com
base                                          | 3.7 kB         00:00
extras                                        | 3.4 kB         00:00
updates                                       | 3.4 kB         00:00
可安装的软件包
Samba4.x86_64                                 4.2.10-15.el6        base
Samba4-client.x86_64                          4.2.10-15.el6        base
Samba4-common.x86_64                          4.2.10-15.el6        base
Samba4-dc.x86_64                              4.2.10-15.el6        base
Samba4-dc-libs.x86_64                         4.2.10-15.el6        base
Samba4-devel.x86_64                           4.2.10-15.el6        base
Samba4-libs.x86_64                            4.2.10-15.el6        base
Samba4-pidl.x86_64                            4.2.10-15.el6        base
Samba4-python.x86_64                          4.2.10-15.el6        base
Samba4-test.x86_64                            4.2.10-15.el6        base
Samba4-winbind.x86_64                         4.2.10-15.el6        base
Samba4-winbind-clients.x86_64                 4.2.10-15.el6        base
Samba4-winbind-krb5-locator.x86_64            4.2.10-15.el6        base
```

3. 安装 Samba 相关软件包

安装 Samba 相关软件包，代码如下。

```
[root@smb_share_152 ~]# yum install Samba4 Samba4-common Samba4-client Samba4-libs
已加载插件：fastestmirror
设置安装进程
（中间代码略）
作为依赖被安装：
  avahi-libs.x86_64 0:0.6.25-17.el6          cups-libs.x86_64 1:1.4.2-81.el6_10
gnutls.x86_64 0:2.12.23-22.el6               libjpeg-turbo.x86_64 0:1.2.1-3.el6_5
  libldb.x86_64 0:1.1.25-2.el6_7             libpng.x86_64 2:1.2.49-2.el6_7
libtalloc.x86_64 0:2.1.5-1.el6_7             libtdb.x86_64 0:1.3.8-3.el6_8.2
  libtevent.x86_64 0:0.9.26-2.el6_7          libtiff.x86_64 0:3.9.4-21.el6_8
pytalloc.x86_64 0:2.1.5-1.el6_7

完毕！
```

4. 确认 Samba 软件包是否被安装成功

确认 Samba 软件包是否被安装成功，代码如下。

```
[root@smb_share_152 ~]# rpm -qa Samba4*
Samba4-libs-4.2.10-15.el6.x86_64
Samba4-4.2.10-15.el6.x86_64
Samba4-common-4.2.10-15.el6.x86_64
Samba4-client-4.2.10-15.el6.x86_64
```

上述代码表示，Samba 相关软件包已经被成功安装。

8.3.3 配置 Samba 服务器

1. 在根目录下建立/data/movie 文件夹并授权

在根目录下建立/data/movie 文件夹并授权,代码如下。

```
[root@smb_share_152 ~]# mkdir -pv /data/movie
mkdir: 已创建目录 "/data"
mkdir: 已创建目录 "/data/movie"
[root@smb_share_152 ~]# ls -ld /data/movie
drwxr-xr-x 2 root root 4096 5月  26 17:15 /data/movie
[root@smb_share_152 ~]# chmod  757 /data/movie/
[root@smb_share_152 ~]# ls -ld /data/movie/
drwxr-xrwx 3 root root 4096 5月  26 17:37 /data/movie/
```

2. 编辑/etc/Samba/smb.conf 配置文件

备份/etc/Samba/smb.conf 文件。

```
[root@smb_share_152 ~]# cp -av /etc/Samba/smb.conf{,_bak}
"/etc/Samba/smb.conf" -> "/etc/Samba/smb.conf_bak"
```

设置/etc/Samba/smb.conf 配置文件,代码如下。

```
[root@smb_share_152 ~]# cat /etc/Samba/smb.conf
[global]
workgroup = workgroup
server string =    movie share
netbios name = CentOS_152_smb_server
interfaces = eth0 192.168.2.152
hosts allow = 192.168.2.0/24
log file = /var/log/Samba/log.%m
max log size = 10000

# Samba 4 已废弃 security = share 此参数
# security = share
security=user
map to guest =Bad User

[movie]
comment = movie share
path = /data/movie
public = yes
writable = yes
browseable = yes
```

3. 注意事项

配置过程中报错,代码如下。

```
[root@smb_share_152 ~]# testparm  -s
Load smb config files from /etc/Samba/smb.conf
rlimit_max: increasing rlimit_max (1024) to minimum Windows limit (16384)
WARNING: Ignoring invalid value 'share' for parameter 'security'
```

Samba 4 较之前的 Samba 3 有一个重大的变化是 security 不再支持 share,参数需要做调整,原来的参数 security=share 需要修改为如下代码。

```
security=user
map to guest =Bad User
```

4. 重启 Samba 服务

重启 Samba 服务,代码如下。

```
[root@smb_share_152 ~]# /etc/init.d/smb restart
关闭 SMB 服务:                                          [确定]
启动 SMB 服务:                                          [确定]
[root@smb_share_152 ~]# /etc/init.d/nmb restart
关闭 NMB 服务:                                          [确定]
启动 NMB 服务:                                          [确定]
```

```
[root@smb_share_152 ~]# netstat -ntpl
Active Internet connections (only servers)
Proto Recv-Q Send-Q Local Address           Foreign Address        State
PID/Program name
tcp        0      0 0.0.0.0:139    0.0.0.0:*              LISTEN      1930/smbd
tcp        0      0 0.0.0.0:22     0.0.0.0:*              LISTEN      1211/sshd
tcp        0      0 127.0.0.1:25   0.0.0.0:*              LISTEN      1290/master
tcp        0      0 0.0.0.0:445    0.0.0.0:*              LISTEN      1930/smbd
tcp        0      0 :::139         :::*                   LISTEN      1930/smbd
tcp        0      0 :::22          :::*                   LISTEN      1211/sshd
tcp        0      0 ::1:25         :::*                   LISTEN      1290/master
tcp        0      0 :::445         :::*                   LISTEN      1930/smbd
```

5. Windows 7 客户端测试

打开运行对话框，输入\\192.168.2.152\movie，测试过程如图 8-18 和图 8-19 所示。

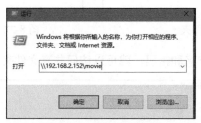

图 8-18　打开共享文件

图 8-19　新建目录

6. 将共享目录映射为 Windows 操作系统驱动器

在 Windows 10 操作系统中，使用 Windows 图标＋R 组合键打开运行对话框，如图 8-20 所示，输入\\192.168.2.151\movie。

图 8-20　输入 Samba 服务器连接地址

右击文件夹，选择"映射网络驱动器"，如图 8-21 和图 8-22 所示。

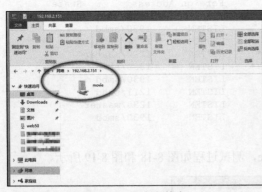

图 8-21 右击文件夹

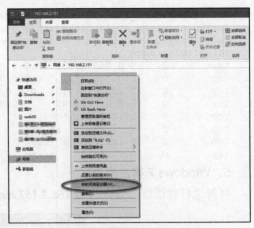

图 8-22 选择"映射网络驱动器"

按照提示输入 Samba 用户名和密码,就可以访问共享文件夹了,如图 8-23 所示。

图 8-23 访问共享文件夹

第 9 章

网站架构之 LAMP

9.1 LAMP 架构安装前基本规划

9.1.1 LAMP 基础知识

1. LAMP 简介

LAMP（Linux-Apache-MySQL-PHP）网站架构是目前国际流行的 Web 架构之一。

该架构包括 Linux 操作系统、Apache 网络服务器、MySQL 数据库、Perl、PHP 或 Python 编程语言。所有组成产品均是开源软件，是成熟的架构，很多流行的商业应用都采用该架构。LAMP 具有通用、跨平台、高性能、低价格的优势，因此无论是从性能、质量，还是从价格上考虑，LAMP 都是企业搭建网站的首选。

2. LAMP 架构概述

LAMP 是一个多 C/S 架构的平台，最初 Web 客户端基于 TCP/IP，通过 HTTP 发起传送，请求可能是动态的，也可能是静态的。

3. LAMP 架构的优缺点

（1）LAMP 架构的优点如下。
- 架构简单，适合建站初期使用。
- 易于构建，适合无基础者或初级 Linux 系统管理员使用。
- PHP 生态成熟，有完整的周边技术支持。

（2）LAMP 架构的缺点如下。
- 本身不利于架构扩展。
- 动态、静态文件都由 Apache 负责调度处理，查询率有限。
- 不适合大流量、高并发网站。

9.1.2 LAMP 架构数据流

1. LAMP 工作流程

- 用户发送 HTTP 请求到 httpd 服务器。
- httpd 解析 URL，获取需要的资源的路径，并通过内核空间读取硬盘资源。如果是静态资源，则构建响应报文，发回给用户；如果是动态资源，将资源地址发给 PHP 解析器，解析 PHP 程序文件，解析完毕后，将内容发回给 httpd，httpd 构建响应报文，发回给用户。
- 如果涉及数据库操作，则利用 PHP-MySQL 驱动，获取数据库数据，返回给 PHP 解析器。

2. "A" "M" 和 "P" 如何联动工作

Apache 与 PHP 结合的方式。

第一种，通过编译时直接把 PHP 编译成 Apache 的模块和模块化的方式进行工作（Apache 的默认方式）。

第二种，CGI。Apache 基于 CGI 与 PHP 通信。

3. PHP 与 MySQL 之间的通信

PHP 与 MySQL 怎么整合起来，PHP 又怎么被 httpd 调用呢？LAMP 架构数据流走向如图 9-1 所示。

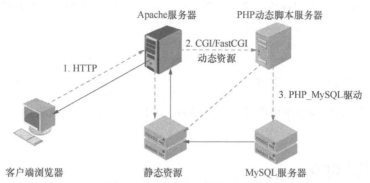

图 9-1　LAMP 架构数据流走向

httpd 并不具备解析代码的能力，而是依赖于 PHP 的解析器。PHP 本身不依赖于 MySQL，它只是一个解析器，那么它什么时候用到 MySQL 呢？是在 MySQL 中存储数据时用到 MySQL，还是当 PHP 中有运行的 MySQL 语句时才用到 MySQL 呢？

PHP 与 MySQL 没有任何关系，只有程序员在 PHP 代码中编写 MySQL 语句时才会连接 MySQL 来执行 MySQL 语句。

基于 PHP-MySQL 连接 MySQL 时只使用一个函数 MySQL_connect()，而 MySQL_connect() 正是 PHP-MySQL 提供的一个 API，只要指明要连接的服务器即可。

当客户端请求的是静态资源时，Web 服务器会直接把静态资源返回给客户端。当客户端请求的是动态资源时，httpd 的 PHP 模块会进行相应的动态资源运算。如果此过程还需要数据库的数据作为运算参数，那么 PHP 会连接 MySQL 取得数据后进行运算，运算结果转换为静态资源，并由 Web 服务器返回到客户端。

9.2　安装 LAMP

9.2.1　环境规划

使用 yum 指令安装 LAMP 的环境规划，如表 9-1 所示。

表 9-1　LAMP 的环境规划

操作系统	安装软件	默认数据目录	运行用户
CentOS 7.6 x86_64	httpd 2.4.7	/var/www/html	Apache
CentOS 7.6 x86_64	MariaDB 1:5.5.60	/data/MySQL	MySQL
CentOS 7.6 x86_64	PHP 5.4.16	/data/PHP	PHP-FPM

确认 SELinux 和防火墙处于关闭状态，代码如下。

```
[root@lamp ~]# sestatus
SELINUX status:            disabled
[root@lamp ~]# iptables -L -n
Chain INPUT (policy ACCEPT)
```

```
target       prot opt source            destination
ACCEPT       all  --  0.0.0.0/0         0.0.0.0/0       state RELATED,ESTABLISHED
ACCEPT       icmp --  0.0.0.0/0         0.0.0.0/0
ACCEPT       all  --  0.0.0.0/0         0.0.0.0/0
ACCEPT       tcp  --  0.0.0.0/0         0.0.0.0/0       state NEW tcp dpt:51518
REJECT       all  --  0.0.0.0/0         0.0.0.0/0       reject-with icmp-host-prohibited

Chain FORWARD (policy ACCEPT)
target       prot opt source            destination
REJECT       all  --  0.0.0.0/0         0.0.0.0/0       reject-with icmp-host-prohibited

Chain OUTPUT (policy ACCEPT)
target       prot opt source            destination
```

从上述代码的返回结果中可以看到，SELinux 处于关闭状态，防火墙开放了 51518 端口，其他端口暂未开放。

9.2.2 安装 httpd

1. 安装 httpd 2.4

安装 httpd 2.4，过程如下。

```
[root@lamp ~]# yum install httpd -y
Loaded plugins: fastestmirror, langpacks
Repository epel is listed more than once in the configuration
Loading mirror speeds from cached hostfile
Resolving Dependencies
（中间代码略）
Installed:
  httpd.x86_64 0:2.4.6-89.el7.centos.1

Dependency Installed:
  apr.x86_64 0:1.4.8-3.el7_4.1              apr-util.x86_64 0:1.5.2-6.el7
  httpd-tools.x86_64 0:2.4.6-89.el7.centos.1   mailcap.noarch 0:2.1.41-2.el7

Complete!
```

上述代码表示 httpd 2.4 已经安装成功。

2. 设置远程访问

安装成功后，在默认情况下，httpd 服务是禁止外部 IP 地址访问的，需要进行如下设置。

```
vim /etc/httpd/conf/httpd.conf
```

进入 httpd 2.4 主配置文件找到如下代码。

```
<Directory />
AllowOverride none
Require all denied
</Directory>
```

将上述代码修改为如下代码，默认设置如图 9-2 所示。

```
<Directory />
AllowOverride none
Require all granted
</Directory>
```

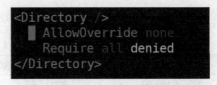

图 9-2　httpd 2.4 默认不允许外部 IP 地址访问

3. 设置防火墙

防火墙开放 80 端口，代码如下。

```
[root@lamp ~]# vim /etc/sysconfig/iptables
[root@lamp ~]# systemctl restart iptables
[root@lamp ~]# iptables -L -n
Chain INPUT (policy ACCEPT)
target       prot opt source            destination
ACCEPT       all  --  0.0.0.0/0         0.0.0.0/0       state RELATED,ESTABLISHED
```

```
ACCEPT     icmp --  0.0.0.0/0           0.0.0.0/0
ACCEPT     all  --  0.0.0.0/0           0.0.0.0/0
ACCEPT     tcp  --  0.0.0.0/0           0.0.0.0/0           state NEW tcp dpt:51518
ACCEPT     tcp  --  0.0.0.0/0           0.0.0.0/0           state NEW tcp dpt:80
REJECT     all  --  0.0.0.0/0           0.0.0.0/0           reject-with
icmp-host-prohibited

Chain FORWARD (policy ACCEPT)
target     prot opt source              destination
REJECT     all  --  0.0.0.0/0           0.0.0.0/0           reject-with
icmp-host-prohibited

Chain OUTPUT (policy ACCEPT)
target     prot opt source              destination
```

4. 浏览器访问

使用浏览器访问服务器的 IP 地址。如果显示测试页面，则表示 httpd 安装并启动成功，如图 9-3 所示。

图 9-3　测试页面

5. 管理 httpd 服务常用指令

启动 httpd 服务，代码如下。

```
[root@lamp ~]# systemctl start httpd
```

重启 httpd 服务，代码如下。

```
[root@lamp ~]# systemctl restart  httpd
```

停止 httpd 服务，代码如下。

```
[root@lamp ~]# systemctl stop httpd
```

开机启动 httpd，代码如下。

```
[root@lamp ~]# systemctl enable  httpd
```

查看 httpd 服务状态，代码如下。

```
[root@lamp ~]# systemctl status  httpd
httpd.service - The Apache HTTP Server
   Loaded: loaded (/usr/lib/systemd/system/httpd.service; disabled; vendor preset: disabled)
   Active: active (running) since Sat 2019-08-24 21:28:31 CST; 56s ago
     Docs: man:httpd(8)
           man:apachectl(8)
 Main PID: 5650 (httpd)
   Status: "Total requests: 0; Current requests/sec: 0; Current traffic:   0 B/sec"
   CGroup: /system.slice/httpd.service
           ├─5650 /usr/sbin/httpd -DFOREGROUND
           ├─5656 /usr/sbin/httpd -DFOREGROUND
           ├─5657 /usr/sbin/httpd -DFOREGROUND
           ├─5658 /usr/sbin/httpd -DFOREGROUND
           ├─5659 /usr/sbin/httpd -DFOREGROUND
           └─5660 /usr/sbin/httpd -DFOREGROUND

Aug 24 21:28:30 lamp systemd[1]: Starting The Apache HTTP Server...
Aug 24 21:28:31 lamp httpd[5650]: AH00558: httpd: Could not reliably determine the
server's fully qualified domain name, using f... message
Aug 24 21:28:31 lamp systemd[1]: Started The Apache HTTP Server.
Hint: Some lines were ellipsized, use -l to show in full.
```

上述代码表示 httpd 服务已经成功启动。

9.2.3　安装 PHP

1. 安装 PHP

安装 PHP，代码如下。

```
[root@lamp ~]# yum install php -y
Loaded plugins: fastestmirror, langpacks
Repository epel is listed more than once in the configuration
Loading mirror speeds from cached hostfile
Resolving Dependencies
（中间代码略）

Installed:
  PHP.x86_64 0:5.4.16-46.el7

Dependency Installed:
  libzip.x86_64 0:0.10.1-8.el7                        PHP-cli.x86_64 0:5.4.16-46.el7
PHP-common.x86_64 0:5.4.16-46.el7

Complete!
```

2. 编写测试文件

在 Apache 的目录下面新建文件 test.php，代码如下。

```
cd /var/www/html
vim test.php
```

可以输入相关 PHP 代码，以输入 hello world 为例，代码如下。

```
<?PHP
    echo "hello world";
?>
```

建议用户使用如下代码进行测试。

```
<?PHP
echo '获取系统类型及版本号：' . PHP_uname();
echo '<br />';
echo '获取PHP运行方式：' . PHP_sapi_name();
echo '<br />';
echo '获取当前进程用户名：' . Get_Current_User();
echo '<br />';
echo '获取PHP版本：' . PHP_VERSION;
echo '<br />';
echo '获取Zend版本：' . Zend_Version();
echo '<br />';
echo '获取PHP安装路径：' . DEFAULT_INCLUDE_PATH;
echo '<br />';
echo '获取当前文件绝对路径：' . __FILE__;
echo '<br />';
echo '接收请求的服务器IP地址：' . GetHostByName($_SERVER['SERVER_NAME']);
echo '<br />';
echo '获取服务器Web端口：' . $_SERVER['SERVER_PORT'];
echo '<br />';
echo '获取服务器IP地址：' . $_SERVER["HTTP_HOST"];
echo '<br />';
echo '获取服务器语言：' . $_SERVER['HTTP_ACCEPT_LANGUAGE'];
echo '<br / >';
echo '当前的系统时间是：' .date('Y-m-d  H:i:s',time());
echo '<br />';
?>
```

3. 访问测试页面

重启 httpd 服务，代码如下。

```
[root@lamp ~]# systemctl restart  httpd
```

使用浏览器访问网站 http://106.52.196.184/index.php。如果正常解析，则说明 PHP 测试完成，如图 9-4 和图 9-5 所示。

图 9-4　浏览器访问 PHP 代码　　　　　图 9-5　PHP 测试代码

9.2.4　安装 MariaDB

1. 安装 MariaDB

在 CentOS 7 中，MariaDB 代替了 MySQL。其实 MariaDB 只是 MySQL 的一个分支，是一个开源数据库。可使用如下代码安装 MariaDB 服务。

```
[root@lamp ~]# yum install mariadb-embedded mariadb-libs mariadb-bench mariadb mariadb-sever mariadb-server  -y
（中间代码略）

Installed:
  mariadb-bench.x86_64 1:5.5.60-1.el7_5
mariadb-embedded.x86_64 1:5.5.60-1.el7_5

Dependency Installed:
  dejavu-fonts-common.noarch 0:2.33-6.el7              dejavu-sans-fonts.noarch 0:2.33-6.el7
                  fontconfig.x86_64 0:2.13.0-4.3.el7
  fontpackages-filesystem.noarch 0:1.44-8.el7          gd.x86_64 0:2.0.35-26.el7
libX11.x86_64 0:1.6.5-2.el7
  libX11-common.noarch 0:1.6.5-2.el7                   libXau.x86_64 0:1.0.8-2.1.el7
libXpm.x86_64 0:3.5.12-1.el7
  libxcb.x86_64 0:1.13-1.el7                           perl-Compress-Raw-Bzip2.x86_64 0:2.061-3.el7             perl-Compress-Raw-Zlib.x86_64 1:2.061-4.el7
  perl-DBI.x86_64 0:1.627-4.el7                        perl-Data-Dumper.x86_64 0:2.145-3.el7                    perl-GD.x86_64 0:2.49-3.el7
  perl-IO-Compress.noarch 0:2.061-2.el7                perl-Net-Daemon.noarch 0:0.48-5.el7                      perl-PlRPC.noarch 0:0.2020-14.el7

Complete!
```

安装成功后，root 用户默认密码为空且仅限本机登录。首先启动 MariaDB 服务，代码如下。

```
[root@lamp ~]# systemctl start mariadb
[root@lamp ~]# ps -ef |grep MySQLd
MySQL     20264     1  0 20:39 ?        00:00:00 /bin/sh /usr/bin/MySQLd_safe --basedir=/usr
MySQL     20426 20264  0 20:39 ?        00:00:00 /usr/libexec/MySQLd --basedir=/usr --datadir=/var/lib/MySQL --plugin-dir=/usr/lib64/MySQL/plugin --log-error=/var/log/mariadb/mariadb.log --pid-file=/var/run/mariadb/mariadb.pid --socket=/var/lib/MySQL/MySQL.sock
root      22339 22274  0 20:57 pts/1    00:00:00 grep --color=auto MySQLd
```

2. 修改 root 用户密码

修改 MariaDB 数据库的 root 用户密码，代码如下。

```
[root@lamp ~]# MySQL -u root -h localhost
Welcome to the MariaDB monitor.  Commands end with ; or \g.
Your MariaDB connection id is 3
Server version: 5.5.60-MariaDB MariaDB Server

Copyright (c) 2000, 2018, Oracle, MariaDB Corporation Ab and others.

Type 'help;' or '\h' for help. Type '\c' to clear the current input statement.

MariaDB [(none)]> \s
```

```
--------------
MySQL  Ver 15.1 Distrib 5.5.60-MariaDB, for Linux (x86_64) using readline 5.1

Connection id:          3
Current database:
Current user:           root@localhost
SSL:                    Not in use
Current pager:          stdout
Using outfile:          ''
Using delimiter:        ;
Server:                 MariaDB
Server version:         5.5.60-MariaDB MariaDB Server
Protocol version:       10
Connection:             Localhost via UNIX socket
Server characterset:    latin1
Db     characterset:    latin1
Client characterset:    utf8
Conn.  characterset:    utf8
UNIX socket:            /var/lib/MySQL/MySQL.sock
Uptime:                 20 min 19 sec

Threads: 1  Questions: 5  Slow queries: 0  Opens: 0  Flush tables: 2  Open tables: 26
Queries per second avg: 0.004
--------------

MariaDB [(none)]> select user,host,password from MySQL.user;
+------+-----------+----------+
| user | host      | password |
+------+-----------+----------+
| root | localhost |          |
| root | lamp      |          |
| root | 127.0.0.1 |          |
| root | ::1       |          |
|      | localhost |          |
|      | lamp      |          |
+------+-----------+----------+
6 rows in set (0.00 sec)


MariaDB [(none)]> GRANT ALL PRIVILEGES ON *.* TO 'root'@'172.16.0.16'IDENTIFIED BY
'123.com' WITH GRANT OPTION;
Query OK, 0 rows affected (0.00 sec)

MariaDB [(none)]> select user,host,password from MySQL.user;
+------+-------------+-------------------------------------------+
| user | host        | password                                  |
+------+-------------+-------------------------------------------+
| root | localhost   |                                           |
| root | lamp        |                                           |
| root | 127.0.0.1   |                                           |
| root | ::1         |                                           |
|      | localhost   |                                           |
|      | lamp        |                                           |
| root | 172.16.0.16 | *AC241830FFDDC8943AB31CBD47D758E79F7953EA |
+------+-------------+-------------------------------------------+
7 rows in set (0.00 sec)
MariaDB [(none)]> select user,host,password from MySQL.user;
+------+-------------+-------------------------------------------+
| user | host        | password                                  |
+------+-------------+-------------------------------------------+
| root | localhost   |                                           |
| root | lamp        |                                           |
| root | 127.0.0.1   |                                           |
| root | ::1         |                                           |
```

```
|       | localhost     |                                                  |
|       | lamp          |                                                  |
| root  | 172.16.0.16   | *AC241830FFDDC8943AB31CBD47D758E79F7953EA        |
+-------+---------------+--------------------------------------------------+
```

上述代码修改 MariaDB 数据库 root 用户密码。

3. MariaDB 数据库账户安全初始化

MariaDB 数据库账户安全初始化，代码如下。

```
MariaDB [(none)]> select user,host,password from MySQL.user;
+------+-------------+--------------------------------------------+
| user | host        | password                                   |
+------+-------------+--------------------------------------------+
| root | localhost   |                                            |
| root | lamp        |                                            |
| root | 127.0.0.1   |                                            |
| root | ::1         |                                            |
|      | localhost   |                                            |
|      | lamp        |                                            |
| root | 172.16.0.16 | *AC241830FFDDC8943AB31CBD47D758E79F7953EA  |
+------+-------------+--------------------------------------------+
7 rows in set (0.00 sec)

MariaDB [(none)]> delete from MySQL.user where host!='localhost';
Query OK, 5 rows affected (0.00 sec)

MariaDB [(none)]> select user,host,password from MySQL.user;
+------+-----------+----------+
| user | host      | password |
+------+-----------+----------+
| root | localhost |          |
|      | localhost |          |
+------+-----------+----------+
2 rows in set (0.00 sec)

MariaDB [(none)]> delete from MySQL.user where user!='root';
Query OK, 1 row affected (0.00 sec)

MariaDB [(none)]> select user,host,password from MySQL.user;
+------+-----------+----------+
| user | host      | password |
+------+-----------+----------+
| root | localhost |          |
+------+-----------+----------+
1 row in set (0.00 sec)

MariaDB [(none)]>  GRANT ALL PRIVILEGES ON *.* TO 'root'@'172.16.0.16'IDENTIFIED BY 
'123.com' WITH GRANT OPTION;
Query OK, 0 rows affected (0.00 sec)

MariaDB [(none)]> select user,host,password from MySQL.user;
+------+-------------+--------------------------------------------+
| user | host        | password                                   |
+------+-------------+--------------------------------------------+
| root | localhost   |                                            |
| root | 172.16.0.16 | *AC241830FFDDC8943AB31CBD47D758E79F7953EA  |
+------+-------------+--------------------------------------------+
2 rows in set (0.00 sec)

MariaDB [(none)]> flush privileges;
Query OK, 0 rows affected (0.00 sec)

MariaDB [(none)]> exit
Bye
[root@lamp ~]# MySQL -h 172.16.0.16 -p123.com
```

```
Welcome to the MariaDB monitor.  Commands end with ; or \g.
Your MariaDB connection id is 6
Server version: 5.5.60-MariaDB MariaDB Server

Copyright (c) 2000, 2018, Oracle, MariaDB Corporation Ab and others.

Type 'help;' or '\h' for help. Type '\c' to clear the current input statement.

MariaDB [(none)]>
```

建议 Linux 系统管理员必须进行登录测试，以免出现授权后无法登录的情况。用户可以通过以下方式进行授权。

（1）配置任意 IP 地址远程访问，代码如下。

```
MySQLadmin -u root --password 'password'
GRANT ALL PRIVILEGES ON *.* TO 'root'@'%'IDENTIFIED BY 'password' WITH GRANT OPTION;
```

（2）刷新权限，重启服务，代码如下。

```
flush privileges; systemctl restart mariadb.service
```

这样任意 IP 地址就可以通过 MySQL -h 192.168.2.100 -u root -p 访问数据库服务器了。当然，为了安全起见，要指定特定主机访问数据库服务器。

4．安装 PHP 扩展

默认情况下，PHP 不支持 MariaDB，需要安装 PHP 扩展，可用指令 yum install php-mysql -y 进行安装，安装过程如下。

```
[root@lamp ~]# yum install php-mysql -y
Loaded plugins: fastestmirror, langpacks
Repository epel is listed more than once in the configuration
Loading mirror speeds from cached hostfile
（中间代码略）

Installed:
  PHP-MySQL.x86_64 0:5.4.16-46.el7

Dependency Installed:
  PHP-pdo.x86_64 0:5.4.16-46.el7

Complete!
```

上述代码表示已成功安装 php-mysql 软件包。

5．安装 PHP 常用模块

安装 PHP 常用模块，代码如下。

```
[root@lamp html]#  yum install -y php-gd php-ldap php-odbc php-pear php-xml
php-xmlrpcphp-mbstring php-snmp php-soap curl curl-devel php-bcmath
Loaded plugins: fastestmirror, langpacks
（中间代码略）

Dependency Installed:
  libtool-ltdl.x86_64 0:2.4.2-22.el7_3          libxslt.x86_64 0:1.1.28-5.el7
lm_sensors-libs.x86_64 0:3.4.0-6.20160601gitf9185e5.el7
  net-snmp.x86_64 1:5.7.2-38.el7_6.2            net-snmp-agent-libs.x86_64
1:5.7.2-38.el7_6.2          net-snmp-libs.x86_64 1:5.7.2-38.el7_6.2
  PHP-process.x86_64 0:5.4.16-46.el7            t1lib.x86_64 0:5.1.2-14.el7
unixODBC.x86_64 0:2.3.1-11.el7

Updated:
  cURL.x86_64 0:7.29.0-51.el7_6.3

Dependency Updated:
  libcURL.x86_64 0:7.29.0-51.el7_6.3

Complete!
```

6. 创建数据库和测试表

创建数据库和测试表,代码如下。

```
MariaDB [(none)]> create database handuoduo;
Query OK, 1 row affected (0.00 sec)

MariaDB [(none)]> use handuoduo;
Database changed
MariaDB [handuoduo]> CREATE TABLE food(
    -> id INT(10) PRIMARY KEY NOT NULL UNIQUE AUTO_INCREMENT,
    -> name VARCHAR(20) NOT NULL,
    -> company VARCHAR(30) NOT NULL,
    -> price FLOAT,
    -> produce_time YEAR,
    -> validity_time INT(4),
    -> address VARCHAR(50)
    -> );
Query OK, 0 rows affected (0.01 sec)

MariaDB [handuoduo]> show tables;
+---------------------+
| Tables_in_handuoduo |
+---------------------+
| food                |
+---------------------+
1 row in set (0.00 sec)

MariaDB [handuoduo]> INSERT INTO food VALUES
    -> (NULL,'EE 果冻','EE 果冻厂', 1.5 ,'2007', 2 ,'北京') ,
    -> (NULL,'FF 咖啡','FF 咖啡厂', 20 ,'2002', 5 ,'天津') ,
    -> (NULL,'GG 奶糖','GG 奶糖', 14 ,'2003', 3 ,'广东');
Query OK, 3 rows affected, 9 warnings (0.00 sec)
Records: 3  Duplicates: 0  Warnings: 9

MariaDB [handuoduo]> select * from food;
+----+------+---------+-------+--------------+---------------+---------+
| id | name | company | price | produce_time | validity_time | address |
+----+------+---------+-------+--------------+---------------+---------+
|  1 | EE?? | EE???   |  1.5  |     2007     |      2        |   ??    |
|  2 | FF?? | FF???   |   20  |     2002     |      5        |   ??    |
|  3 | GG?? | GG??    |   14  |     2003     |      3        |   ??    |
+----+------+---------+-------+--------------+---------------+---------+
3 rows in set (0.00 sec)
```

7. 测试 PHP 连接 MariaDB

测试整个 Web 环境是否可用,重点检测 PHP 调用数据库,在网站目录下新建测试数据库文件,代码如下。

```
[root@lamp html]# cat conn.php
<?PHP
$con = MySQLi_connect("172.16.0.16","root","123.com","handuoduo");
if(!$con){
        die("failed".MySQLi_error());
        }else{
             echo "db connect success";
                }
```

测试结果如图 9-6 所示。

8. 打开 PHP 调试

默认情况下,PHP 没有打开错误调试,需要在/etc/PHP.ini 中将错误调试打开,代码如下。

```
1 ;display_errors
2 ;error_reporting
```

上述代码第 1~2 行前面的 ";" 表示注释，注释后的代码不会生效，打开 PHP 调试，如图 9-7 所示。

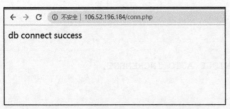

图 9-6　PHP 连接 MariaDB 测试结果

图 9-7　打开 PHP 调试

至此，简单的 LAMP 环境搭建成功！

9．MySQL 指令精讲

```
mysql -h localhost -P 3306 -u root -proot
```

（1）mysql 可以理解成一个关键字或者一个固定的指令，是固定写法，类似于 Java、JDK 中的 javac 指令或 Java 指令。

（2）-h 表示 host，即主机的 IP 地址。

（3）-P 表示端口，MySQL 数据库的默认端口号是 3306，用户可以更改端口号（注意是大写字母 P）。

（4）-u 表示用户名。

（5）-p 表示密码，注意是小写字母 p。

MySQL 连接数据库指令的基本注意事项。

- P 表示端口号，p 表示密码。
- -p 和密码之间一定不能有空格，-u、-h、-P 是可以有空格的，也可以没有空格。

> 注意：如果是本机，那么可以不写主机 IP 地址和端口号（即可以省略主机 IP 地址和端口号），直接写成 mysql -u root -proot。

语法：mysql -h 主机 IP 地址 -P 端口号 -u 用户名 -p 密码（-h 和主机 IP 地址之间有空格，-P 和端口号之间有空格，-u 和用户名之间有空格，-p 和密码之间一定不能有空格）。

```
mysql -h localhost -P 3306 -u root -proot
```

如果是连接本机 MySQL 服务，可以省略-h localhost -P 3306，直接写成 mysql -u root -proot 或者 mysql -uroot -proot 即可。

10．管理 MariaDB 服务常用指令

MariaDB 服务常用管理指令如下。

（1）启动 MariaDB 服务，代码如下。

```
systemctl start mariadb
```

（2）停止 MariaDB 服务，代码如下。

```
systemctl stop mariadb
```

（3）重启 MariaDB 服务，代码如下。

```
systemctl restart mariadb
```

（4）设置 MariaDB 服务开机启动，代码如下。

```
systemctl enable mariadb
```

9.2.5 LAMP 常用运维指令

1. LAMP 运维管理常用指令

（1）查看哪些端口被打开，代码如下。
```
netstat -ntlp
```
（2）查看监听的端口，代码如下。
```
netstat -lntp
```
（3）查看端口被哪个进程占用，代码如下。
```
netstat -lnp|grep 8080
```
（4）查看进程 PID，代码如下。
```
ps -ef | grep httpd
```
（5）查看本机 IP 地址，代码如下。
```
ifconfig -a
```
（6）重启网络服务，代码如下。
```
sudo service network restart
```
（7）查看 CentOS 版本，代码如下。
```
cat /etc/redhat-release
```

2. Apache 运维管理常用指令

Apache 运维管理常用指令如下。

（1）查看 Apache 版本，代码如下。
```
httpd -v
```
（2）启动 Apache 服务，代码如下。
```
systemctl start httpd.service
```
（3）停止 Apache 服务，代码如下。
```
systemctl stop httpd.service
```
（4）重启 Apache 服务，代码如下。
```
systemctl restart httpd.service
```

3. MySQL 运维管理常用指令

MySQL 运维管理常用指令如下。

（1）查看 MySQL 版本，代码如下。
```
rpm -qa | grep mysql*
```
（2）启动 MySQL 服务，代码如下。
```
systemctl start mariadb.service
systemctl start mysqld.service
```
（3）停止 MySQL 服务，代码如下。
```
systemctl stop mariadb.service
systemctl stop mysqld.service
```
（4）重启 MySQL 服务，代码如下。
```
systemctl restart mariadb.service
systemctl restart mysqld.service
```
（5）设置 MySQL 服务开机自动启动，代码如下。
```
systemctl enable mariadb.service
systemctl enable mysqld.service
```

4. PHP 运维管理常用指令

PHP 运维管理常用指令如下。

（1）查看 PHP 编译参数，代码如下。
```
PHP -i |less
```

(2) 查看 PHP 加载的模块扩展，代码如下。
```
PHP -m |less
```
(3) 找到实际加载的 ini 文件路径，代码如下。
```
PHP --ini
```

5. 防火墙常用指令

防火墙常用指令如下。

(1) 查看防火墙规则，代码如下。
```
iptables -L -n -v
```
(2) 查看防火墙状态，代码如下。
```
service iptables status
```
(3) 重启防火墙，代码如下。
```
service iptables restart
```
(4) 禁用防火墙，代码如下。
```
service iptables stop
```
(5) 添加端口 8080，代码如下。
```
iptables -I INPUT 4 -p tcp -m state --state NEW -m tcp --dport 8080 -j ACCEPT
```
(6) 保存防火墙规则，代码如下。
```
service iptables save
```
上述代码保存防火墙规则到/etc/sysconfig/iptables 配置文件中。

9.3 优化编译安装 LAMP 架构

9.3.1 配置 LAMP 运行环境

LAMP 安装规划如表 9-2 所示，编译 LAMP 软件包如图 9-8 所示。

表 9-2　　　　　　　　　　　LAMP 安装规划

操作系统	安装软件	安装目录	运行用户
CentOS 6.5	Apache 2.4.17	/data/Apache	Apache
CentOS 6.5	Percona-Server-5.6.28	/data/MySQL	MySQL
CentOS 6.5	PHP 5.6.14	/data/PHP	PHP-FPM

图 9-8　编译 LAMP 软件包

安装软件包之前，要先安装 epel 源。
```
yum -y install epel-release
```
采用 laohan.booxin*.vip 作为域名，网络可以直接访问，方便测试。

9.3.2 为什么要编译 LAMP

在 CentOS 7 上可以直接使用 yum 指令安装 LAMP，比手动编译 LAMP 要简单得多，但在实际的生产环境中都会手动编译 LAMP，手动编译 LAMP 有以下几个优点。

- 方便扩展模块，如添加 PHP 的扩展模块、HTTP 扩展模块等，可根据实际需求进行配置。
- 可以自由选择安装较适合的版本，系统自带的 httpd、MySQL 或者 PHP 的版本可能较低，不能满足某些应用需求。
- 可以以 FPM 的方式运行 PHP，一般系统自带的 PHP 是以 httpd 模块的形式运行的，若需要让 PHP 以 FPM 的方式运行，必须手动编译安装 httpd 和 PHP。

Apache 早期是一个提供 Web 软件的小组，现在已演变成 Apache 基金会。httpd 就是 Apache 基金会提供的一个 Web 服务器，通常 LAMP 中的 Apache 就是指 httpd。

9.4 高标准编译安装 Apache

9.4.1 彻底隐藏 Apache 版本

一般情况下，软件的漏洞信息和特定版本是相关的，因此，软件的版本号对攻击者来说是很有价值的。在默认情况下，系统会把 Apache 版本模块都显示出来（HTTP 返回头信息）。如果列举目录的话，会显示域名信息（文件列表正文），代码如下。

```
root@bashedu.com 20:48:43 opt #cURL -I 127.0.0.1
HTTP/1.1 403 Forbidden
Date: Mon, 10 Oct 2016 12:48:50 GMT
Server: Apache
Accept-Ranges: bytes
Content-Length: 4961
Connection: close
Content-Type: text/html; charset=UTF-8
```

隐藏 Apache 版本方法如下。

1. 修改配置文件

隐藏 Apache 版本的方法是修改 Apache 的配置文件，代码如下。

```
vim /etc/httpd/conf/httpd.conf
```

分别搜索关键字 ServerTokens 和 ServerSignature 并对相应内容进行修改：

ServerTokens OS 修改为 ServerTokensProductOnly。

ServerSignature On 修改为 ServerSignature Off。

CentOS 6.8 可以安装 httpd 版本进行测试，代码如下。

```
yum -y install httpd
root@bashedu.com 20:44:31 opt #grep "ServerTokens OS" /etc/httpd/conf/httpd.conf
ServerTokens OS
root@bashedu.com 20:44:32 opt #grep "ServerSignature On" /etc/httpd/conf/httpd.conf
ServerSignature On
root@bashedu.com 20:44:49 opt #sed -i 's/ServerTokens OS/ServerTokensProductOnly/' /etc/httpd/conf/httpd.conf
root@bashedu.com 20:46:37 opt #sed -i 's/ServerSignature On/ServerSignature Off/' /etc/httpd/conf/httpd.conf
root@bashedu.com 20:47:05 opt #egrep "ServerSignature|ServerTokens" /etc/httpd/conf/httpd.conf
ServerTokensProductOnly
ServerSignature Off
root@bashedu.com 20:47:36 opt #/etc/init.d/httpd restart
Stopping httpd:                                            [  OK  ]
Starting httpd: httpd: Could not reliably determine the server's fully qualified domain name, using bashedu.com for ServerName
[  OK  ]
```

```
root@bashedu.com 20:48:43 opt #curl -I 127.0.0.1
HTTP/1.1 403 Forbidden
Date: Mon, 10 Oct 2016 12:48:50 GMT
Server: Apache
Accept-Ranges: bytes
Content-Length: 4961
Connection: close
Content-Type: text/html; charset=UTF-8
```

2. 重新加载配置文件

重启或重新加载Apache，代码如下。

```
Apachectl restart
```

测试一下，代码如下。

```
root@bashedu.com 20:48:43 opt #cURL -I 127.0.0.1
HTTP/1.1 403 Forbidden
Date: Mon, 10 Oct 2016 12:48:50 GMT
Server: Apache
Accept-Ranges: bytes
Content-Length: 4961
Connection: close
Content-Type: text/html; charset=UTF-8
```

版本号与操作系统信息已经隐藏了。

3. 修改源代码

（1）手动修改Apache版本信息。

上面的方法适用于默认情况下安装的Apache。如果是编译安装的，还可以用修改源代码编译的方法：进入Apache的源代码目录下的include目录，然后编辑ap_release.h文件，会看到一些变量文件。代码如下。

```
#define AP_SERVER_BASEVENDOR "Apache Software Foundation"
#define AP_SERVER_BASEPROJECT "Apache HTTP Server"
#define AP_SERVER_BASEPRODUCT "Apache"

#define AP_SERVER_MAJORVERSION_NUMBER 2
#define AP_SERVER_MINORVERSION_NUMBER 2
#define AP_SERVER_PATCHLEVEL_NUMBER 15
#define AP_SERVER_DEVBUILD_BOOLEAN 0
```

Nginx版本描述信息在默认文件的第40～47行，如图9-9所示。

```
40 #define AP_SERVER_BASEVENDOR "NGINX Software Foundation"
41 #define AP_SERVER_BASEPROJECT "NGINX HTTP Server"
42 #define AP_SERVER_BASEPRODUCT "NGINX"
43
44 #define AP_SERVER_MAJORVERSION_NUMBER  1
45 #define AP_SERVER_MINORVERSION_NUMBER  10
46 #define AP_SERVER_PATCHLEVEL_NUMBER    1
47 #define AP_SERVER_DEVBUILD_BOOLEAN     0
```

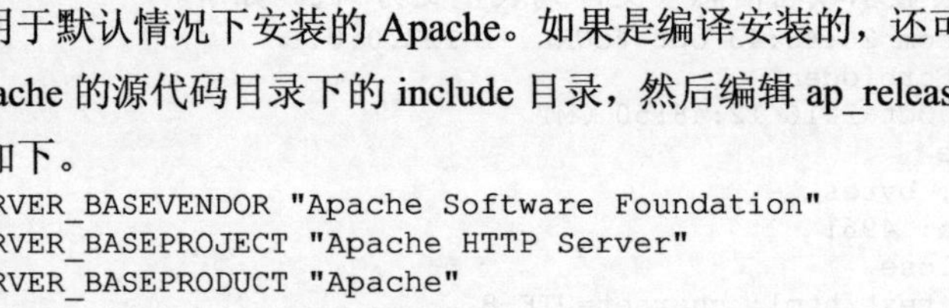
40 #define AP_SERVER_BASEVENDOR "NGINX Software Foundation"
41 #define AP_SERVER_BASEPROJECT "NGINX HTTP Server"
42 #define AP_SERVER_BASEPRODUCT "NGINX"
43
44 #define AP_SERVER_MAJORVERSION_NUMBER 1
45 #define AP_SERVER_MINORVERSION_NUMBER 10
46 #define AP_SERVER_PATCHLEVEL_NUMBER 1
47 #define AP_SERVER_DEVBUILD_BOOLEAN 0

图9-9 修改Apache默认版本

这里是结合Nginx官网的Nginx 1.10.1版本修改，系统管理员可根据公司实际业务场景修改或隐藏版本号与名字，手动修改版本，代码如下。

```
# vim include/ap_release.h
```

服务器供应商名称为 Apache 基金会
```
#define AP_SERVER_BASEVENDOR "Apache Software Foundation"
服务的项目名称
#define AP_SERVER_BASEPROJECT "Apache HTTP Server"
服务的产品名称
#define AP_SERVER_BASEPRODUCT "Apache"
#define AP_SERVER_MAJORVERSION_NUMBER 2    主版本号
#define AP_SERVER_MINORVERSION_NUMBER 4    次版本号
#define AP_SERVER_PATCHLEVEL_NUMBER    6    修正号
```

（2）使用 sed 指令自动替换版本信息。

sed 指令可实现自动替换版本信息，避免手动替换时出现未知问题，代码如下。

```
sed -i 's/Apache/BASHEDU/' /opt/httpd-2.4.17/include/ap_release.h
sed -i 's/#define AP_SERVER_MAJORVERSION_NUMBER 2/#define
AP_SERVER_MAJORVERSION_NUMBER 1/' /opt/httpd-2.4.17/include/ap_release.h
sed -i 's/#define AP_SERVER_MINORVERSION_NUMBER 4/#define
AP_SERVER_MINORVERSION_NUMBER 11/' /opt/httpd-2.4.17/include/ap_release.h
sed -i 's/#define AP_SERVER_PATCHLEVEL_NUMBER   17/#define
AP_SERVER_PATCHLEVEL_NUMBER    1/' /opt/httpd-2.4.17/include/ap_release.h
sed -i 's/#define AP_SERVER_DEVBUILD_BOOLEAN  0/#define AP_SERVER_DEVBUILD_BOOLEAN
0/' /opt/httpd-2.4.17/include/ap_release.h
```

9.4.2　安装 httpd 依赖包

Apache 2.4 和 Apache 2.2 的编译安装方式有所不同，编译 Apache 2.4 时需要先安装两个高版本的依赖包 apr 和 apr-util，而编译 Apache 2.2 则无此要求。

Apache 可移植运行时（Apache Portable Runtime，APR）是 Apache HTTP 服务器的支持库，提供了一组映射到下层操作系统的 API。如果操作系统不支持某个特定的功能，APR 将提供一个模拟的实现。

1. 创建项目目录

创建项目目录，代码如下。

```
[root@lamp ~]# mkdir -pv /data/app
mkdir: created directory '/data/app'
```

2. 安装 apr

安装 apr 依赖包，代码如下。

```
yum -y install gcc gcc-c++ pcre-devel zlib-devel openssl-devel openssl libtool
libtool-ltdl-devel  perl perl-devel
cd /opt
```

安装过程的代码如下。

```
[root@lamp ~]# yum -y install gcc gcc-c++ pcre-devel zlib-devel openssl-devel
openssl libtool libtool-ltdl-devel  perl perl-devel
Loaded plugins: fastestmirror, langpacks
Repository epel is listed more than once in the configuration
Loading mirror speeds from cached hostfile
epel
| 5.3 kB  00:00:00
extras
| 3.4 kB  00:00:00
os
| 3.6 kB  00:00:00
updates
| 3.4 kB  00:00:00

6/28
（中间代码略）

Installed:
  gcc-c++.x86_64 0:4.8.5-36.el7_6.2 libtool.x86_64 0:2.4.2-22.el7_3
```

```
  libtool-ltdl-devel.x86_64 0:2.4.2-22.el7_3  openssl-devel.x86_64 1:1.0.2k-16.el7_6.1
    pcre-devel.x86_64 0:8.32-17.el7    perl-devel.x86_64 4:5.16.3-294.el7_6
  zlib-devel.x86_64 0:1.2.7-18.el7

Dependency Installed:
  autoconf.noarch 0:2.69-11.el7              automake.noarch 0:1.13.4-3.el7
gdbm-devel.x86_64 0:1.10-8.el7
    keyutils-libs-devel.x86_64 0:1.5.8-3.el7    krb5-devel.x86_64
0:1.15.1-37.el7_6              libcom_err-devel.x86_64 0:1.42.9-13.el7
    libdb-devel.x86_64 0:5.3.21-24.el7       libkadm5.x86_64
0:1.15.1-37.el7_6              libSELINUX-devel.x86_64 0:2.5-14.1.el7
    libsepol-devel.x86_64 0:2.5-10.el7       libstdc++-devel.x86_64
0:4.8.5-36.el7_6.2              libverto-devel.x86_64 0:0.2.5-4.el7
    m4.x86_64 0:1.4.16-10.el7              perl-ExtUtils-Install.noarch
0:1.58-294.el7_6             perl-ExtUtils-MakeMaker.noarch 0:6.68-3.el7
    perl-ExtUtils-Manifest.noarch 0:1.61-244.el7    perl-ExtUtils-ParseXS.noarch
1:3.18-3.el7              perl-Test-Harness.noarch 0:3.28-3.el7
    perl-Thread-Queue.noarch 0:3.02-2.el7     pyparsing.noarch 0:1.5.6-9.el7
systemtap-sdt-devel.x86_64 0:3.3-3.el7

Complete!
[root@lamp apr-1.5.2]# rpm -q gcc gcc-c++ pcre-devel zlib-devel openssl-devel
openssl libtool libtool-ltdl-devel  perl perl-devel
gcc-4.8.5-36.el7_6.2.x86_64
gcc-c++-4.8.5-36.el7_6.2.x86_64
pcre-devel-8.32-17.el7.x86_64
zlib-devel-1.2.7-18.el7.x86_64
openssl-devel-1.0.2k-16.el7_6.1.x86_64
openssl-1.0.2k-16.el7_6.1.x86_64
libtool-2.4.2-22.el7_3.x86_64
libtool-ltdl-devel-2.4.2-22.el7_3.x86_64
perl-5.16.3-294.el7_6.x86_64
perl-devel-5.16.3-294.el7_6.x86_64
```

上述代码表示 apr 依赖包已经安装完成。

3. 编译安装 apr

（1）生成 configure 文件，代码如下。

```
[root@lamp opt]# tar xf apr-1.5.2.tar.gz
[root@lamp opt]# cd apr-1.5.2
[root@lamp apr-1.5.2]# ./configure -prefix=/data/app/apr ;echo $?
```

输出代码如下。

```
[root@lamp apr-1.5.2]# ./configure -prefix=/data/app/apr ;echo $?
Checking build system type… x86_64-unknown-linux-gnu
checking host system type… x86_64-unknown-linux-gnu
checking target system type… x86_64-unknown-linux-gnu
Configuring APR library
Platform: x86_64-unknown-linux-gnu
checking for working mkdir -p… yes
APR Version: 1.5.2
checking for chosen layout… apr
checking for gcc… gcc
（中间代码略）
config.status: executing default commands
0
```

（2）编译安装 apr，代码如下。

```
make&& make install && echo $?
cd ..
```

上述指令输出代码如下。

```
[root@lamp apr-1.5.2]# make&& make install && echo $?
（中间代码略）
Libraries have been installed in:
   /data/app/apr/lib
```

```
If you ever happen to want to link against installed libraries
in a given directory, LIBDIR, you must either use libtool, and
specify the full pathname of the library, or use the '-LLIBDIR'
flag during linking and do at least one of the following:
   - add LIBDIR to the 'LD_LIBRARY_PATH' environment variable
     during execution
   - add LIBDIR to the 'LD_RUN_PATH' environment variable
     during linking
   - use the '-Wl,-rpath -Wl,LIBDIR' linker flag
   - have your system administrator add LIBDIR to '/etc/ld.so.conf'

See any operating system documentation about shared libraries for
more information, such as the ld(1) and ld.so(8) manual pages.
/usr/bin/install -c -m 644 apr.exp /data/app/apr/lib/apr.exp
/usr/bin/install -c -m 644 apr.pc /data/app/apr/lib/pkgconfig/apr-1.pc
for f in libtool shlibtool; do \
if test -f ${f}; then /usr/bin/install -c -m 755 ${f} /data/app/apr/build-1; fi; \
done
/usr/bin/install -c -m 755 /opt/apr-1.5.2/build/mkdir.sh /data/app/apr/build-1
for f in make_exports.awk make_var_export.awk; do \
    /usr/bin/install -c -m 644 /opt/apr-1.5.2/build/${f} /data/app/apr/build-1; \
done
/usr/bin/install -c -m 644 build/apr_rules.out /data/app/apr/build-1/apr_rules.mk
/usr/bin/install -c -m 755 apr-config.out /data/app/apr/bin/apr-1-config
0
```

上述指令返回结果状态码为 0，表示 apr 已经编译安装成功。

4．安装 apr-util

安装 apr-util 依赖包，代码如下。

```
[root@lamp apr-1.5.2]# cd ..
[root@lamp opt]# tar xf apr-util-1.5.4.tar.bz2 && cd apr-util-1.5.4
```

开始编译 apr-util，代码如下。

```
./configure -prefix=/data/app/apr-util -with-apr=/data/app/apr/&& echo $?
```

输出代码如下。

```
[root@lamp apr-util-1.5.4]# ./configure -prefix=/data/app/apr-util -
with-apr=/data/app/apr/&& echo $?
Checking build system type… x86_64-unknown-linux-gnu
checking host system type… x86_64-unknown-linux-gnu
checking target system type… x86_64-unknown-linux-gnu
checking for a BSD-compatible install… /usr/bin/install -c
checking for working mkdir -p… yes
APR-util Version: 1.5.4
checking for chosen layout… apr-util
checking for gcc… gcc
checking whether the C compiler works… yes
checking for C compiler default output file name… a.out
checking for suffix of executables…
checking whether we are cross compiling… no
checking for suffix of object files… o
checking whether we are using the GNU C compiler… yes
checking whether gcc accepts -g… yes
checking for gcc option to accept ISO C89… none needed
Applying apr-util hints file rules for x86_64-unknown-linux-gnu
checking for APR… yes
  setting CPP to "gcc -E"
  adding "-pthread" to CFLAGS
  setting CPPFLAGS to " -DLINUX -D_REENTRANT -D_GNU_SOURCE"
(中间代码略)
configure: creating ./config.status
config.status: creating Makefile
config.status: creating export_vars.sh
config.status: creating build/pkg/pkginfo
```

```
config.status: creating apr-util.pc
config.status: creating apu-1-config
config.status: creating include/private/apu_select_dbm.h
config.status: creating include/apr_ldap.h
config.status: creating include/apu.h
config.status: creating include/apu_want.h
config.status: creating test/Makefile
config.status: creating include/private/apu_config.h
config.status: executing default commands
0
```

上述代码返回结果为 0，表示指令已经成功执行。

```
Make&& make install && echo $?
```

代码输出如下。

```
[root@lamp apr-util-1.5.4]# make&& make install && echo $?
(中间代码略)
Libraries have been installed in:
   /data/app/apr-util/lib

If you ever happen to want to link against installed libraries
in a given directory, LIBDIR, you must either use libtool, and
specify the full pathname of the library, or use the '-LLIBDIR'
flag during linking and do at least one of the following:
   - add LIBDIR to the 'LD_LIBRARY_PATH' environment variable
     during execution
   - add LIBDIR to the 'LD_RUN_PATH' environment variable
     during linking
   - use the '-Wl,-rpath -Wl,LIBDIR' linker flag
   - have your system administrator add LIBDIR to '/etc/ld.so.conf'

See any operating system documentation about shared libraries for
more information, such as the ld(1) and ld.so(8) manual pages.
/usr/bin/install -c -m 644 aprutil.exp /data/app/apr-util/lib
/usr/bin/install -c -m 755 apu-config.out /data/app/apr-util/bin/apu-1-config
0
```

上述代码返回结果为 0，表示 apr-util 编译安装成功。

```
[root@lamp apr-util-1.5.4]# cd ..
```

5．安装 ngHTTP2

Apache 的 mod_http2.so 模块需要 ngHTTP2 的支持，而 ngHTTP2 是 HTTP/2 的 C 运行库，所以接下来我们要安装 ngHTTP2。

```
[root@lamp ngHTTP2-1.39.2]# tar xf ngHTTP2-1.39.2.tar.bz2
[root@lamp ngHTTP2-1.39.2]# cd ngHTTP2-1.39.2/
[root@lamp ngHTTP2-1.39.2]# ./configure  --prefix=/data/app/lamp/ngHTTP2 ; echo $?
(开头及中间代码略)
   Features:
     Applications:    no
     HPACK tools:     no
     LibngHTTP2_asio:no
     Examples:        no
     Python bindings:no
     Threading:       no

0
```

安装 ngHTTP2。

```
[root@lamp ngHTTP2-1.39.2]# make && make install ; echo $?
// 开头代码略
make[2]: Nothing to be done for 'install-exec-am'.
 /usr/bin/mkdir -p '/data/app/lamp/ngHTTP2/share/doc/ngHTTP2'
 /usr/bin/install -c -m 644 README.rst '/data/app/lamp/ngHTTP2/share/doc/ngHTTP2'
make[2]: Leaving directory '/opt/ngHTTP2-1.39.2'
make[1]: Leaving directory '/opt/ngHTTP2-1.39.2'
0
```

上述安装指令返回值为 0，表示安装成功。

```
[root@lamp ngHTTP2-1.39.2]# ll /data/app/lamp/nghttp2/
total 16
drwxr-xr-x 2 root root 4096 Aug 28 09:03 bin
drwxr-xr-x 3 root root 4096 Aug 28 09:03 include
drwxr-xr-x 3 root root 4096 Aug 28 09:03 lib
drwxr-xr-x 5 root root 4096 Aug 28 09:03 share
```

上述代码查看 httpd 编译目录。

9.4.3　Apache 2.4 编译参数详解

1. 创建项目目录

使用 mdkir 指令创建项目目录，代码如下。

```
[root@lamp opt]# mkdir -pv /data/app/lamp/
mkdir: created directory '/data/app/lamp/
```

2. 解压 Apache 2.4

使用 tar 指令解压 Apache 压缩包，代码如下。

```
tar xf httpd-2.4.17.tar.bz2 &&  cd httpd-2.4.17
[root@lamp httpd-2.4.17]# ./configure --prefix=/data/app/lamp/Apache
--with-apr=/data/app/apr --with-apr-util=/data/app/apr-util --with-pcre
--with-zlib --enable-deflate --enable-expires --enable-headers --enable-HTTP2
--with-ngHTTP2=/data/app/lamp/ngHTTP2/ --enable-so --enable-deflate=shared
--enable-expires=shared --enable-ssl=shared --enable-rewrite --enable-headers=shared
--enable-rewrite=shared --enable-static-support --enable-cgi --with-mpm=event
--enable-modules-shared=most --enable-mpms-share=all;echo $?
```

输出代码如下。

```
[root@lamp httpd-2.4.17]# ./configure --prefix=/data/app/lamp/Apache
--with-apr=/data/app/apr --with-apr-util=/data/app/apr-util --with-pcre
--with-zlib --enable-deflate --enable-expires --enable-headers --enable-HTTP2
--with-ngHTTP2=/data/app/lamp/ngHTTP2/ --enable-so --enable-deflate=shared
--enable-expires=shared --enable-ssl=shared --enable-rewrite --enable-headers=
shared --enable-rewrite=shared --enable-static-support --enable-cgi
--with-mpm=event --enable-modules-shared=most --enable-mpms-share=all;echo $?
checking for chosen layout... Apache
checking for working mkdir -p... yes
checking for grep that handles long lines and -e... /usr/bin/grep
checking for egrep... /usr/bin/grep -E
checking build system type... x86_64-unknown-linux-gnu
checking host system type... x86_64-unknown-linux-gnu
checking target system type... x86_64-unknown-linux-gnu
configure:
configure: Configuring Apache Portable Runtime library...
configure:
checking for APR... yes
  setting CC to "gcc"
  setting CPP to "gcc -E"
  setting CFLAGS to " -g -O2 -pthread"
  setting CPPFLAGS to " -DLINUX -D_REENTRANT -D_GNU_SOURCE"
  setting LDFLAGS to " "
configure:
configure: Configuring Apache Portable Runtime Utility library...
configure:
checking for APR-util... yes
checking for gcc... gcc
checking whether the C compiler works... yes
checking for C compiler default output file name... a.out
checking for suffix of executables...
checking whether we are cross compiling... no
checking for suffix of object files... o
checking whether we are using the GNU C compiler... yes
```

```
checking whether gcc accepts -g... yes
checking for gcc option to accept ISO C89... none needed
// 中间代码略
creating support/Makefile
creating test/Makefile
config.status: creating docs/conf/httpd.conf
config.status: creating docs/conf/extra/httpd-autoindex.conf
config.status: creating docs/conf/extra/httpd-dav.conf
config.status: creating docs/conf/extra/httpd-default.conf
config.status: creating docs/conf/extra/httpd-info.conf
config.status: creating docs/conf/extra/httpd-languages.conf
config.status: creating docs/conf/extra/httpd-manual.conf
config.status: creating docs/conf/extra/httpd-mpm.conf
config.status: creating docs/conf/extra/httpd-multilang-errordoc.conf
config.status: creating docs/conf/extra/httpd-ssl.conf
config.status: creating docs/conf/extra/httpd-userdir.conf
config.status: creating docs/conf/extra/httpd-vhosts.conf
config.status: creating docs/conf/extra/proxy-html.conf
config.status: creating include/ap_config_layout.h
config.status: creating support/apxs
config.status: creating support/Apachectl
config.status: creating support/dbmmanage
config.status: creating support/envvars-std
config.status: creating support/log_server_status
config.status: creating support/logresolve.pl
config.status: creating support/phf_abuse_log.cgi
config.status: creating support/split-logfile
config.status: creating build/rules.mk
config.status: creating build/pkg/pkginfo
config.status: creating build/config_vars.sh
config.status: creating include/ap_config_auto.h
config.status: executing default commands
0
```

上述代码表示已经成功生成 makefile 文件，Apache 常用编译选项如表 9-3 所示。

表 9-3　　　　　　　　　　　　Apache 常用编译选项

选项	基本含义
--prefix	指定 Apache 安装目录
--sysconfdir	指定 Apache 配置文件目录
--enable-so	允许运行时加载 DSO 模块
--with-zlib	启用 zlib 库文件
--enable-deflate	压缩传输编码支持
--enable-expires	Expires 头控制
--enable-headers	HTTP 头控制
--enable-HTTP2	支持 HTTP/2
--enable-ssl	启动 SSL 加密功能，SSL/TLS 支持（mod_ssl）
--enable-rewrite	基于规则的 URL 操作，启用 URL 重写功能
--with-pcre	指定 pcre 的安装路径
--with-apr	指定 apr 路径
--with-apr-util	指定 apr-util 路径
--enable-mpms-shared	支持动态安装、卸载所有 mpm
--with-mpm=event	Apache 进程模型，mpm 默认使用 event
--enable-modules	all 为安装所有模块，most 为安装常用模块（默认设置）

使用 ./configure --help | grep disable 指令可以查看哪些模块是默认会加载的，默认加载的可

以不用在 ./configure 中指定，代码如下。

```
[root@lamp ~]# cd /opt/httpd-2.4.17/
[root@lamp httpd-2.4.17]# ./configure --help | grep disable
  --cache-file=FILE       cache test results in FILE [disabled]
  --disable-option-checking  ignore unrecognized --enable/--with options
  --disable-FEATURE       do not include FEATURE (same as --enable-FEATURE=no)
  --disable-authn-file    file-based authentication control
  --disable-authn-core    core authentication module
  --disable-authz-host    host-based authorization control
  --disable-authz-groupfile
  --disable-authz-user    'require user' authorization control
  --disable-authz-core    core authorization provider vector module
  --disable-access-compat mod_access compatibility
  --disable-auth-basic    basic authentication
  --disable-reqtimeout    Limit time waiting for request from client
  --disable-filter        Smart Filtering
  --disable-charset-lite  character set translation. Enabled by default only
                          server. It is only useful to disable it if you want
                          disable this module unless you are really sure what
  --disable-mime          mapping of file-extension to MIME. Disabling this
  --disable-log-config    logging configuration. You won't be able to log
  --disable-env           clearing/setting of ENV vars
  --disable-headers       HTTP header control
  --disable-setenvif      basing ENV vars on headers
  --disable-version       determining httpd version in config files
  --disable-status        process/thread monitoring
  --disable-autoindex     directory listing
  --disable-dir           directory request handling
  --disable-alias         mapping of requests to different filesystem parts
```

用户可以使用如下指令查看编译选项，代码如下。

```
[root@lamp httpd-2.4.17]# ./configure  --help
'configure' configures this package to adapt to many kinds of systems.

Usage: ./configure [OPTION]... [VAR=VALUE]...

To assign environment variables (e.g., CC, CFLAGS...), specify them as
VAR=VALUE.  See below for descriptions of some of the useful variables.

Defaults for the options are specified in brackets.

Configuration:
  -h, --help              display this help and exit
      --help=short        display options specific to this package
      --help=recursive    display the short help of all the included packages
  -v, --version           display version information and exit
  -q, --quiet, --silent   do not print 'checking ...' messages
      --cache-file=FILE   cache test results in FILE [disabled]
  -c, --config-cache      alias for '--cache-file=config.cache'
  -n, --no-create         do not create output files
      --srcdir=DIR        find the sources in DIR [configure dir or '..']

Installation directories:
  --prefix=PREFIX         install architecture-independent files in PREFIX
                          [/usr/local/apache2]
  --exec-prefix=EPREFIX   install architecture-dependent files in EPREFIX
                          [PREFIX]

By default, 'make install' will install all the files in
'/usr/local/apache2/bin', '/usr/local/apache2/lib' etc.  You can specify
an installation prefix other than '/usr/local/apache2' using '--prefix',
for instance '--prefix=$HOME'.

For better control, use the options below.
```

```
Fine tuning of the installation directories:
  --bindir=DIR            user executables [EPREFIX/bin]
  --sbindir=DIR           system admin executables [EPREFIX/sbin]
  --libexecdir=DIR        program executables [EPREFIX/libexec]
  --sysconfdir=DIR        read-only single-machine data [PREFIX/etc]
  --sharedstatedir=DIR    modifiable architecture-independent data [PREFIX/com]
  --localstatedir=DIR     modifiable single-machine data [PREFIX/var]
  --libdir=DIR            object code libraries [EPREFIX/lib]
  --includedir=DIR        C header files [PREFIX/include]
  --oldincludedir=DIR     C header files for non-gcc [/usr/include]
  --datarootdir=DIR       read-only arch.-independent data root [PREFIX/share]
  --datadir=DIR           read-only architecture-independent data [DATAROOTDIR]
  --infodir=DIR           info documentation [DATAROOTDIR/info]
  --localedir=DIR         locale-dependent data [DATAROOTDIR/locale]
  --mandir=DIR            man documentation [DATAROOTDIR/man]
  --docdir=DIR            documentation root [DATAROOTDIR/doc/PACKAGE]
  --htmldir=DIR           html documentation [DOCDIR]
  --dvidir=DIR            dvi documentation [DOCDIR]
  --pdfdir=DIR            pdf documentation [DOCDIR]
  --psdir=DIR             ps documentation [DOCDIR]
（中间代码略）
  --with-program-name     alternate executable name
  --with-suexec-bin       Path to suexec binary
  --with-suexec-caller    User allowed to call SuExec
  --with-suexec-userdir   User subdirectory
  --with-suexec-docroot   SuExec root directory
  --with-suexec-uidmin    Minimal allowed UID
  --with-suexec-gidmin    Minimal allowed GID
  --with-suexec-logfile   Set the logfile
  --with-suexec-safepath  Set the safepath
  --with-suexec-umask     umask for suexec'd process

Some influential environment variables:
  CC          C compiler command
  CFLAGS      C compiler flags
  LDFLAGS     linker flags, e.g. -L<lib dir> if you have libraries in a
              nonstandard directory <lib dir>
  LIBS        libraries to pass to the linker, e.g. -l<library>
  CPPFLAGS    (Objective) C/C++ preprocessor flags, e.g. -I<include dir> if
              you have headers in a nonstandard directory <include dir>
  CPP         C preprocessor

Use these variables to override the choices made by 'configure' or to help
it to find libraries and programs with nonstandard names/locations.

Report bugs to the package provider.
```

实际生产环境中，建议读者根据场景需要增加编译选项。

9.4.4 编译安装 Apache 2.4

1. 编译安装 Apache 2.4

编译安装 Apache 2.4，代码如下：

```
[root@lamp httpd-2.4.17]# make && make install ; echo $?
（中间代码略）
make[4]: Leaving directory '/opt/httpd-2.4.17/modules/mappers'
make[3]: Leaving directory '/opt/httpd-2.4.17/modules/mappers'
make[2]: Leaving directory '/opt/httpd-2.4.17/modules'
make[2]: Entering directory '/opt/httpd-2.4.17/support'
make[2]: Leaving directory '/opt/httpd-2.4.17/support'

Installing configuration files
mkdir /data/app/lamp/Apache/conf
mkdir /data/app/lamp/Apache/conf/extra
```

```
mkdir /data/app/lamp/Apache/conf/original
mkdir /data/app/lamp/Apache/conf/original/extra
Installing HTML documents
mkdir /data/app/lamp/Apache/htdocs
Installing error documents
mkdir /data/app/lamp/Apache/error
Installing icons
mkdir /data/app/lamp/Apache/icons
mkdir /data/app/lamp/Apache/logs
Installing CGIs
mkdir /data/app/lamp/Apache/cgi-bin
Installing header files
mkdir /data/app/lamp/Apache/include
Installing build system files
mkdir /data/app/lamp/Apache/build
Installing man pages and online manual
mkdir /data/app/lamp/Apache/man
mkdir /data/app/lamp/Apache/man/man1
mkdir /data/app/lamp/Apache/man/man8
mkdir /data/app/lamp/Apache/manual
make[1]: Leaving directory '/opt/httpd-2.4.17'
0
```

上述结果表示 Apache 2.4 已经安装成功。

2. 可能遇到的问题

预编译生成 configure 文件后，可以使用 make 指令进行编译，代码如下。

```
make && make install && echo $?
cd ..
```

Apache 编译结果如图 9-10 所示。

图 9-10　编译结果

Apache 编译过程中安装出现错误，如图 9-11 和图 9-12 所示。

图 9-11　编译过程错误 1

图 9-12　编译过程错误 2

错误代码如下。

```
***
[exports.lo] 错误 1
[all-recursive] 错误 1
[all-recursive] 错误 1
***
```

解决办法如下。

```
cd /opt
cp -rv /opt/apr-1.5.2 /opt/httpd-2.4.17/srclib/apr
cp -rv /opt/apr-util-1.5.4 /opt/httpd-2.4.17/srclib/apr-util
```

3. 编译后目录结构

```
[root@lamp httpd-2.4.17]# ll /data/app/lamp/Apache/
total 56
drwxr-xr-x  2 root root  4096 Aug 26 22:01 bin
drwxr-xr-x  2 root root  4096 Aug 26 22:01 build
drwxr-xr-x  2 root root  4096 Aug 26 22:01 cgi-bin
drwxr-xr-x  4 root root  4096 Aug 26 22:01 conf
drwxr-xr-x  3 root root  4096 Aug 26 22:01 error
drwxr-xr-x  2 root root  4096 Aug 26 21:55 htdocs
drwxr-xr-x  3 root root  4096 Aug 26 22:01 icons
drwxr-xr-x  2 root root  4096 Aug 26 22:01 include
drwxr-xr-x  2 root root  4096 Aug 26 22:01 logs
drwxr-xr-x  4 root root  4096 Aug 26 22:01 man
drwxr-xr-x 14 root root 12288 Oct 10  2015 manual
drwxr-xr-x  2 root root  4096 Aug 26 22:01 modules
```

建议 Linux 系统管理员熟练掌握 Apache 目录结构，方便日后系统维护。查看 Apache 安装信息，代码如下。

```
[root@lamp httpd-2.4.17]# du -sh /data/app/lamp/Apache/
35M     /data/app/lamp/Apache/
```

查看编译后 Apache 目录结构，代码如下。

```
[root@lamp httpd-2.4.17]# tree -L 1 /data/app/lamp/Apache/
/data/app/lamp/Apache/
|-- bin
|-- build
|-- cgi-bin
|-- conf
|-- error
|-- htdocs
|-- icons
|-- include
|-- logs
|-- man
|-- manual
'-- modules

12 directories, 0 files
```

使用 tree 指令查看 Apache 目录结构更直观。

4. 启动 Apache 服务

启动 Apache 服务，代码如下。

```
[root@lamp httpd-2.4.17]# /data/app/lamp/apache/bin/apachectl -t
AH00558: httpd: Could not reliably determine the server's fully qualified domain
name, using fe80::5054:ff:fe65:96bd. Set the 'ServerName' directive globally
to suppress this message
Syntax OK
[root@lamp httpd-2.4.17]# /data/app/lamp/apache/bin/apachectl -k start
AH00558: httpd: Could not reliably determine the server's fully qualified domain
name, using fe80::5054:ff:fe65:96bd. Set the 'ServerName' directive globally
to suppress this message
[root@lamp httpd-2.4.17]# ps -ef |grep httpd
```

```
root         1586     1  0 09:50 ?        00:00:00 /data/app/lamp/apache/bin/httpd -k start
daemon       1587  1586  0 09:50 ?        00:00:00 /data/app/lamp/apache/bin/httpd -k start
daemon       1588  1586  0 09:50 ?        00:00:00 /data/app/lamp/apache/bin/httpd -k start
daemon       1589  1586  0 09:50 ?        00:00:00 /data/app/lamp/apache/bin/httpd -k start
root         1707  1076  0 09:51 pts/0    00:00:00 grep --color=auto httpd
[root@lamp httpd-2.4.17]# /data/app/lamp/apache/bin/apachectl  -k stop
AH00558: httpd: Could not reliably determine the server's fully qualified domain
name, using fe80::5054:ff:fe65:96bd. Set the 'ServerName' directive globally to
suppress this message
[root@lamp httpd-2.4.17]# ps -ef |grep httpd
root         1725  1076  0 09:51 pts/0    00:00:00 grep --color=auto httpd
```

9.5 高标准安装 MySQL Percona

9.5.1 为什么要使用 Percona 版本

1. Percona 版本的优点

下面是 Percona 版本的优点。

- 可扩展性：处理更多事务；在强大的服务器上进行扩展。
- 性能：使用了 XtraDB 的 Percona 版本速度非常快。
- 可靠性：避免损坏，提供崩溃安全（Crash-safe）复制。
- 管理：在线备份，在线表格导入/导出。
- 诊断：高级分析和检测。
- 灵活性：可变的页面大小，改进的缓冲池管理。

2. 如何下载 Percona

建议读者到 Percona 官网下载对应的安装版本，如图 9-13 所示。

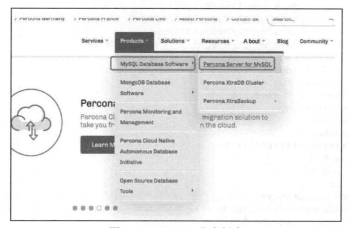

图 9-13　Percona 分支版本

读者可以根据需要下载对应的稳定版本，目前生产环境中使用的比较稳定的版本是 Percona 5.6 和 Percona 5.7，如图 9-14 所示。

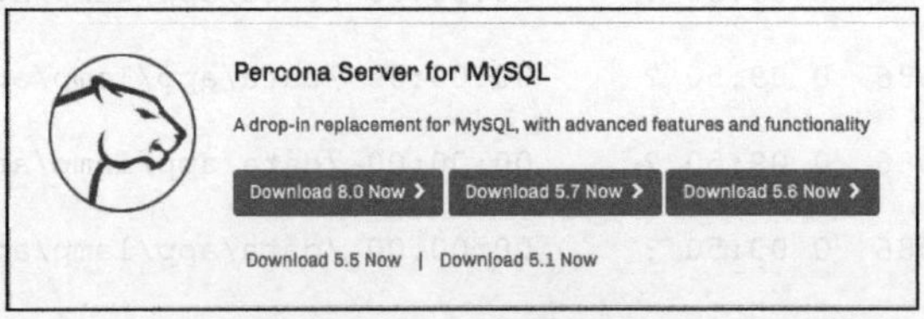

图 9-14 Percona 稳定版本

9.5.2 优化 Percona 5.6.28 运行环境

首先，下载 percona-server 版本的 MySQL，理由是它支持线程池，这对高并发的环境是非常重要的。

然后，为 Percona 二进制安装包建立软链接，代码如下。

```
tar xf Percona-Server-5.6.28-rel76.1-Linux.x86_64.ssl101.tar.gz
ln -sv /opt/Percona-Server-5.6.28-rel76.1-Linux.x86_64.ssl101 /data/app/lamp/MySQL
```

接下来，创建 MySQL 项目运行环境，代码如下。

```
groupadd MySQL&&useradd MySQL -g MySQL -s /sbin/nologin -M
mkdir -pv /data/MySQL/MySQL3316/{data,logs,tmp,conf,run}
mkdir -pv /data/sh
chown -R MySQL. /data/app/lamp/MySQL/*
chown -R MySQL. /data/MySQL/
```

接下来，推送 MySQL 5.6 标准模板配置文件到/data/MySQL/MySQL3316/目录下，代码如下。

```
cat>/data/MySQL/MySQL3316/conf/my3316.cnf<<E
[client]
port           = 3316
socket         = /data/MySQL/MySQL3316/run/MySQL3316.sock
prompt                        = '(mystql-3316)\u@\h [\d]> '

# The MySQL server
[MySQLd]
# Basic
port           = 3316
user           = MySQL
basedir        = /data/app/lamp/MySQL/
datadir        = /data/MySQL/MySQL3316/data
tmpdir         = /data/MySQL/MySQL3316/tmp
socket         = /data/MySQL/MySQL3316/run/MySQL3316.sock

log-bin        = /data/MySQL/MySQL3316/logs/MySQL-bin
log-error      = error.log
slow-query-log-file = slow.log
skip-external-locking
skip-name-resolve
log-slave-updates
###############################
# FOR Percona 5.6
#extra_port = 3345
gtid-mode = on
enforce-gtid-consistency=1
#thread_handling=pool-of-threads
#thread_pool_oversubscribe=8
explicit_defaults_for_timestamp
###############################
server-id       =763316
character-set-server = utf8
slow-query-log
binlog_format = row
```

```
max_binlog_size = 128M
binlog_cache_size = 1M
expire-logs-days = 5
back_log = 500
long_query_time=1
max_connections=1100
max_user_connections=1000
max_connect_errors=1000

wait_timeout=100
interactive_timeout=100

connect_timeout = 20
slave-net-timeout=30

max-relay-log-size = 256M
relay-log = relay-bin
transaction_isolation = READ-COMMITTED

performance_schema=0
#myisam_recover
key_buffer_size = 64M
max_allowed_packet = 16M
#table_cache = 3096
table_open_cache = 6144
table_definition_cache = 4096

sort_buffer_size = 128K
read_buffer_size = 1M
read_rnd_buffer_size = 1M
join_buffer_size = 128K

myisam_sort_buffer_size = 32M
tmp_table_size = 32M
max_heap_table_size = 64M
query_cache_type=0
query_cache_size = 0
bulk_insert_buffer_size = 32M

thread_cache_size = 64
#thread_concurrency = 32
thread_stack = 192K
skip-slave-start

# InnoDB
innodb_data_home_dir = /data/MySQL/MySQL3316/data
innodb_log_group_home_dir = /data/MySQL/MySQL3316/logs
innodb_data_file_path = ibdata1:100M:autoextend

innodb_buffer_pool_size = 100M

#innodb_buffer_pool_instances      = 8
#innodb_additional_mem_pool_size = 16M

innodb_log_file_size = 100M
innodb_log_buffer_size = 16M
innodb_log_files_in_group = 3
innodb_flush_log_at_trx_commit = 0
innodb_lock_wait_timeout = 10
innodb_sync_spin_loops = 40
innodb_max_dirty_pages_pct = 90
innodb_support_xa = 0
innodb_thread_concurrency = 0
innodb_thread_sleep_delay = 500
innodb_file_io_threads       = 4
innodb_concurrency_tickets = 1000
```

```
log_bin_trust_function_creators = 1
#innodb_flush_method = O_DIRECT
innodb_file_per_table
innodb_read_io_threads = 16
innodb_write_io_threads = 16
innodb_io_capacity = 2000
innodb_file_format = Barracuda
innodb_purge_threads=1
innodb_purge_batch_size = 32
innodb_old_blocks_pct=75
innodb_change_buffering=all
innodb_stats_on_metadata=OFF

[MySQLdump]
quick
max_allowed_packet = 128M
#myisam_max_sort_file_size = 10G

[MySQL]
no-auto-rehash
max_allowed_packet = 128M
prompt                       = '(product)\u@\h [\d]> '
default_character_set        = utf8

[myisamchk]
key_buffer_size = 64M
sort_buffer_size = 512k
read_buffer = 2M
write_buffer = 2M

[MySQLhotcopy]
interactive-timeout

[MySQLd_safe]
#malloc-lib= /usr/local/MySQL/lib/MySQL/libjemalloc.so
```

上述代码中的核心参数，如内存缓存等，建议读者根据实际硬件参数进行设定。

最后，安装依赖包。

```
yum -y install numactl-devel perl perl-devel
```

9.5.3 初始化 MySQL

初始化 MySQL 时主要是生成 MySQL 的字典文件和自身的管理文件等，代码如下。

```
[root@lamp opt]# chown -R MySQL. /data/app/lamp/MySQL/*
[root@lamp opt]# chown -R MySQL. /data/MySQL/
cd /data/app/lamp/MySQL/
./scripts/MySQL_install_db
--defaults-file=/data/MySQL/MySQL3316/conf/my3316.cnf;echo $?
```

完成后，初始化数据库报错如下。

```
./scripts/MySQL_install_db  --user=MySQL
-bash: ./scripts/MySQL_install_db: /usr/bin/perl: bad interpreter: No such file or directory
```

提示注释器错误，没有 /usr/bin/perl 文件或者档案，解决办法是安装 perl 与 perl-devel。

执行安装程序，代码如下。

```
yum -y install perl perl-devel
```

初始化数据库，初始化成功的标志就是在屏幕的终端会先输出两个 OK，MySQL 初始化成功和启动成功如图 9-15～图 9-17 所示。

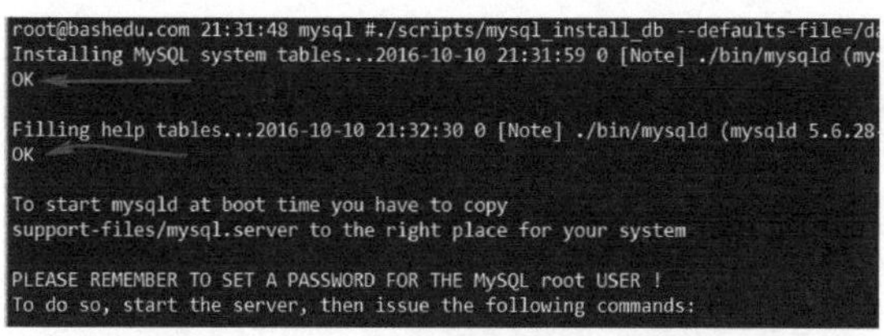

图 9-15　MySQL 初始化成功提示信息 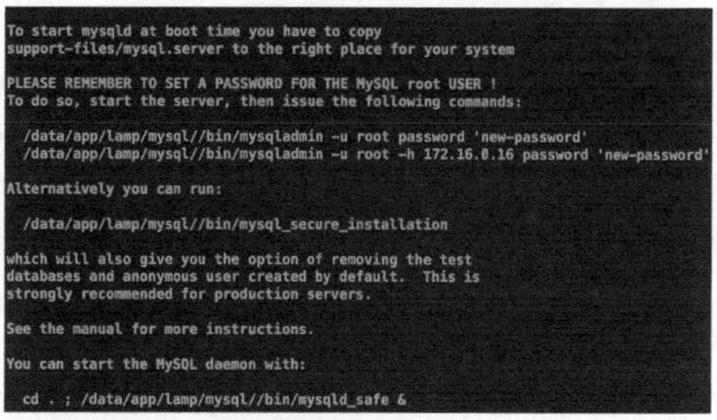　　　　图 9-16　MySQL 启动成功提示信息 1

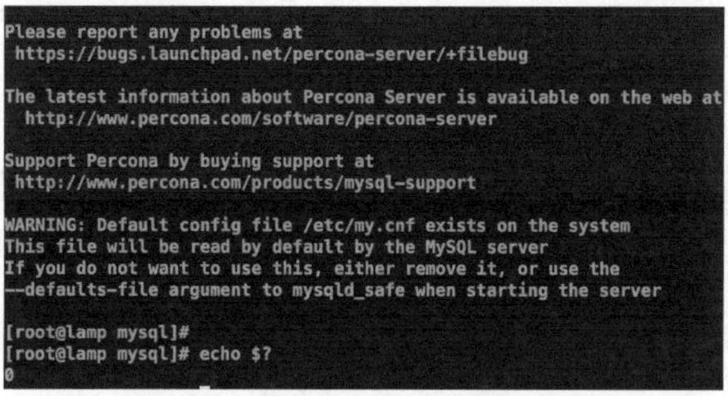

图 9-17　MySQL 启动成功提示信息 2

9.5.4　导出 MySQL 头文件和库文件

导出 MySQL 头文件和库文件，主要是为了方便后面的 PHP 程序使用（如果仅是单纯的 MySQL 安装部署，则不需要此步骤），代码如下。

```
echo " /data/app/lamp/MySQL/lib/" >/etc/ld.so.conf.d/MySQL.conf
ldconfig
ldconfig -p |grep MySQL
cd  /data/app/lamp/MySQL/lib/
ln -sv libperconaserverclient.so.18.1.0 libMySQLclient_r.so
```

9.5.5　安装 MySQL 总结

MySQL 安装目录结构，代码如下。

```
[root@lamp lib]# ll /data/app/lamp/MySQL/
total 140
drwxr-xr-x   2 MySQL MySQL  4096 Jan   9  2016 bin
-rw-r--r--   1 MySQL MySQL 17987 Jan   9  2016 COPYING
-rw-r--r--   1 MySQL MySQL 34520 Jan   9  2016 COPYING.AGPLv3
-rw-r--r--   1 MySQL MySQL 17987 Jan   9  2016 COPYING.GPLv2
drwxr-xr-x   3 MySQL MySQL  4096 Jan   9  2016 data
drwxr-xr-x   2 MySQL MySQL  4096 Jan   9  2016 docs
drwxr-xr-x   3 MySQL MySQL  4096 Jan   9  2016 include
-rw-r--r--   1 MySQL MySQL   301 Jan   9  2016 INSTALL-BINARY
drwxr-xr-x   3 MySQL MySQL  4096 Aug  28 10:32 lib
drwxr-xr-x   4 MySQL MySQL  4096 Jan   9  2016 man
drwxr-xr-x  10 MySQL MySQL  4096 Jan   9  2016 MySQL-test
-rw-r--r--   1 MySQL MySQL  2211 Jan   9  2016 PATENTS
-rw-r--r--   1 MySQL MySQL  4442 Jan   9  2016 README.md
-rw-r--r--   1 MySQL MySQL  2496 Jan   9  2016 README.MySQL
drwxr-xr-x   2 MySQL MySQL  4096 Jan   9  2016 scripts
drwxr-xr-x  28 MySQL MySQL  4096 Jan   9  2016 share
drwxr-xr-x   4 MySQL MySQL  4096 Jan   9  2016 sql-bench
drwxr-xr-x   2 MySQL MySQL  4096 Jan   9  2016 support-files
```

MySQL 项目目录相关信息，代码如下。

```
[root@lamp lib]# ll /data/MySQL/MySQL3316/
total 20
drwxr-xr-x 2 MySQL MySQL 4096 Aug 28 10:28 conf
drwxr-xr-x 5 MySQL MySQL 4096 Aug 28 10:30 data
drwxr-xr-x 2 MySQL MySQL 4096 Aug 28 10:30 logs
drwxr-xr-x 2 MySQL MySQL 4096 Aug 28 10:28 run
drwxr-xr-x 2 MySQL MySQL 4096 Aug 28 10:30 tmp
```

启动 MySQL 服务，代码如下。

```
[root@lamp lib]# /data/app/lamp/MySQL/bin/MySQLd
--defaults-file=/data/MySQL/MySQL3316/conf/my3316.cnf &
[1] 7922
[root@lamp lib]# 2019-08-28 10:47:00 0 [Note] /data/app/lamp/MySQL/bin/MySQLd
(MySQLd 5.6.28-76.1-log) starting as process 7922 ...
[root@lamp lib]#
[root@lamp lib]#
[root@lamp lib]# ps -ef |grep MySQLd
MySQL     7922  5107  0 10:47 pts/0    00:00:00 /data/app/lamp/MySQL/bin/MySQLd
--defaults-file=/data/MySQL/MySQL3316/conf/my3316.cnf
root      8000  5107  0 10:47 pts/0    00:00:00 grep --color=auto MySQLd
[root@lamp lib]# netstat -ntpl
Active Internet connections (only servers)
Proto Recv-Q Send-Q Local Address          Foreign Address         State       PID/Program name
tcp        0      0 0.0.0.0:51518          0.0.0.0:*               LISTEN      13743/sshd
tcp6       0      0 :::3316                :::*                    LISTEN      7922/MySQLd
```

1. 初始化 MySQL 3316 实例

初始化 MySQL 3316 实例，代码如下。

```
cd /opt/MySQL-5.5.53-linux2.6-x86_64 && ./scripts/MySQL_install_db
--defaults-file=/data/MySQL/MySQL3316/conf/my3316.cnf

[root@lamp lib]# mkdir -pv /data/scripts/shell/
mkdir: created directory '/data/scripts'
mkdir: created directory '/data/scripts/shell/'
```

2. 启动、连接及关闭 MySQL 服务脚本测试

启动 MySQL 服务脚本，代码如下。

```
cat >/data/scripts/shell/MySQL3316-start.sh<<EOF
#start MySQL server
/data/app/lamp/MySQL/bin/MySQLd
--defaults-file=/data/MySQL/MySQL3316/conf/my3316.cnf &
EOF
```

关闭 MySQL 服务脚本，代码如下。

```
cat >/data/scripts/shell/MySQL3316-stop.sh<<EOF
#stop MySQL server
/data/app/lamp/MySQL/bin/MySQLadmin -S /data/MySQL/MySQL3316/run/MySQL3316.sock
shutdown &
EOF
```

连接 MySQL 服务脚本，代码如下。

```
cat >/data/scripts/shell/MySQL3316-conn.sh<<EOF
#link MySQL server
/data/app/lamp/MySQL/bin/MySQL -S /data/MySQL/MySQL3316/run/MySQL3316.sock
--prompt="MySQL3316>"
EOF
```

MySQL 启动、关闭及连接服务脚本，代码如下。

```
[root@lamp lib]# ll /data/scripts/shell/
total 12
-rw-r--r-- 1 root root 117 Aug 28 10:56 MySQL3316-conn.sh
-rw-r--r-- 1 root root 109 Aug 28 10:55 MySQL3316-start.sh
```

```
-rw-r--r-- 1 root root 281 Aug 28 10:55 MySQL3316-stop.sh
```

3. MySQL 5.5.48 安装部署小结

二进制安装是大批量部署的前提，所以最好使用基于二进制发行版的安装。

4. 大规模批量部署 MySQL 实例思路

- 采用统一的操作系统平台，并安装好相关的依赖包。
- 然后采用 Ansible 推送相关的配置文件到各个服务节点上面。
- 修改每个实例的 server-id，与端口号不同即可。

5. 查看 MySQL 错误日志

查看 MySQL 错误日志，结果显示如下。

```
root@www.MySQL55m.com 15:02:06 ~ #head /data/MySQL/MySQL3316/MySQL-error.log
2020-01-12 23:09:25 10471 [Note] InnoDB: Using atomics to ref count buffer pool pages
2020-01-12 23:09:25 10471 [Note] InnoDB: The InnoDB memory heap is disabled
2020-01-12 23:09:25 10471 [Note] InnoDB: Mutexes and rw_locks use GCC atomic builtins
2020-01-12 23:09:25 10471 [Note] InnoDB: Memory barrier is not used
2020-01-12 23:09:25 10471 [Note] InnoDB: Compressed tables use zlib 1.2.3
2020-01-12 23:09:25 10471 [Note] InnoDB: Using Linux native AIO
2020-01-12 23:09:25 10471 [Note] InnoDB: Using CPU crc32 instructions
2020-01-12 23:09:25 10471 [Note] InnoDB: Initializing buffer pool, size = 5.0G
2020-01-12 23:09:25 10471 [Note] InnoDB: Completed initialization of buffer pool
```

查看 MySQL 服务器字符集设置，代码如下。

```
[root@lamp shell]# bash  /data/scripts/shell/MySQL3316-conn.sh
Welcome to the MySQL monitor.  Commands end with ; or \g.
Your MySQL connection id is 4
Server version: 5.6.28-76.1-log Percona Server (GPL), Release 76.1, Revision 5759e76

Copyright (c) 2009-2015 Percona LLC and/or its affiliates
Copyright (c) 2000, 2015, Oracle and/or its affiliates. All rights reserved.

Oracle is a registered trademark of Oracle Corporation and/or its
affiliates. Other names may be trademarks of their respective
owners.

Type 'help;' or '\h' for help. Type '\c' to clear the current input statement.

MySQL3316>show global variables like 'char%';
+--------------------------+----------------------------------------------------------------------+
| Variable_name            | Value                                                                |
+--------------------------+----------------------------------------------------------------------+
| character_set_client     | utf8                                                                 |
| character_set_connection | utf8                                                                 |
| character_set_database   | utf8                                                                 |
| character_set_filesystem | binary                                                               |
| character_set_results    | utf8                                                                 |
| character_set_server     | utf8                                                                 |
| character_set_system     | utf8                                                                 |
| character_sets_dir       | /opt/Percona-Server-5.6.28-rel76.1-Linux.x86_64.ssl101/share/charsets/ |
+--------------------------+----------------------------------------------------------------------+
8 rows in set (0.00 sec)
```

9.6 高标准编译安装 PHP

熟练编译 PHP 是 Linux 系统管理员最基本的专业技能，维护企业级 PHP 应用时，经常需要编译第三方模块，建议 Linux 系统管理员熟练掌握该项技能。

9.6.1 构建 PHP 基础环境

1. 安装 PHP 依赖库

安装 PHP 依赖库，代码如下：

```
yum -y install gcc gcc-c++ \
cURL-devel bzip2-devel \
openssl-devel libxml2-devel \
libjpeg-devel libmcrypt-devel \
libpng-devel freetype-devel \
PHP-mcrypt libmcrypt libmcrypt-devel\
autoconf freetype gd jpegsrc libmcrypt \
libpng libpng-devel libjpeg \
libxml2 libxml2-devel zlib\
cURL cURL-devel \
bison \
re2c
[root@lamp shell]# yum -y install gcc gcc-c++ curl-devel bzip2-devel openssl-devel
libxml2-devel libjpeg-devel libmcrypt-devel libpng-devel freetype-devel PHP-mcrypt
libmcrypt libmcrypt-develautoconf freetype gd jpegsrc libmcrypt libpng libpng-devel
libjpeg libxml2 libxml2-devel zlibcurl curl-devel bison re2c
```

2. 下载 PHP

下载 PHP 稳定版本并解压，代码如下。

```
cd /opt/
wget -c HTTP://cn2.PHP.net/distributions/PHP-5.6.14.tar.bz2
tar xf PHP-5.6.14.tar.bz2
cd PHP-5.6.14
```

3. 编译安装 PHP

编译安装 PHP，代码如下。

```
./configure --prefix=/data/app/lamp/PHP \
--with-MySQL=/data/app/lamp/MySQL \
--with-apxs2=/data/app/lamp/Apache/bin/apxs \
--with-config-file-path=/data/app/lamp/PHP/etc \
--with-config-file-scan-dir=/data/app/lamp/PHP/etc.d \
--enable-mbstring \
--enable-sockets \
--enable-bcmath \
--enable-xml \
--enable-zip \
--enable-gd-native-ttf \
--enable-pdo \
--enable-fpm \
--enable-maintainer-zts \
--enable-opcache \
--with-gd \
--with-zlib \
--with-bz2 \
--with-mcrypt \
--with-openssl \
--with-MySQLi \
--with-MySQL-sock \
--enable-maintainer-zts \
```

```
--with-pdo-MySQL \
--with-gettext \
--with-cURL \
--with-pdo-MySQL \
--with-freetype-dir \
--with-jpeg-dir \
--with-png-dir;echo $?
[root@lamp PHP-5.6.14]# ./configure --prefix=/data/app/lamp/PHP \
> --with-MySQL=/data/app/lamp/MySQL \
> --with-apxs2=/data/app/lamp/Apache/bin/apxs \
> --with-config-file-path=/data/app/lamp/PHP/etc \
> --with-config-file-scan-dir=/data/app/lamp/PHP/etc.d \
> --enable-mbstring \
> --enable-sockets \
> --enable-bcmath \
> --enable-xml \
> --enable-zip \
> --enable-gd-native-ttf \
> --enable-pdo \
> --enable-fpm \
> --enable-maintainer-zts \
> --enable-opcache \
> --with-gd \
> --with-zlib \
> --with-bz2 \
> --with-mcrypt \
> --with-openssl \
> --with-MySQLi \
> --with-MySQL-sock \
> --enable-maintainer-zts \
> --with-pdo-MySQL \
> --with-gettext \
> --with-cURL \
> --with-pdo-MySQL \
> --with-freetype-dir \
> --with-jpeg-dir \
> --with-png-dir;echo $?
checking for grep that handles long lines and -e... /usr/bin/grep
checking for egrep... /usr/bin/grep -E
checking for a sed that does not truncate output... /usr/bin/sed
checking build system type... x86_64-unknown-linux-gnu
checking host system type... x86_64-unknown-linux-gnu
checking target system type... x86_64-unknown-linux-gnu
checking for cc... cc
checking whether the C compiler works... yes
checking for C compiler default output file name... a.out
checking for suffix of executables...
checking whether we are cross compiling... no
checking for suffix of object files... o
checking whether we are using the GNU C compiler... yes
checking whether cc accepts -g... yes
checking for cc option to accept ISO C89... none needed
checking how to run the C preprocessor... cc -E
// 中间代码略
Generating files
configure: creating ./config.status
creating main/internal_functions.c
creating main/internal_functions_cli.c
+--------------------------------------------------------------------+
| License:                                                           |
| This software is subject to the PHP License, available in this     |
| distribution in the file LICENSE.  By continuing this installation |
| process, you are bound by the terms of this license agreement.     |
| If you do not agree with the terms of this license, you must abort |
| the installation process at this point.                            |
+--------------------------------------------------------------------+
```

```
Thank you for using PHP.

config.status: creating PHP5.spec
config.status: creating main/build-defs.h
config.status: creating scripts/PHPize
config.status: creating scripts/man1/PHPize.1
config.status: creating scripts/PHP-config
config.status: creating scripts/man1/PHP-config.1
config.status: creating sapi/cli/PHP.1
config.status: creating sapi/fpm/PHP-FPM.conf
config.status: creating sapi/fpm/init.d.PHP-FPM
config.status: creating sapi/fpm/PHP-FPM.service
config.status: creating sapi/fpm/PHP-FPM.8
config.status: creating sapi/fpm/status.html
config.status: creating sapi/cgi/PHP-cgi.1
config.status: creating ext/phar/phar.1
config.status: creating ext/phar/phar.phar.1
config.status: creating main/PHP_config.h
config.status: executing default commands
0
```

编译安装 PHP，代码如下。
make&& make install;echo $?

```
[root@lamp PHP-5.6.14]# make && make install;echo $?
[root@lamp PHP-5.6.14]# screen -S compile_PHP
[root@lamp PHP-5.6.14]# time make && make install;echo $?
/bin/sh /opt/PHP-5.6.14/libtool --silent --preserve-dup-deps --mode=compile
/opt/PHP-5.6.14/meta_ccld -Iext/date/lib -D HAVE_TIMELIB_CONFIG_H=1 -Iext/date/
-I/opt/PHP-5.6.14/ext/date/ -DPHP_ATOM_INC -I/opt/PHP-5.6.14/include
-I/opt/PHP-5.6.14/main -I/opt/PHP-5.6.14 -I/opt/PHP-5.6.14/ext/date/lib
-I/opt/PHP-5.6.14/ext/ereg/regex -I/usr/include/libxml2 -I/usr/include/freetype2
-I/usr/include/libpng15 -I/opt/PHP-5.6.14/ext/mbstring/oniguruma -I/opt/PHP-5.6.14/
ext/mbstring/libmbfl -I/opt/PHP-5.6.14/ext/mbstring/libmbfl/mbfl -I/data/app
/lamp/MySQL/include -I/opt/PHP-5.6.14/ext/sqlite3/libsqlite -I/opt/PHP-5.6.14
/ext/zip/lib -I/opt/PHP-5.6.14/TSRM -I/opt/PHP-5.6.14/Zend  -D_REENTRANT
-I/usr/include -g -O2 -fvisibility=hidden -pthread -DZTS  -c /opt/PHP-5.6.14
/ext/date/lib/parse_date.c -o ext/date/lib/parse_date.lo
// 中间代码略
```

查看编译 PHP 后台执行任务进度，代码如下。

```
[root@lamp PHP-5.6.14]# screen -ls
There is a screen on:
        1310.compile_PHP        (Detached)
1 Socket in /var/run/screen/S-root.


[root@lamp ~]# screen -r 1310
/data/app/lamp/Apache/build/instdso.sh SH_LIBTOOL='/data/app/apr/build-1/libtool'
libPHP5.la /data/app/lamp/Apache/modules
/data/app/apr/build-1/libtool --mode=install install libPHP5.la
/data/app/lamp/Apache/modules/
libtool: install: install .libs/libPHP5.so /data/app/lamp/Apache/modules/libPHP5.so
libtool: install: install .libs/libPHP5.lai /data/app/lamp/Apache/modules/libPHP5.la
libtool: install: warning: remember to run 'libtool --finish /opt/PHP-5.6.14/libs'
chmod 755 /data/app/lamp/Apache/modules/libPHP5.so
[activating module 'PHP5' in /data/app/lamp/Apache/conf/httpd.conf]
Installing shared extensions:
/data/app/lamp/PHP/lib/PHP/extensions/no-debug-zts-20131226/
Installing PHP CLI binary:        /data/app/lamp/PHP/bin/
Installing PHP CLI man page:      /data/app/lamp/PHP/PHP/man/man1/
Installing PHP FPM binary:        /data/app/lamp/PHP/sbin/
Installing PHP FPM config:        /data/app/lamp/PHP/etc/
Installing PHP FPM man page:      /data/app/lamp/PHP/PHP/man/man8/
Installing PHP FPM status page:   /data/app/lamp/PHP/PHP/PHP/fpm/
```

```
Installing PHP CGI binary:          /data/app/lamp/PHP/bin/
Installing PHP CGI man page:        /data/app/lamp/PHP/PHP/man/man1/
Installing build environment:       /data/app/lamp/PHP/lib/PHP/build/
Installing header files:            /data/app/lamp/PHP/include/PHP/
Installing helper programs:         /data/app/lamp/PHP/bin/
  program: PHPize
  program: PHP-config
Installing man pages:               /data/app/lamp/PHP/PHP/man/man1/
  page: PHPize.1
  page: PHP-config.1
Installing PEAR environment:        /data/app/lamp/PHP/lib/PHP/
[PEAR] Archive_Tar      - installed: 1.3.12
[PEAR] Console_Getopt   - installed: 1.3.1
[PEAR] Structures_Graph - installed: 1.0.4
[PEAR] XML_Util         - installed: 1.2.3
[PEAR] PEAR             - installed: 1.9.5
Wrote PEAR system config file at: /data/app/lamp/PHP/etc/pear.conf
You may want to add: /data/app/lamp/PHP/lib/PHP to your PHP.ini include_path
/opt/PHP-5.6.14/build/shtool install -c ext/phar/phar.phar /data/app/lamp/PHP/bin
ln -s -f phar.phar /data/app/lamp/PHP/bin/phar
Installing PDO headers:             /data/app/lamp/PHP/include/PHP/ext/pdo/
0
```

4．PHP 安装提示信息

PHP 编译成功后安装提示信息如图 9-18～图 9-20 所示。

图 9-18　安装提示信息 1

图 9-19　安装提示信息 2

图 9-20　安装提示信息 3

到此，基本的 LAMP 编译安装工作已经完成。

9.6.2　配置 PHP

复制 PHP 源代码下的模板配置文件到指定的 PHP 编译配置文件路径，代码如下。

```
/bin/cp -av /opt/PHP-5.6.14/PHP.ini-production  /data/app/lamp/PHP/etc/PHP.ini
[root@lamp PHP-5.6.14]# /bin/cp -av /opt/PHP-5.6.14/PHP.ini-production
/data/app/lamp/PHP/etc/PHP.ini
```

1. 配置 Apache 支持 PHP

配置 Apache 支持 PHP，代码如下。
```
vim /data/app/lamp/Apache/conf/httpd.conf
```
查找字符串 AddType，并增加代码如下。
```
####AddType application/x-httpd-PHP .PHP
####AddType application/x-httpd-PHP-source .PHPs
```
执行代码如下。
```
sed -i -e '/    AddType application\/x-compress .Z/a\AddType
application/x-httpd-PHP .PHP\nAddType application/x-httpd-PHP-source .PHPs'
/data/app/lamp/Apache/conf/httpd.conf
```
这里用到了 sed 的多行追加，注意以下几点。
- 不用加-g 选项就可默认把符合条件的内容都加入 a\ 后面。
- 加入一行默认是自动换行；最后一行不用加 \n。
- 有时我们会遇到在指定多行之后添加多行，其实这样的需求用 sed 很难实现，可以尝试把文件组装成 JSON 或者 XML 的格式，然后用 Python 来处理这个 JSON 或 XML 文件，并添加代码支持 PHP，如图 9-21 所示。

```
sed -i 's/    DirectoryIndex index.html/    DirectoryIndex   index.PHP index.html/g'
/data/app/lamp/Apache/conf/httpd.conf
```

```
    AddType application/x-compress .Z
AddType application/x-httpd-php .php
AddType application/x-httpd-php-source .phps
    AddType application/x-gzip .gz .tgz
```

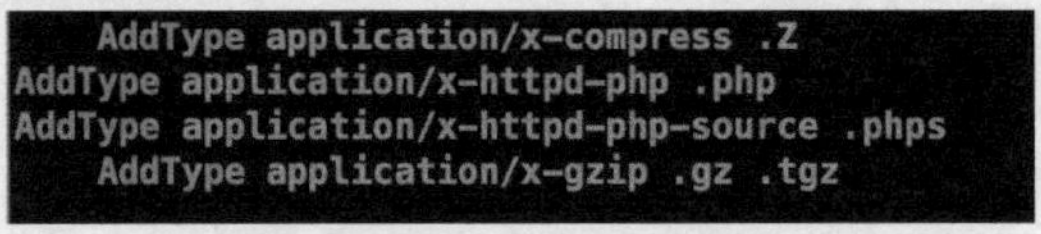

图 9-21　添加代码支持 PHP

默认首页支持文件如图 9-22 所示。

```
<IfModule dir_module>
    DirectoryIndex   index.php index.html
</IfModule>
```

图 9-22　默认首页支持文件

2. 开启 PHP OPcache

开启、查看 PHP OPcache 如图 9-23 和图 9-24 所示，代码如下。
```
sed -i 's/;opcache.enable=0/;opcache.enable=1/g' /data/app/lamp/PHP/etc/PHP.ini
[opcache]
opcache.fast_shutdown=1
opcache.enable_cli=1
opcache.memory_consumption=128
opcache.interned_strings_buffer=8
opcache.max_accelerated_files=4000
opcache.revalidate_freq=60
opcache.fast_shutdown=1
opcache.enable= 1
opcache.enable_cli=0

zend_extension = opcache.so
[root@lamp PHP-5.6.14]# grep opcache /data/app/lamp/PHP/etc/PHP.ini
[opcache]
opcache.enable=1
opcache.enable_cli=0
opcache.memory_consumption=64
```

```
[opcache]
opcache.fast_shutdown=1
opcache.enable_cli=1
opcache.memory_consumption=128
opcache.interned_strings_buffer=8
opcache.max_accelerated_files=4000
opcache.revalidate_freq=60
opcache.fast_shutdown=1
opcache.enable= 1
opcache.enable_cli=0

zend_extension = opcache.so
```

图 9-23 开启 OPcache

PHP OPcache

Opcode Caching	Up and Running
Optimization	Enabled
Startup	OK
Shared memory model	mmap
Cache hits	1
Cache misses	1
Used memory	450104
Free memory	133767624
Wasted memory	0
Cached scripts	1
Cached keys	1
Max keys	7963
OOM restarts	0
Hash keys restarts	0
Manual restarts	0

图 9-24 查看 PHP OPcache

```
opcache.interned_strings_buffer=4
opcache.max_accelerated_files=2000
opcache.max_wasted_percentage=5
opcache.use_cwd=1
opcache.validate_timestamps=1
opcache.revalidate_freq=2
opcache.revalidate_path=0
opcache.save_comments=1
opcache.load_comments=1
opcache.fast_shutdown=0
opcache.enable_file_override=0
opcache.optimization_level=0xffffffff
opcache.inherited_hack=1
opcache.dups_fix=0
opcache.max_file_size=0
opcache.consistency_checks=0
opcache.force_restart_timeout=180
;opcache.error_log=
;opcache.log_verbosity_level=1
;opcache.preferred_memory_model=
;opcache.protect_memory=0
```

3. PHP 连接 MySQL

创建连接 MySQL 数据库的测试账号并授权，代码如下。

```
grant all privileges on *.* to 'ceshi'@'%' identified by '123.com';
flush privileges;
```

代码输出结果如下。

```
MySQL3316>grant all privileges on *.* to 'ceshi'@'%' identified by '123.com';
Query OK, 0 rows affected (0.00 sec)
```

```
MySQL3316>flush privileges;
Query OK, 0 rows affected (0.00 sec)

MySQL3316>select user,host,password from MySQL.user;
+-------+-----------+-------------------------------------------+
| user  | host      | password                                  |
+-------+-----------+-------------------------------------------+
| root  | localhost |                                           |
| root  | lamp      |                                           |
| root  | 127.0.0.1 |                                           |
| root  | ::1       |                                           |
|       | localhost |                                           |
|       | lamp      |                                           |
| ceshi | %         | *AC241830FFDDC8943AB31CBD47D758E79F7953EA |
+-------+-----------+-------------------------------------------+
7 rows in set (0.00 sec)

MySQL3316>flush privileges;
Query OK, 0 rows affected (0.00 sec)


[root@lamp htdocs]# bash  /data/scripts/shell/MySQL3316-conn.sh
Welcome to the MySQL monitor.  Commands end with ; or \g.
Your MySQL connection id is 4
Server version: 5.6.28-76.1-log Percona Server (GPL), Release 76.1, Revision 5759e76

Copyright (c) 2009-2015 Percona LLC and/or its affiliates
Copyright (c) 2000, 2015, Oracle and/or its affiliates. All rights reserved.

Oracle is a registered trademark of Oracle Corporation and/or its
affiliates. Other names may be trademarks of their respective
owners.

Type 'help;' or '\h' for help. Type '\c' to clear the current input statement.

MySQL3316>grant all privileges on *.* to 'ceshi'@'172.16.%' identified by '123.com';
Query OK, 0 rows affected (0.00 sec)

MySQL3316>selece user,host,password from MySQL.user;
ERROR 1064 (42000): You have an error in your SQL syntax; check the manual that
corresponds to your MySQL server version for the right syntax to use near 'selece
user,host,password from MySQL.user' at line 1
MySQL3316>
MySQL3316>
MySQL3316>selecet user,host,password from MySQL.user;
ERROR 1064 (42000): You have an error in your SQL syntax; check the manual that
corresponds to your MySQL server version for the right syntax to use near 'selecet
user,host,password from MySQL.user' at line 1
MySQL3316>
MySQL3316>
MySQL3316>select user,host,password from MySQL.user;
+-------+-----------+-------------------------------------------+
| user  | host      | password                                  |
+-------+-----------+-------------------------------------------+
| root  | localhost |                                           |
| root  | lamp      |                                           |
| root  | 127.0.0.1 |                                           |
| root  | ::1       |                                           |
|       | localhost |                                           |
|       | lamp      |                                           |
| ceshi | %         | *AC241830FFDDC8943AB31CBD47D758E79F7953EA |
| ceshi | 172.16.%  | *AC241830FFDDC8943AB31CBD47D758E79F7953EA |
+-------+-----------+-------------------------------------------+
```

```
8 rows in set (0.00 sec)

MySQL3316>delete from MySQL.user where user = 'root' and host!='localhost';
ERROR 2006 (HY000): MySQL server has gone away
No connection. Trying to reconnect...
Connection id:    5
Current database: *** NONE ***

Query OK, 3 rows affected (0.00 sec)

MySQL3316>select user,host,password from MySQL.user;
+-------+-----------+-------------------------------------------+
| user  | host      | password                                  |
+-------+-----------+-------------------------------------------+
| root  | localhost |                                           |
|       | localhost |                                           |
|       | lamp      |                                           |
| ceshi | %         | *AC241830FFDDC8943AB31CBD47D758E79F7953EA |
| ceshi | 172.16.%  | *AC241830FFDDC8943AB31CBD47D758E79F7953EA |
+-------+-----------+-------------------------------------------+
5 rows in set (0.00 sec)

MySQL3316>delete from MySQL.user where user='';
Query OK, 2 rows affected (0.00 sec)

MySQL3316>select user,host,password from MySQL.user;
+-------+-----------+-------------------------------------------+
| user  | host      | password                                  |
+-------+-----------+-------------------------------------------+
| root  | localhost |                                           |
| ceshi | %         | *AC241830FFDDC8943AB31CBD47D758E79F7953EA |
| ceshi | 172.16.%  | *AC241830FFDDC8943AB31CBD47D758E79F7953EA |
+-------+-----------+-------------------------------------------+
3 rows in set (0.00 sec)

MySQL3316>delete from MySQL.user where user='ceshi' and host='%';
Query OK, 1 row affected (0.00 sec)

MySQL3316>flush privileges;
Query OK, 0 rows affected (0.00 sec)

MySQL3316>exit
Bye
```

编写 PHP 程序连接 MySQL 代码，测试 PHP 连接 MySQL 结果。

```
[root@lamp htdocs]# cat conn.PHP
 <?PHP
$link=MySQL_connect("172.16.0.16:3316","ceshi","123.com");
if ($link)

        echo   "<h1><font color='red'>succ..</font></h1>";
else
        echo "fail..";
MySQL_close($link);
?>
<?PHP
$link=MySQL_connect("192.168.1.111:3316","ceshi","123.com");
if ($link)

    echo   "<h1><font color='red'>succ..</font></h1>";
else
    echo "fail..";
MySQL_close($link);
?>
```

目前为止我们已经构建了属于自己的单机 LAMP。

9.7 使用 WordPress 搭建企业级站点

优化建站环境

下载并解压 WordPress，代码如下。WordPress 版本信息如图 9-25 所示。

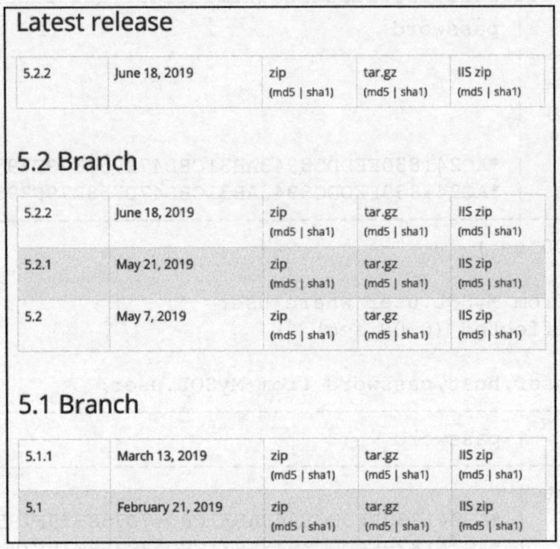

Latest release				
5.2.2	June 18, 2019	zip (md5 | sha1)	tar.gz (md5 | sha1)	IIS zip (md5 | sha1)

5.2 Branch				
5.2.2	June 18, 2019	zip (md5 | sha1)	tar.gz (md5 | sha1)	IIS zip (md5 | sha1)
5.2.1	May 21, 2019	zip (md5 | sha1)	tar.gz (md5 | sha1)	IIS zip (md5 | sha1)
5.2	May 7, 2019	zip (md5 | sha1)	tar.gz (md5 | sha1)	IIS zip (md5 | sha1)

5.1 Branch				
5.1.1	March 13, 2019	zip (md5 | sha1)	tar.gz (md5 | sha1)	IIS zip (md5 | sha1)
5.1	February 21, 2019	zip (md5 | sha1)	tar.gz (md5 | sha1)	IIS zip (md5 | sha1)

图 9-25　WordPress 版本信息

```
[root@lamp opt]# ls -lh latest.tar.gz
-rw-r--r-- 1 root root 11M Jun 19 01:52 latest.tar.gz
[root@lamp opt]# tar xf latest.tar.gz
[root@lamp opt]# ls -lh
total 161M
drwxr-xr-x  28   1000  1000   4.0K Aug 26 21:46 apr-1.5.2
-rw-r--r--   1   root  root  1008K Aug 25 22:10 apr-1.5.2.tar.gz
drwxr-xr-x  20   1000  1000   4.0K Aug 26 21:51 apr-util-1.5.4
-rw-r--r--   1   root  root   679K Aug 25 22:10 apr-util-1.5.4.tar.bz2
drwxr-xr-x  12    501 games   4.0K Aug 28 09:47 httpd-2.4.17
-rw-r--r--   1   root  root   5.0M Aug 25 22:10 httpd-2.4.17.tar.bz2
-rw-r--r--   1   root  root    11M Jun 19 01:52 latest.tar.gz
-rw-r--r--   1   root  root   3.3K Aug 25 22:11 my3316.cnf
drwxr-xr-x  14   1000  1000   4.0K Aug 28 09:02 ngHTTP2-1.39.2
-rw-r--r--   1   root  root   2.0M Aug 14 07:44 ngHTTP2-1.39.2.tar.bz2
drwxr-xr-x  13   root  root   4.0K Jan  9  2016 Percona-Server-5.6.28-rel76.1-Linux.x86_64.ssl101
-rw-r--r--   1   root  root   128M Aug 25 22:11 Percona-Server-5.6.28-rel76.1-Linux.x86_64.ssl101.tar.gz
drwxr-xr-x  18   1000  1000   4.0K Aug 28 11:35 PHP-5.6.14
-rw-r--r--   1   root  root    14M Aug 25 22:11 PHP-5.6.14.tar.bz2
drwxr-xr-x.  2   root  root   4.0K Oct 31  2018 rh
drwxr-xr-x   5 nobody 65534   4.0K Jun 19 01:50 wordpress
[root@lamp opt]# du -sh wordpress/
46M     wordpress/
```

出现图 9-26 所示的提示信息一般是 PHP 版本"低于"WordPress 版本导致的，有两种解决方法。第一种方法是升级 PIIP 到指定版本（或者高于指定版本）。第二种方法是降低 WordPress 版本，本书选择降低 WordPress 版本，读者可根据实际需要选择适合自己场景的方案。

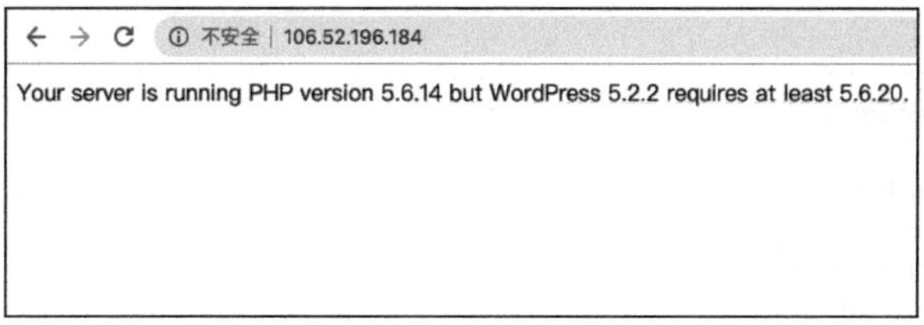

图 9-26　PHP 版本提示

解压 latest.zip 文件，并将其复制到/var/www/html/目录下，代码如下。

```
# unzip -q latest.zip
# cp -rf wordpress/* /var/www/html/

[root@lamp opt]# du -sh wordpress/
46M     wordpress/
[root@lamp opt]# cp -rp wordpress/* /data/app/lamp/Apache/htdocs/
[root@lamp opt]# du -sh /data/app/lamp/Apache/htdocs/
46M     /data/app/lamp/Apache/htdocs/
[root@lamp opt]# ls -lh /data/app/lamp/Apache/htdocs/
total 208K
-rw-r--r--  1 root   root    173 Aug 28 16:00 conn.PHP
-rw-r--r--  1 root   root     45 Jun 12  2007 index.html
-rw-r--r--  1 nobody 65534   420 Dec  1  2017 index.PHP
-rw-r--r--  1 nobody 65534   20K Jan  2  2019 license.txt
-rw-r--r--  1 nobody 65534  7.3K Apr  9 06:59 readme.html
-rw-r--r--  1 nobody 65534  6.8K Jan 12  2019 wp-activate.PHP
drwxr-xr-x  9 nobody 65534  4.0K Jun 19 01:50 wp-admin
-rw-r--r--  1 nobody 65534   369 Dec  1  2017 wp-blog-header.PHP
-rw-r--r--  1 nobody 65534  2.3K Jan 21  2019 wp-comments-post.PHP
-rw-r--r--  1 nobody 65534  2.9K Jan  8  2019 wp-config-sample.PHP
drwxr-xr-x  4 nobody 65534  4.0K Jun 19 01:50 wp-content
-rw-r--r--  1 nobody 65534  3.8K Jan  9  2019 wp-cron.PHP
drwxr-xr-x 20 nobody 65534   12K Jun 19 01:50 wp-includes
-rw-r--r--  1 nobody 65534  2.5K Jan 16  2019 wp-links-opml.PHP
-rw-r--r--  1 nobody 65534  3.3K Dec  1  2017 wp-load.PHP
-rw-r--r--  1 nobody 65534   39K Jun 10 21:34 wp-login.PHP
-rw-r--r--  1 nobody 65534  8.3K Dec  1  2017 wp-mail.PHP
-rw-r--r--  1 nobody 65534   19K Mar 29 03:04 wp-settings.PHP
-rw-r--r--  1 nobody 65534   31K Jan 17  2019 wp-signup.PHP
-rw-r--r--  1 nobody 65534  4.7K Dec  1  2017 wp-trackback.PHP
-rw-r--r--  1 nobody 65534  3.0K Aug 17  2018 xmlrpc.PHP
```

编辑配置文件，代码如下。

```
# cd /var/www/html
# cp wp-config-sample.PHP wp-config.PHP
# vim wp-config.PHP
```

新建库和用户并授权，代码如下。

```
[root@lamp ~]# bash   /data/scripts/shell/MySQL3316-conn.sh
Welcome to the MySQL monitor.  Commands end with ; or \g.
Your MySQL connection id is 10
Server version: 5.6.28-76.1-log Percona Server (GPL), Release 76.1, Revision 5759e76

Copyright (c) 2009-2015 Percona LLC and/or its affiliates
Copyright (c) 2000, 2015, Oracle and/or its affiliates. All rights reserved.

Oracle is a registered trademark of Oracle Corporation and/or its
affiliates. Other names may be trademarks of their respective
owners.

Type 'help;' or '\h' for help. Type '\c' to clear the current input statement.

MySQL3316>show databases;
```

```
+--------------------+
| Database           |
+--------------------+
| information_schema |
| MySQL              |
| performance_schema |
| test               |
+--------------------+
4 rows in set (0.00 sec)

MySQL3316>create database wordpress;
Query OK, 1 row affected (0.00 sec)

MySQL3316>grant all privileges on wordpress.* to 'wordpress_user'@'172.16.0.16'
identified by 'wordpress_!!!!!!123.com';
ERROR 2006 (HY000): MySQL server has gone away
No connection. Trying to reconnect...
Connection id:    11
Current database: *** NONE ***

Query OK, 0 rows affected (0.00 sec)

MySQL3316>select user,host,password from MySQL.user;
+----------------+-------------+-------------------------------------------+
| user           | host        | password                                  |
+----------------+-------------+-------------------------------------------+
| root           | localhost   |                                           |
| wordpress_user | 172.16.0.16 | *BD6BB7B044725B153F59412ED00BCCC4743C5FB8 |
| ceshi          | 172.16.%    | *AC241830FFDDC8943AB31CBD47D758E79F7953EA |
+----------------+-------------+-------------------------------------------+
3 rows in set (0.00 sec)

MySQL3316>show grants for wordpress_user@172.16.0.16;
+-----------------------------------------------------------------------------
---------------------------------------------+
| Grants for wordpress_user@172.16.0.16
                                             |
+-----------------------------------------------------------------------------
---------------------------------------------+
| GRANT USAGE ON *.* TO 'wordpress_user'@'172.16.0.16' IDENTIFIED BY PASSWORD '
*BD6BB7B044725B153F59412ED00BCCC4743C5FB8' |
| GRANT ALL PRIVILEGES ON 'wordpress'.* TO 'wordpress_user'@'172.16.0.16'
                                             |
+-----------------------------------------------------------------------------
---------------------------------------------+
2 rows in set (0.00 sec)

MySQL3316>show databases;
+--------------------+
| Database           |
+--------------------+
| information_schema |
| MySQL              |
| performance_schema |
| test               |
| wordpress          |
+--------------------+
5 rows in set (0.00 sec)
MySQL3316>exit
Bye
[root@lamp htdocs]# cat /data/app/lamp/Apache/htdocs/wp-config.PHP
<?PHP
define( 'DB_NAME', 'wordpress' );
define( 'DB_USER', 'wordpress_user' );
define( 'DB_PASSWORD', 'wordpress_!!!!!!123.com' );
```

```
define( 'DB_HOST', '172.16.0.16:3316' );
define( 'DB_CHARSET', 'utf8' );
define( 'DB_COLLATE', '' );
define( 'AUTH_KEY',         'put your unique phrase here' );
define( 'SECURE_AUTH_KEY',  'put your unique phrase here' );
define( 'LOGGED_IN_KEY',    'put your unique phrase here' );
define( 'NONCE_KEY',        'put your unique phrase here' );
define( 'AUTH_SALT',        'put your unique phrase here' );
define( 'SECURE_AUTH_SALT', 'put your unique phrase here' );
define( 'LOGGED_IN_SALT',   'put your unique phrase here' );
define( 'NONCE_SALT',       'put your unique phrase here' );
$table_prefix = 'wp_';
define( 'WP_DEBUG', false );
if ( ! defined( 'ABSPATH' ) ) {
        define( 'ABSPATH', dirname( __FILE__ ) . '/' );
}
require_once( ABSPATH . 'wp-settings.PHP' );
```

记得复制管理员密码。